Proceedings
of the Steklov Institute of Mathematics

edited by
I. G. Petrovskiĭ and S. M. Nikol'skiĭ

Number 114 (1970)

Some Questions in Constructive Functional Analysis

by

Fan Din'-Zieu [Phan Dinh Diêu]

AMERICAN MATHEMATICAL SOCIETY
Providence, Rhode Island
1974

Академия Наук

Союза Советских Социалистических Республик

ТРУДЫ

ордена Ленина

МАТЕМАТИЧЕСКОГО ИНСТИТУТА

имени В. А. СТЕКЛОВА

CXIV

НЕКОТОРЫЕ ВОПРОСЫ

КОНСТРУКТИВНОГО ФУНКЦИОНАЛЬНОГО АНАЛИЗА

ФАН ДИНЬ ЗИЕУ

Ответственный редактор

академик И. Г. ПЕТРОВСКИЙ

Зам. ответственного редактора

член-корреспондент АН СССР С. М. НИКОЛЬСКИЙ

Издательство „Наука"

Москва 1970

Translated by J. M. Danskin

AMS (MOS) subject classifications (1970).

Primary 02E99, 46A05, 46F05.

Library of Congress Cataloging in Publication Data CIP

Phan Dình Diêu.

Some questions in constructive functional analysis.

(Proceedings of the Steklov Institute of Mathematics, no. 114 (1970))

Translation of Nekotorye voprosy konstruktivnogo funkt͡sional'nogo analiza by Fan Din'-Zieu.

Bibliography: p.

1. Locally convex spaces. 2. Distributions, Theory of (Functional analysis) 3. Constructive mathematics. I. Title. II. Series: Akademii͡a nauk SSSR. Matematicheskiĭ institut. Proceedings, no. 114.

QA1.A4 t.114,1974 [QA322] 510'.8s [515'.7] 73-21929

ISBN 0-8218-3014-7

PROCEEDINGS OF THE STEKLOV INSTITUTE OF MATHEMATICS IN THE ACADEMY OF SCIENCES OF THE USSR

(Труды математического института им. В. А. Стеклова, т. CXIV, 1970)

TABLE OF CONTENTS

INTRODUCTION

1. The present monograph belongs to the constructive direction in mathematics. It is devoted to a number of questions of constructive functional analysis. By constructive mathematics (the constructive tendency in mathematics) we understand the variant of constructive mathematics developed by A. A. Markov, N. A. Šanin and their school. Constructive mathematics is characterized by the following general features, for which see [**11**–**14**] and [**24**–**26**]:

a) As objects of study one considers only constructively defined objects. In applications to these objects one admits the abstraction of potential realizability, but absolutely rejects the use of the abstraction of actual infinity.

b) Mathematical inferences are understood in the constructive sense, on the basis of rules of constructive evaluation of inferences. Here we must note that certain logical axioms and rules of deduction of classical mathematics are inconsistent with the constructive understanding of inferences.

c) One admits the principle of constructive choice introduced by A. A. Markov.

In the course of the last decade constructive mathematics, in particular constructive mathematical analysis, has been the object of study by many authors and achieved a considerable development. In a series of papers of A. A. Markov [**12**], N. A. Šanin [**26**], G. S. Ceĭtin [**27**], I. D. Zaslavskiĭ [**4**], and Zaslavskiĭ and Ceĭtin [**6**], the foundations were laid down for the theory of constructive functions of a real variable, and for the theory of constructive metric spaces and constructive linear normed spaces. The present work is devoted to the construction of the foundations of theories of constructive locally convex linear topological spaces and constructive generalized functions.

The monograph consists of six chapters. The first three chapters are devoted to general questions of the theory of constructive locally convex linear spaces, and the last three to questions of the theory of constructive generalized functions. In formulating the constructive analogues of concepts and results of the corresponding theories of classical mathematics the author has endeavored to explain the peculiarities of constructive mathematics within the confines of the theories in question.

2. In classical mathematics a locally compact space is defined as a linear topological space in which there exists a fundamental system of convex neighborhoods of the origin. The topology of a locally convex space is defined by such a fundamental system of neighborhoods of the origin (see for example [**7**, **38**, **53**]). The fundamental system of neighborhoods in its turn is some system of sets. Thus, in order to define the constructive analogue of the

concept of a locally convex space, we need first to define the method of giving a "system of sets" in constructive mathematics. We note that constructive mathematics does not use one and the same precise concept of set for all of its various theories. The term "set" in constructive mathematics means the same as "type of constructive object". Within the limits of the concrete logico-mathematical language intended for the formation of a concrete theory of constructive mathematics a set is given by a condition with one parameter (a one-parameter formula) in that language (see [**24**]). In connection with this understanding of the term "set", we adopt the following method of giving a system of sets. We fix on some logico-mathematical language. Within the limits of this language a system of sets is defined by a list, consisting of the following objects: a) a one-parameter formula $\mathscr{A}$, defining a set of words in some alphabet whose elements are regarded as indices of sets of the system; b) a two-parameter formula $\mathfrak{P}$, one of whose parameters is a variable for the indices just mentioned. On substituting into $\mathfrak{P}$, in place of the index variable, an element U of the set $\mathscr{A}$, the formula $\mathfrak{P}$ turns into a one-parameter formula, defining some set. This set is called the set with index U of the system of sets $(\mathscr{A}, \mathfrak{P})$. A detailed exposition of this question is given in [**35**].

Basing ourselves on the method just described of giving a system of sets, we introduce in Chapter I the concept of constructive locally convex space.

A constructive locally convex space is defined by a list of constructive objects satisfying a system of conditions which correspond to conditions imposed on the concept of locally convex space in classical mathematics.

Using the method used in the present work to define the concept of constructive locally convex space, we may introduce also the concepts of constructive uniform space and of constructive topological space. The definitions of these concepts may be found in [**30, 36**].

In classical mathematics a generalized function is, in the definition of S. L. Sobolev and L. Schwartz, a continuous linear functional in the linear topological space of all infinitely differentiable functions with compact support (see [**1, 2, 51**]). Along with this most frequently used definition there are also other definitions of the concept of generalized function. Among these we note the definition of J. Mikusiński and R. Sikorski [**48**], according to which a generalized function is defined by a sequence of continuous functions fundamental in a certain sense. The definition of Mikusiński and Sikorski will be adopted in the present monograph as the initial point for the formation of the concept of constructive generalized function—the constructive analogue of the concept of generalized function in classical mathematics. By the definition of Mikusiński and Sikorski, the concept of generalized function is a direct extension of the concept of continuous function. Generalized functions are constructed from continuous functions by means of a method close to the method of completion of topological linear spaces. However, one should note that in the definition of Mikusiński and Sikorski a "fundamental sequence" is still not a fundamental sequence in the usual sense in an arbitrary linear topological space. We shall seek to overcome this "defect" in the definition of the concept of a constructive generalized function.

As the constructive analogue of the classical concept of an everywhere defined continuous function, we adopt the constructive notion of complete cipher of an almost uniformly continuous function. This concept will be defined in Chapter IV. The space of constructive generalized functions of a given order is defined as the completion of the space of complete ciphers of almost uniformly continuous functions relative to some appropriate metrizable topology. Thus, the space of constructive generalized functions of a given order is a complete metrizable constructive locally convex space. The union of all such spaces will be the space of all constructive generalized functions.

We note that in the application of generalized functions to concrete problems we frequently have to do not with all generalized functions, but rather only with generalized functions of some definite order. For the constructive generalized functions defined in our monograph, this means that in each concrete case of their application we will have to do only with points of a metrizable constructive locally convex space.

Along with the concept of constructive generalized function, defined by the method indicated above, we consider also the constructive analogue of the concept of generalized function defined according to Sobolev and Schwartz. Chapter VI is devoted to this question.

3. Along with the concepts of locally convex space and generalized function, for which constructive analogues are defined, we define also constructive analogues of a number of other concepts from the classical theories of locally convex spaces and generalized functions. We note the following peculiarity: when they are carried over to constructive mathematics, certain concepts of classical mathematics split. In other words, for one and the same concept of classical mathematics there may be several different constructive analogues. These splittings can arise for several reasons. In certain cases the reason for the splitting is that the constructive interpretation of inferences differs from the classical interpretation. In particular, certain classically equivalent combinations of logical connectives turn out to be nonequivalent in the constructive interpretation. Thus, for example, in classical mathematics the following two conditions are equivalent for a locally convex space: a) the intersection of all neighborhoods of zero consists only of zero; and b) every point different from the origin may be separated from it by some neighborhood. Either of these two conditions defines the "separation" of the space. In the "constructive reading" these conditions are not equivalent. Therefore in place of the classical concept of separated space there arise two constructive concepts, corresponding to these conditions: a T'-separated space and a T-

separated space (see §2.1 of Chapter I). One proves that it is possible to realize a constructive locally convex T'-separated space which is not T-separated (Theorem 2.1.1, Chapter I). For the same reason the concepts of closed set and open set split on being carried over into constructive mathematics. In place of the concept of closed set there arise two constructive analogues: a З-closed set and a З′-closed set.* In place of the concept of open set there arise two constructive analogues: an O-open set and an O'-open set. In Chapter I we construct in the space R of all real duplexes an example of a З-closed set which is not З′-closed, and in the space $\mathscr{S}$ (see 1.4) an example of a З′-closed set which is not З-closed and an example of an O'-open set which not O-open (Theorems 2.2.6–2.2.8 of Chapter I).

In other cases the splitting arises from the difference in the "quantity of information" contained on the objects being defined. As an example we select the concepts of F-generalized function and FR-generalized function, defined in §3 of Chapter IV. By definition, an FR-generalized function is a word of the type[1] $U \tau P \odot Q$, where U is the "index" of an algorithm of the type (н → н), P is the transcription of a sequence of complete ciphers of almost uniformly continuous functions, and Q is the transcription of the regulator of convergence into itself of this sequence relative to the topology corresponding to U. For a detailed definition see §3 of Chapter IV. On the other hand, an F-generalized function is a word of the type $P \odot$, where P is the transcription of a sequence of complete ciphers of almost uniformly continuous functions, for which there is potentially realizable an "index" U of an algorithm of the type (н → н) such that $\langle P \rangle$ is a fundamental sequence relative to the topology corresponding to U. Both of these concepts may be regarded as constructive analogues of the concept of generalized function defined in the Mikusiński-Sikorski sense. We prove that it is not possible to have an algorithm constructing relative to each F-generalized function an FR-generalized function equal to it (Theorem 3.3.1, Chapter IV). For the same reason certain other concepts whose constructive analogues are considered in the present monograph split.

4. In this monograph we consider a number of concrete types of constructive objects (i.e. concrete sets). Each such set is characterized by some one-parameter formula in a definite logico-mathematical language of the type described in §3 of [**24**]. We shall say that a set is *normal* if the formula characterizing that set is a normal formula or is equivalent to some normal formula. Here we recall that by a normal formula in a logico-mathematical language of the type described in §3 of [**24**] we understand a formula not

---

* *Editor's note.* The notation used here and below is motivated by the Russian words for closed (замкнутое) and open (открытое).

[1] For simplicity we drop the special case when the FR-generalized function has the form $U \tau P$, where P is the complete cipher of an almost uniformly continuous function.

containing the logical connectives $\vee$ and $\exists$. In a certain sense normal sets have a simpler logical structure than nonnormal sets. Indeed, for justifying an assertion of the type "the object P belongs to the normal set W" it suffices to prove an inference in which the logical signs $\vee$ and $\exists$ do not occur. For the proof of such an inference there is no necessity to construct any kind of additional object. It is possible to say that the objects belonging to a normal set contain "complete information" in themselves as to the objects included in that set. Sets which cannot be expressed by means of normal formulas will be called, following I. D. Zaslavskiĭ [4], *essentially problematic sets*. The objects belonging to an essentially problematic set contain in themselves "incomplete information". If, for example, W is such a set, then for the justification of an inference of the type "the object P belongs to W" it is necessary in principle to solve a constructive problem of formation of certain other objects. It is because of the merits just mentioned that normal sets take on a preeminent place in the formation of concrete theories of constructive mathematics.

In reinterpreting the concepts of classical mathematics on a constructive foundation, in most cases we shall try to find constructive analogues for them which are characterized by normal formulas. For example, the following sets turn out to be normal: the set of all complete ciphers of almost uniformly continuous constructive functions, the set of all FR-generalized functions, the set of all complete transcriptions of infinitely differentiable constructive functions. For constructive locally convex spaces we introduce the hypothesis of the normality of the set of indices of a fundamental system of neighborhoods of zero in each such space (see the Remark in Chapter I, §1.1). This hypothesis makes it possible in a fairly easy way to prove constructive analogues of a number of basic results of the classical theory of locally convex spaces.

As we noted above, that a set W is essentially problematic means that for justifying an inference of the type "the object P is an element of the set W" it is necessary to solve a constructive problem on the formation of certain additional objects. Here an algorithm constructing, relative to any element of the set W, the corresponding additional objects is not possible. In the present monograph, along with the introduction of certain normal sets playing an important role, we will make clear at the same the essentially problematic character of certain other sets closely connected with them. For example, the set of F-generalized functions turns out to be essentially problematic. The proof that a set is essentially problematic usually comes down to the proof of the impossibility of an algorithm connected with some mass problem. In the paper we establish the following theorems:

1) It is impossible to have an algorithm constructing relative to each F-generalized function an FR-algorithm equal to it (Theorem 3.3.1, Chapter IV).

2) It is impossible to have an algorithm which transforms each F-generalized function into its "formal order" (Theorem 3.3.3 of Chapter IV).

3) It is impossible to have an algorithm which transforms each FR-generalized function d_1 of finite order into a natural number k such that d_1 is an FR-generalized function of order $\leqq k$ (Theorem 2.1.1, Chapter V).

4) For any natural number k, it is impossible to have an algorithm which transforms each FR-generalized function of order $\leqq k$ into a complete cipher of an almost uniformly continuous function f such that d_1 is the generalized derivative of order k of f (Theorem 2.1.6, Chapter V).

5) It is not possible to have an algorithm which transforms each FR-generalized function with bounded support into that support (Theorem 2.2.3, Chapter V).

6) It is not possible to have an algorithm which transforms the transcription of each infinitely differentiable constructive function with compact support into a natural number m such that f vanishes outside the segment $-m\Delta m$ (Theorem 1.3.1, Chapter VI).

From these theorems follows the essentially problematic character of certain sets. For example, in view of Theorems 3) –5) the following sets turn out to be essentially problematic: the set of FR-generalized functions of finite order, the set of FR-generalized functions of order not exceeding some given natural number k, and the set of FR-generalized functions with bounded support.

5. In working out the foundations of the theory of constructive locally convex spaces and the theory of constructive generalized functions in this monograph, we on the one side prove the constructive analogues of a number of results of the corresponding theories of classical mathematics, and on the other side explain points of essential divergence between classical and constructive mathematics.

It turns out that many basic theorems of the classical theory of locally convex spaces and the majority of theorems of the classical theory of generalized functions, in the form developed by Mikusiński and Sikorski, may be reinterpreted on constructive bases. For theorems of constructive mathematics which are constructive analogues of the corresponding theorems of classical mathematics, the proofs are frequently obtained simply by copying (of course on constructive bases) the proofs of the corresponding classical theorems. In the present work we will not present the proofs of such theorems in detail, rather simply outlining the proof or referring to the proofs of the corresponding classical theorems.

However, there are also cases when the constructive analogue of some theorem of classical mathematics turns out to be true in constructive mathematics, but its proof and the proof of the corresponding classical theorem as given in the literature are substantially different. We give some examples.

1) In the classical theory one proves a theorem stating that the space adjoint to the space of functions infinitely differentiable and vanishing outside

a given segment is complete. The proof of this theorem in classical mathematics is based on considerations which are inadmissible in constructive mathematics. The constructive analogue of this theorem is Theorem 2.2.2 of Chapter VI, which asserts that the space K_M^s is complete. The proof of Theorem 2.2.2 has a completely different character. It is based on the theorem that every linear constructive functional in K_M is continuous (Theorem 2.2.1, Chapter VI), a theorem having no analogue in classical mathematics.

2) In the classical theory one proves a theorem stating that in any locally convex space the strong topology majorizes the initial topology (see, for example, [7], Chapter XI, §4, Theorem 1). Hence it follows directly that in the countable-normed space of all the functions which are infinitely differentiable and vanish outside a given segment, the strong topology majorizes the initial topology. The constructive analogue of this last assertion may be proved (Theorem 2.4.1, Chapter VI) in constructive mathematics. However, for this a special proof is required, since the general proposition "in any constructive locally convex space the strong topology majorizes the initial topology" is contradicted by an example (Theorem 2.5.1, Chapter III).

3) In classical mathematics one proves a theorem stating that each separable locally convex space is multinormable. The constructive analogue of this theorem does not hold in constructive mathematics (Theorem 4.2.4, Chapter I). Therefore, in constructive mathematics, in order to prove the multinormability of some concrete constructive locally convex space, generally speaking, special arguments are needed. An example of this is Theorem 1.5.2 of Chapter VI, which asserts that the space K is multinormable. For the proof of this theorem we need to construct an algorithm which defines a multinorm in K.

In the monograph there are sections which give explanations of essential differences between the classical theory and the constructive theory of locally convex spaces and generalized functions. In particular, the following theorems are established.

1) It is possible to construct a separable constructive locally convex space which has a countable fundamental system of neighborhoods of the origin, and has a bounded neighborhood of the origin, but is neither metrizable nor normable (Theorems 4.1.2 and 4.2.2 of Chapter I).

2) For any natural number $n \geqq 1$, it is possible to construct a T'-separated, n-dimensional, constructive, locally convex space which is not isomorphic to the space $R^{(n)}$, where $R^{(n)}$ is the n-times product of the space of all real duplexes (Theorem 4.2.2, Chapter II).

3) It is possible to construct a strictly expanding sequence of constructive locally convex spaces such that in its strict inductive limit a bounded set may be realized which is not contained in any space of that sequence. Accordingly, this strict inductive limit is not regular (Theorem 1.3.4, Chapter III).

4) It is possible to construct a constructive locally convex space such that in it the strong topology does not majorize the initial topology (Theorem 2.5.1, Chapter III).

5) It is possible to construct a constructive normed space whose adjoint is not normable (Theorem 3.2.2, Chapter III).

6) For any natural numbers m and n with $0 \leqq m \leqq n$ it is possible to construct a T'-separated, n-dimensional, constructive, locally convex space whose adjoint is an m-dimensional normable space (Theorem 4.1.1, Chapter III).

7) For any natural number m, it is possible to construct a T'-separated, infinite-dimensional, constructive, locally convex space whose adjoint is an m-dimensional normable space (Theorem 4.2.1, Chapter III).

In the following results we denote by K_M the constructive countable normed space of all complete transcriptions of infinitely differentiable constructive functions vanishing outside the segment $-M\Delta M$. We denote by $\overset{s}{K}_M$ the space adjoint to K_M.

8) The space K_M is not bornological. In fact it is possible to construct a convex set in K_M which absorbs any bounded set in K_M but does not contain any of its neighborhoods of the origin (Theorem 1.2.2, Chapter VI).

9) It is possible to construct a constructive locally convex space $\mathfrak{M}$ such a constructive linear operator λ mapping K_M into $\mathfrak{M}$ such that λ is bounded and п-continuous,* but not continuous (Theorem 2.1.1, Chapter VI).

10) Every linear functional in K_M is continuous (Theorem 2.2.1, Chapter VI).

11) It is possible to construct a bounded set in $\overset{s}{K}_M$ which is not bounded in any neighborhood of the origin of K_M (Theorem 2.3.3, Chapter VI).

12) In the space K_M the initial topology and the strong topology are not equivalent (Theorem 2.4.2, Chapter VI).

13) As M runs through the natural numbers, the spaces K_M form an expanding sequence of constructive locally convex spaces such that the space adjoint to the strict inductive limit of this sequence is not isomorphic to the projective limit of the sequence of adjoint spaces $\overset{s}{K}_M$ under the canonical mapping [Part b) of Theorem 3.2.2, Chapter VI].

These results show that the constructive analogues of certain theorems in the classical theory of locally convex spaces and generalized functions are contradicted by examples.

To these results we may adjoint the following two theorems, which elucidate the connection between the concepts of constructive function and constructive generalized function.

1) It is possible to construct a constructive continuous function f for which there does not exist a constructive generalized function whose values are calculated by the function f (Theorem 3.2.1, Chapter V).

* *Translator's note.* The definition of "п-continuous" is in Chapter III, §2.1.

2) It is possible to construct a constructive generalized function for which there does not exist a constructive function which computes its values (Theorem 3.2.2, Chapter V).

6. This monograph is only the first step on the path of reinterpretation of the classical theories of locally convex spaces and generalized functions on a constructive basis. A number of sections of these theories, important for the applications of mathematics, still do not have constructive analogues in the present monograph.

One of the most important sections of the classical theory of generalized functions is the theory of the Fourier transform of generalized functions. The constructive analogue of the concept "Fourier transform" is not defined in this monograph. The concept of "constructive generalized function", to whose study Chapters IV and V are devoted, is a constructive analogue of the generalized function defined in the sense of Mikusiński and Sikorski. In connection with this, in the formation of the constructive analogue of the concept of "Fourier transform", an appropriate initial point is the classical concept of "ultrageneralized function",[2] introduced, for example, in [**55**].

7. The monograph consists of six chapters and an Appendix. Each chapter is divided into sections, and each section into subsections.

The enumeration of sections in each chapter is carried out independently of the enumeration in the other chapters. The subsections are denoted by pairs of natural numbers, where the first number coincides with the index of the section and the second number indicates the position of the given subsection within its section.

Theorems are denoted by triples of natural numbers, where the first two numbers denote the subsection in which the theorem occurs, and the third number indicates the position of the given theorem within its subsection.

Certain lines of the text are provided with indices, appearing at the left of the line in question and included in parentheses. Within the confines of each section the numeration of the lines thus distinguished is carried out independently of the numeration in other sections.

We will use without explanation the terminology and notation introduced in the papers [**11, 24, 26**]. In every case when we employ the term "algorithm" we understand a normal algorithm in the corresponding alphabet [**11**]. However, we almost nowhere construct concrete schemas of the algorithms in question, simply referring to the possibility of such constructions.

Concerning the terminology and notation used below, we make the following remarks.

1) For the logico-mathematical languages we keep to the following assumption: in the list of the basic alphabets one has the alphabets $Ч_0$, $Ч_1$,

[2] "Ultradistribution" in the context.

$Ч_2$, $Ч_3$ (see [**26**]), and in the list Σ of characteristic formulas of these languages one has the formulas defining respectively the set of natural numbers, the set of rational numbers, the set of F-numbers and the set of FR-numbers. The subordinated generic letters assigned to these sets are denoted respectively by н, р, в and д. We denote by i, j, k, l, m, n variables for natural numbers, and by x, y, z variables for real duplexes (FR-numbers).

2) Suppose that $Я_\Sigma$ is some logico-mathematical language of the type described in §8 of [**26**]. In principle, when we adjoin to the list of basic alphabets (to the list of characteristic formulas) of this language new alphabets (new formulas), we obtain a new list of alphabets (new formulas), on the basis of which a new language is defined. For this new language we should strictly speaking use a new notation. However, for simplification of the notation we will not do this, preserving for the new language the original notation $Я_\Sigma$. In this case we shall simply say that we are adjoining to the list of basic alphabets (characteristic formulas) of the language $Я_\Sigma$ these new alphabets (new formulas).

3) In certain places we also employ notation and terminology for extended alphabets of the type described in [**35**]. These extended languages are introduced in order to simplify the symbolic transcription of mathematical inferences connected with systems of sets. In [**35**] such extended languages were described in detail. However, the use of some notation and terminology for these languages in the present work is of a very simple character. In every case the meaning of the mathematical inferences expressed in terms of the notation of these extended languages will be directly understood from their transcriptions.

4) Suppose that $\mathfrak{A}$ is an algorithm in the alphabet A. We denote by $\mathcal{E}\mathfrak{A}, \mathrm{A}\mathfrak{Z}$ the transcription of the algorithm $\mathfrak{A}$ relative to the alphabet A (see [**27**]). Suppose further that P is a word in A. We denote by $(\mathfrak{A} \uparrow P)$ an algorithm such that for any word Q in the alphabet A

$$(\mathfrak{A} \uparrow P)(Q) \cong \mathfrak{A}(P \square Q).$$

In [**27**] the notation $\widetilde{\mathfrak{A}}_P$ was introduced for this algorithm. However in certain cases this notation turns out to be inconvenient. Therefore, instead of this notation, we will use the notation $(\mathfrak{A} \uparrow P)$.

The basic results of this monograph were published in the notes [**29**–**34**].

The author expresses his deep gratitude to A. A. Markov and to N. A. Šanin for useful advice and constant attention to the work.

CHAPTER I

CONSTRUCTIVE LOCALLY CONVEX LINEAR TOPOLOGICAL SPACES

§1. Basic concepts and properties

1.1. We suppose that we are given a logico-mathematical language Я of the type described in [24], §3, defined on the basis of a list of alphabets

(1) $$\mathrm{A}_0, \mathrm{A}_1, \cdots, \mathrm{A}_h.$$

We suppose further that we are given some list Σ of basis one-parameter formulas of the language Я and that a definite subordinate* generic letter has been put into correspondence with each formula of the list Σ. We denote by $Я_\Sigma$ a logico-mathematical language of the type described in [24], §8, defined on the basis of the list (1) of alphabets and the list Σ of formulas.

Suppose that A_{m} and A_{n} are alphabets from the list (1). Denote by $\mathfrak{P}$ a one-parameter formula of the language Я whose parameter, denoted by α, is a basis variable of genus t_{m}. Let $\mathfrak{J}$ be a normal one-parameter formula of the language Я whose parameter, denoted by β, is a basis variable of genus t_{n}. We shall suppose that each of the sets $\mathfrak{P}$ and $\mathfrak{J}$ contains at least one element, and that they are formulas of the list Σ. It follows from this last supposition that subordinate generic letters, which we shall denote by θ and ι respectively, have been assigned to the formulas $\mathfrak{P}$ and $\mathfrak{J}$ in the language $Я_\Sigma$. Variables of genus θ (genus ι) will be denoted by $\theta_1, \theta_2, \cdots$ (by $\iota_1, \iota_2, \cdots$).

We suppose we are given the following constructive objects:

a) An algorithm $\underset{\theta}{+}$ in the alphabet $\mathrm{A}^{\mathrm{sa}}_{\mathrm{m}}$ of the type $(\theta\theta \to \theta)$, called the *addition algorithm* for words of type θ.

b) An algorithm $\underset{\theta}{\cdot}$ in the alphabet $\mathrm{A}^{\mathrm{sa}}_{\mathrm{m}} \cup Ч_3$ of the type $(Д\theta \to \theta)$, called the *multiplication algorithm* for words of type θ by real duplexes.

c) An element $\mathfrak{O}$ of the set $\mathfrak{P}$, called the *standard zero element* of type θ.

d) A two-parameter formula $\mathfrak{E}$, whose parameters are variables of genus t_{m}, and are denoted by γ and δ.

e) A two-parameter formula $\mathfrak{Q}$, whose parameters are variables of genus t_{n}, denoted by ζ, and parameters of genus t_{m}, denoted by η.

We will make use of the following notation:

$$(P \underset{\theta}{+} Q) \rightleftharpoons \underset{\theta}{+}(P \square Q); \qquad (P \underset{\theta}{\cdot} Q) \rightleftharpoons \underset{\theta}{\cdot}(P \square Q);$$

---

* *Editor's note*. The Russian is подчиненная, which literally translates as "subordinate" (cf. the English translation of [24], §8.2). It has been rendered as "restricted" in the English translation of [26], §1.3.

$$(P \underset{\theta}{=} Q) \rightleftharpoons F^{\gamma,\,\delta}_{P,\,Q}\llcorner \mathfrak{E} \lrcorner; \qquad \mathfrak{Q}\,(P,\, Q) \rightleftharpoons F^{\zeta,\,\eta}_{P,\,Q}\llcorner \mathfrak{Q} \lrcorner,$$

where P and Q are arbitrary object terms.

DEFINITION. The nine-letter list

$$\text{(2)} \qquad \mathrm{A}_{\mathfrak{m}},\ \mathfrak{P},\ \mathfrak{E},\ \underset{\theta}{+},\ \underset{\theta}{\cdot},\ \mathfrak{O},\ \mathrm{A}_{\mathfrak{n}},\ \mathfrak{J},\ \mathfrak{Q}$$

will be called a *constructive locally convex linear topological space*, if the following conditions are satisfied:

I. Axioms of equality:

I1. $\forall \theta_1 (\theta_1 \underset{\theta}{=} \theta_1)$;

I2. $\forall \theta_1\theta_2\theta_3 (\theta_1 \underset{\theta}{=} \theta_2 \,\&\, \theta_1 \underset{\theta}{=} \theta_3 \supset \theta_2 \underset{\theta}{=} \theta_3)$.

II. Axioms of linearity:

II1. $\forall \theta_1\theta_2\theta_3\theta_4 (\theta_1 \underset{\theta}{=} \theta_2 \,\&\, \theta_3 \underset{\theta}{=} \theta_4 \supset \theta_1 \underset{\theta}{+} \theta_3 \underset{\theta}{=} \theta_2 \underset{\theta}{+} \theta_4)$;

II2. $\forall xy\theta_1\theta_2 (x \underset{\mathrm{B}}{=} y \,\&\, \theta_1 \underset{\theta}{=} \theta_2 \supset x \underset{\theta}{\cdot} \theta_1 \underset{\theta}{=} y \underset{\theta}{\cdot} \theta_2)$;

II3. $\forall \theta_1\theta_2 (\theta_1 \underset{\theta}{+} \theta_2 \underset{\theta}{=} \theta_2 \underset{\theta}{+} \theta_1)$;

II4. $\forall \theta_1\theta_2\theta_3 (\theta_1 \underset{\theta}{+} (\theta_2 \underset{\theta}{+} \theta_3) \underset{\theta}{=} (\theta_1 \underset{\theta}{+} \theta_2) \underset{\theta}{+} \theta_3)$;

II5. $\forall \theta_1 (0 \underset{\theta}{\cdot} \theta_1 \underset{\theta}{=} \mathfrak{O})$;

II6. $\forall \theta_1 (1 \underset{\theta}{\cdot} \theta_1 \underset{\theta}{=} \theta_1)$;

II7. $\forall xy\theta_1 (x \underset{\theta}{\cdot} (y \underset{\theta}{\cdot} \theta_1) \underset{\theta}{=} (x \underset{\mathrm{B}}{\cdot} y) \underset{\theta}{\cdot} \theta_1)$;

II8. $\forall xy\theta_1 ((x \underset{\mathrm{B}}{+} y) \underset{\theta}{\cdot} \theta_1 \underset{\theta}{=} x \underset{\theta}{\cdot} \theta_1 \underset{\theta}{+} y \underset{\theta}{\cdot} \theta_1)$;

II9. $\forall x\theta_1\theta_2 (x \underset{\theta}{\cdot} (\theta_1 \underset{\theta}{+} \theta_2) \underset{\theta}{=} x \underset{\theta}{\cdot} \theta_1 \underset{\theta}{+} x \underset{\theta}{\cdot} \theta_2)$.

III. Topological axioms:

III1. $\forall \iota_1\theta_1\theta_2 (\theta_1 \underset{\theta}{=} \theta_2 \,\&\, \mathfrak{Q}(\iota_1, \theta_1) \supset \mathfrak{Q}(\iota_1, \theta_2))$;

III2. $\forall \iota_1\iota_2 \exists \iota_3 \forall \theta_1 (\mathfrak{Q}(\iota_3, \theta_1) \supset \mathfrak{Q}(\iota_1, \theta_1) \,\&\, \mathfrak{Q}(\iota_2, \theta_1))$;

III3. $\forall \iota_1\theta_1\theta_2 xy (|x| + |y| = 1 \,\&\, \mathfrak{Q}(\iota_1, \theta_1) \,\&\, \mathfrak{Q}(\iota_1, \theta_2)$
$\supset \mathfrak{Q}(\iota_1, x \underset{\theta}{\cdot} \theta_1 \underset{\theta}{+} y \underset{\theta}{\cdot} \theta_2))$;[1)]

III4. $\forall \iota_1\theta_1 \exists x (x > 0 \,\&\, \mathfrak{Q}(\iota_1, x \underset{\theta}{\cdot} \theta_1))$;

III5. $\forall \iota_1 \exists \iota_2 \forall \theta_1\theta_2 (\mathfrak{Q}(\iota_2, \theta_1) \,\&\, \mathfrak{Q}(\iota_2, \theta_2) \supset \mathfrak{Q}(\iota_1, \theta_1 \underset{\theta}{+} \theta_2))$.

[1)] By $|x|$ we mean $\mathrm{M}(x)$, where M is the algorithm of computation of absolute values of real duplexes.

In what follows a constructive locally convex linear topological space will be called for short a locally convex space.

We denote by $\mathfrak{M}$ a locally convex space given by the list (2). A word of the type θ, i.e. an element of the set $\mathfrak{M}$, will be said to be a *point* of the space $\mathfrak{M}$. Suppose that θ_1 and θ_2 are points of the space $\mathfrak{M}$. If $\theta_1 \underset{\theta}{=} \theta_2$ then we will say that θ_1 and θ_2 are equal to $\mathfrak{M}$. Words of the type ι, i.e. elements of the set $\mathfrak{J}$, will be called *indices of the neighborhoods of the origin* in the space $\mathfrak{M}$. For each fixed word U of type ι the formula $\mathfrak{Q}(U,\theta_1)$ defines a set of points of $\mathfrak{M}$, which we will call a *neighborhood of the origin with the index U*. If the point X satisfies the condition $\mathfrak{Q}(U,X)$, then we will say that X belongs to a neighborhood of the origin with the index U. A sublist $(\mathrm{A}_{\mathfrak{n}}, \mathfrak{J}, \mathfrak{Q})$ of the list (2), satisfying III1–III5, will be called a *locally convex topology,* or a *fundamental system of neighborhoods of the origin,* in the space $\mathfrak{M}$.

The list

$$\mathrm{A}_{\mathfrak{m}},\ \mathfrak{P},\ \mathfrak{E},\ \underset{\theta}{+},\ \underset{\theta}{\cdot},\ \mathfrak{D},$$

satisfying the conditions I1–I2 and II1–II9, is called a *constructive linear space.*

Thus we may say that a constructive locally convex space is a constructive linear space in which a locally convex topology is defined.

Remark. In the definition of a locally convex space we have not laid any restrictions on the formulas $\mathfrak{P}$, $\mathfrak{E}$ and $\mathfrak{Q}$. However, concerning the formula $\mathfrak{J}$ we suppose that *it is a normal formula.* On the one hand, this assumption makes it possible for us to prove the constructive analogues of many basic results of the classical theory of locally convex spaces. On the other hand, this assumption does not basically restrict us either in the development of the theory itself or in the application of the theory to the investigation of functional spaces playing an important role in mathematical analysis.

1.2. We define a subtraction algorithm $\underset{\theta}{-}$ as follows:

$$\forall\theta_1\,(\underset{\theta}{-}(\theta_1) \simeq (-1) \underset{\theta}{\cdot} \theta_1);$$

$$\forall\theta_1\theta_2\,(\underset{\theta}{-}(\theta_2 \,\square\, \theta_2) \simeq \theta_1 \underset{\theta}{+} (-1) \underset{\theta}{\cdot} \theta_2).$$

We will write

$$\underset{\theta}{-} P \rightleftharpoons \underset{\theta}{-}(P); \qquad (P \underset{\theta}{-} Q) \rightleftharpoons \underset{\theta}{-}(P \,\square\, Q),$$

where P and Q are arbitrary object terms.

In what follows, in place of $\underset{\theta}{+}$, $\underset{\theta}{\cdot}$, $\underset{\theta}{-}$ and $\underset{\theta}{=}$ we shall adopt the corresponding shorter notation $+$, $\cdot$, $-$ and $=$, if doing so does not lead to confusion.

The axioms of groups I and II imply the following properties:

1.2.1. $\forall\theta_1\theta_2\,(\theta_1 = \theta_2 \supset \theta_2 = \theta_1);$

1.2.2. $\forall\theta_1(\theta_1+\mathfrak{O}=\theta_1)$;

1.2.3. $\forall\theta_1(\theta_1+(-\theta_1)=\mathfrak{O})$;

1.2.4. $\forall\theta_1\theta_2(\theta_1=\theta_2\supset-\theta_1=-\theta_2)$;

1.2.5. $\forall\theta_1\theta_2\theta_3\theta_4(\theta_1=\theta_2\,\&\,\theta_3=\theta_4\supset\theta_1-\theta_3=\theta_2-\theta_4)$;

1.2.6. $\forall\theta_1\theta_2(\theta_1+\theta_2=\mathfrak{O}\supset\theta_1=-\theta_2)$;

1.2.7. $\forall x\theta_1(-(x\cdot\theta_1)=(-x)\cdot\theta_1)$; $\forall x\theta_1(-(x\cdot\theta_1)=x\cdot(-\theta_1))$;

1.2.8. $\forall\theta_1\theta_2((\theta_1-\theta_2)+\theta_2=\theta_1)$;

1.2.9. $\forall\theta_1\theta_2\theta_3(\theta_3+\theta_2=\theta_1\supset\theta_3=\theta_1-\theta_2)$;

1.2.10. $\forall\theta_1\theta_2(\theta_1=\theta_2\equiv\theta_1-\theta_2=\mathfrak{O})$;

1.2.11. $\forall x\theta_1\theta_2(x\cdot(\theta_1-\theta_2)=x\cdot\theta_1-x\cdot\theta_2)$;

1.2.12. $\forall xy\theta_1((x-y)\cdot\theta_1=x\cdot\theta_1-y\cdot\theta_1)$;

1.2.13. $\forall x(x\cdot\mathfrak{O}=\mathfrak{O})$;

1.2.14. $\forall x\theta_1(x\cdot\theta_1=\mathfrak{O}\,\&\,x\neq0\supset\theta_1=\mathfrak{O})$;

1.2.15. $\forall x\theta_1\theta_2(x\cdot\theta_1=x\cdot\theta_2\,\&\,x\neq0\supset\theta_1=\theta_2)$;

1.2.16. $\forall x\theta_1(x\cdot\theta_1=\mathfrak{O}\,\&\,\neg(\theta_1=\mathfrak{O})\supset x=0)$;

1.2.17. $\forall xy\theta_1(x\cdot\theta_1=y\cdot\theta_1\,\&\,\neg(\theta_1=\mathfrak{O})\supset x=y)$.

1.3. From the topological axioms we may deduce the following properties:

1.3.1. $\forall\iota_1\mathfrak{Q}(\iota_1,\mathfrak{O})$;

1.3.2. $\forall\iota_1\theta_1(\mathfrak{Q}(\iota_1,\theta_1)\supset\mathfrak{Q}(\iota_1,-\theta_1))$;

1.3.3. $\forall\iota_1\theta_1 x(|x|\leqslant1\,\&\,\mathfrak{Q}(\iota_1,\theta_1)\supset\mathfrak{Q}(\iota_1,x\cdot\theta_1))$;

1.3.4. $\forall\iota_1\theta_1 xy(x\neq0\,\&\,|y|\leqslant|x|\,\&\,\mathfrak{Q}(\iota_1,x\cdot\theta_1)\supset\mathfrak{Q}(\iota_1,y\cdot\theta_1$;

1.3.5. $\forall\iota_1\theta_1\exists x(x>0\,\&\,\forall y(|y|\leqslant x\supset\mathfrak{Q}(\iota_1,y\cdot\theta_1)))$;

1.3.6. For any natural number $n\geqq1$:

$$\forall\iota_1\exists\iota_2\forall\theta_1\ldots\theta_n(\mathfrak{Q}(\iota_2,\theta_1)\,\&\ldots\&\,\mathfrak{Q}(\iota_2,\theta_n)\supset\mathfrak{Q}(\iota_1,\theta_1+\ldots+\theta_n))$$;

1.3.7. $\forall\iota_1 x\exists\iota_2\forall\theta_1(\mathfrak{Q}(\iota_2,\theta_1)\supset\mathfrak{Q}(\iota_1,x\cdot\theta_1))$;

1.3.8. $\forall\iota_1\exists\iota_2\forall\theta_1\theta_2\theta_3\theta_4(\mathfrak{Q}(\iota_2,\theta_3-\theta_1)\,\&\,\mathfrak{Q}(\iota_2,\theta_4-\theta_2)\supset\mathfrak{Q}(\iota_1,(\theta_3+\theta_4)-(\theta_1+\theta_2)))$;

1.3.9. $\forall x\theta_1\iota_1\exists n\iota_2\forall y\theta_2(\mathfrak{Q}(\iota_2,\theta_2-\theta_1)\,\&\,|y-x|<2^{-n}\supset\mathfrak{Q}(\iota_1,y\cdot\theta_2-x\cdot\theta_1))$;

1.3.10. $\forall x\iota_1\exists\iota_2\forall\theta_1\theta_2(\mathfrak{Q}(\iota_2,\theta_2-\theta_1)\supset\mathfrak{Q}(\iota_1,x\cdot\theta_2-x\cdot\theta_1))$.

Suppose that $\mathfrak{A}$ is a set of points of the space $\mathfrak{M}$. This means that $\mathfrak{A}$ is a one-parameter formula whose parameter is a variable of genus t_m, and that $\mathfrak{A}$ satisfies the condition

$$\forall \nu (\mathfrak{A} \supset F_\nu^\alpha \llcorner \mathfrak{P} \lrcorner),$$

where ν is a parameter of the formula $\mathfrak{A}$ and α a parameter of the formula $\mathfrak{P}$. We will use the symbol $(P \in \mathfrak{A})$ to denote the statement that the word P is an element of the set $\mathfrak{A}$.

We shall say that $\mathfrak{A}$ *is a convex set in the space* $\mathfrak{M}$, if

$$\forall \theta_1 \theta_2 x (0 \leqslant x \leqslant 1 \,\&\, \theta_2 \in \mathfrak{A} \,\&\, \theta_2 \in \mathfrak{A} \supset x \cdot \theta_1 + (1 - x) \cdot \theta_2 \in \mathfrak{A});$$

We shall say that $\mathfrak{A}$ is an *absolutely convex set in* $\mathfrak{M}$, if

$$\forall \theta_1 \theta_2 x y (| x | + | y | = 1 \,\&\, \theta_1 \in \mathfrak{A} \,\&\, \theta_2 \in \mathfrak{A} \supset x \cdot \theta_1 + y \cdot \theta_2 \in \mathfrak{A}).$$

Obviously any absolutely convex set is a convex set. Axiom III3 shows that any neighborhood of the origin of the fundamental system is an absolutely convex set.

Suppose that θ_1 is a point of the set $\mathfrak{M}$. The set of points θ_2 satisfying the condition $\mathfrak{Q}(\iota_1, \theta_2 - \theta_1)$, where ι_1 is some word of type ι, is called a *neighborhood of the point* θ_1 *with index* ι_1.

Property 1.3.8 means that the addition of points in the space $\mathfrak{M}$ is "uniformly continuous" relative to both arguments. Property 1.3.9 means that the multiplication of points of $\mathfrak{M}$ by real duplexes is "continuous" relative to both arguments. Property 1.3.10 means that multiplication by a fixed real duplex is "uniformly continuous" in the space $\mathfrak{M}$.

1.4. We present a number of examples of constructive locally convex spaces. In the formulas below, $c_{3,1}$ and $c_{3,2}$ denote variables for words in the alphabet $Ч_3$, and $c_{0,1}$ a variable for words in the alphabet $Ч_0$.

EXAMPLE 1. We write

$$\mathfrak{P}_д \leftrightharpoons (c_{3,1} \in д); \qquad \mathfrak{E}_д \leftrightharpoons (c_{3,1} \underset{в}{=} c_{3,2});$$

$$\mathfrak{J} \leftrightharpoons (c_{0,1} \in н); \qquad \mathfrak{Q}_д \leftrightharpoons (| c_{3,1} | \leqslant 2^{-c_{0,1}}).$$

It is easy to verify that the list

$$Ч_3,\ \mathfrak{P}_д,\ \mathfrak{E}_д,\ \underset{в}{+},\ \underset{в}{\cdot},\ 0,\ Ч_0,\ \mathfrak{J},\ \mathfrak{Q}_д$$

is a locally convex space: the space of all real duplexes with the usual topology. In what follows we shall denote this by R.

EXAMPLE 2. We construct a formula $\mathfrak{Q}_S$ as follows:

$$\mathfrak{Q}_S \leftrightharpoons \exists j \forall k l (k, l \geqslant j \supset | c_{3,1} | \cdot | S(k) - S(l) | \leqslant 2^{-c_{0,1}}),$$

where S is the Specker algorithm constructed in **[54]** (see also **[26]**, §8).

We may verify that the list

$$Ч_3,\ \mathfrak{P}_д,\ \mathfrak{E}_д,\ \underset{в}{+},\ \underset{в}{\cdot},\ 0,\ Ч_0,\ \mathfrak{J},\ \mathfrak{Q}_S$$

is a locally convex space. In order to verify that this list satisfies Axioms III1–III5, we need only to note that the sequence of rational numbers S is bounded by the number 1. The space defined by this list is denoted by $\mathscr{S}$.

1.5. We introduce some further concepts. Suppose that $\mathfrak{M}$ is a constructive locally convex space given by the list (2). Suppose that $\mathfrak{P}^*$ is the set of points of $\mathfrak{M}$ for which

a) $\mathfrak{O} \in \mathfrak{P}^*$;

b) $\forall \theta_1\theta_2 (\theta_1 \in \mathfrak{P}^* \,\&\, \theta_2 \in \mathfrak{P}^* \supset \theta_1 + \theta_2 \in \mathfrak{P}^*)$;

c) $\forall x\theta_1 (\theta_1 \in \mathfrak{P}^* \supset x \cdot \theta_1 \in \mathfrak{P}^*)$.

Under these hypotheses the list

$$A_{\mathrm{m}},\ \mathfrak{P}^*,\ \mathfrak{E},\ \underset{\theta}{+},\ \underset{\theta}{\cdot},\ \mathfrak{O}$$

defines a constructive linear space, which we shall call a *linear subspace* of the space $\mathfrak{M}$. In addition, the list

$$A_{\mathrm{m}},\ \mathfrak{P}^*,\ \mathfrak{E},\ \underset{\theta}{+},\ \underset{\theta}{\cdot},\ \mathfrak{O},\ A_{\mathrm{n}},\ \mathfrak{J},\ \mathfrak{Q}$$

defines a locally convex space, which we shall call a subspace of $\mathfrak{M}$.

Suppose that $\mathfrak{P}^*$ is a set of points of the space $\mathfrak{M}$. We will say that the set $\mathfrak{P}^*$ is *almost dense* in the space $\mathfrak{M}$, if

$$\forall \theta_1\iota_1 \exists \theta_2 (\theta_2 \in \mathfrak{P}^* \,\&\, \mathfrak{Q}(\iota_1, \theta_1 - \theta_2)).$$

We shall say that the set $\mathfrak{P}^*$ is *dense* in the space $\mathfrak{M}$, if an algorithm $\mathfrak{A}$ of the type $(\theta\iota \to \theta)$ with the property that

$$\forall \theta_1\iota_1 (\mathfrak{A}(\theta_1 \square \iota_1) \in \mathfrak{P}^* \,\&\, \mathfrak{Q}(\iota_1, \theta_1 - \mathfrak{A}(\theta_1 \square \iota_1)))$$

is potentially realizable. Evidently $\mathfrak{P}$ is a normal formula. Hence if $\mathfrak{P}^*$ is almost dense in $\mathfrak{M}$ it follows that $\mathfrak{P}^*$ is dense in $\mathfrak{M}$.

The space $\mathfrak{M}$ is said to be *separable* if an enumerable set $\mathfrak{P}^*$, dense in $\mathfrak{M}$, is realizable.

From the definition one may prove that the space $\mathfrak{M}$ is separable if and only if two algorithms are realizable: an algorithm α of the type $(\mathrm{H} \to \theta)$ and an algorithm β of the type $(\theta\iota \to \mathrm{H})$ such that

$$\forall \theta_1\iota_1 \mathfrak{Q}(\iota_1, \theta_1 - \alpha(\beta(\theta_1 \square \iota_1))).$$

For the proof of this assertion one may use the theorem on inversion of an algorithm [21].

Suppose two locally convex topologies $(A_{\mathrm{n}}, \mathfrak{J}, \mathfrak{Q})$ and $(A_{\tilde{\mathrm{n}}}, \tilde{\mathfrak{J}}, \tilde{\mathfrak{Q}})$ are defined on the constructive linear space

$$A_{\mathrm{m}},\ \mathfrak{P},\ \mathfrak{E},\ \underset{\theta}{+},\ \underset{\theta}{\cdot},\ \mathfrak{O}.$$

We will say that the topology $(A_{\mathrm{n}}, \mathfrak{J}, \mathfrak{Q})$ *majorizes* the topology $(A_{\tilde{\mathrm{n}}}, \tilde{\mathfrak{J}}, \tilde{\mathfrak{Q}})$, or that the topology $(A_{\tilde{n}}, \tilde{\mathfrak{J}}, \tilde{\mathfrak{Q}})$ is *majorized* by the topology $(A_n, \mathfrak{J}, \mathfrak{Q})$, if

$$\forall \tilde{\iota}_1 \exists \iota_1 \forall \theta_1 (\mathfrak{Q}(\iota_1, \theta_1) \supset \tilde{\mathfrak{Q}}(\tilde{\iota}_1, \theta_1)),$$

where $\tilde{\iota}_1$ is a variable for elements of the set $\tilde{\mathfrak{J}}$.

Two locally convex topologies are said to be *equivalent* if each of them majorizes the other.

Along with the space $\mathfrak{M}$ given by the list (2), we consider further the space $\tilde{\mathfrak{M}}$ given by the list

$$A_{\tilde{m}},\ \tilde{\mathfrak{P}},\ \tilde{\mathfrak{E}},\ \tilde{+},\ \tilde{\cdot},\ \tilde{\mathfrak{D}},\ A_{\tilde{n}},\ \tilde{\mathfrak{J}},\ \tilde{\mathfrak{Q}}.$$

Suppose that all the notation with superscript $\sim$ for $\tilde{\mathfrak{M}}$ have values analogous to the values of these notations for $\mathfrak{M}$ without that sign. We denote by η the index of the alphabet $A_m \cup A_{\tilde{m}}$ in the list of alphabets of the language $\mathfrak{A}_\Sigma$, i.e. $A_\eta = A_m \cup A_{\tilde{m}}$.

Suppose that λ is an algorithm of the type $(\theta \underset{\rightarrow}{\cdot} \tilde{\theta})$ in the alphabet A_η^{sa}. We will say that λ is an *operator of the type* $(\theta \underset{\rightarrow}{\cdot} \tilde{\theta})$, if

$$\forall \theta_1 \theta_2 (!\lambda(\theta_1) \,\&\, \theta_1 = \theta_2 \supset !\lambda(\theta_2) \,\&\, \lambda(\theta_1) \cong \lambda(\theta_2)).$$

We shall say that λ is an *operator of the type* $(\theta \rightarrow \tilde{\theta})$, if it is an operator of the type $(\theta \underset{\rightarrow}{\cdot} \tilde{\theta})$, and $\forall \theta_1 !\lambda(\theta_1)$. Every operator of the type $(\theta \rightarrow \tilde{\theta})$ is said to be a *mapping of the space* $\mathfrak{M}$ *into the space* $\tilde{\mathfrak{M}}$.

The operator λ of type $(\theta \rightarrow \tilde{\theta})$ is said to be a *mapping of the space* $\mathfrak{M}$ *onto the space* $\tilde{\mathfrak{M}}$ if

$$\forall \tilde{\theta}_1 \exists \theta_1 (\lambda(\theta_1) \cong \tilde{\theta}_1).$$

Suppose that λ is an operator of the type $(\theta \rightarrow \tilde{\theta})$ and $\tilde{\lambda}$ an operator of the type $(\tilde{\theta} \underset{\rightarrow}{\cdot} \theta)$. We will say that $\tilde{\lambda}$ *is an operator inverse to the operator* λ, and write $\tilde{\lambda}\,\underline{\mathrm{inv}}\,\lambda$, if

$$\forall \theta_1 (!\tilde{\lambda}(\lambda(\theta_1)) \,\&\, \lambda(\tilde{\lambda}(\lambda(\theta_1))) \cong \lambda(\theta_1)).$$

An operator λ of the type $(\theta \rightarrow \tilde{\theta})$ is said to be *one-to-one* if

$$\forall \theta_1 \theta_2 (\lambda(\theta_1) \cong \lambda(\theta_2) \supset \theta_1 = \theta_2);$$

It is said to be *linear* if

$$\forall \theta_1 \theta_2 (\lambda(\theta_1 + \theta_2) \cong \lambda(\theta_1) \tilde{+} \lambda(\theta_2));$$

$$\forall x \theta_1 (\lambda(x \cdot \theta_1) \cong x \tilde{\cdot} \lambda(\theta_1)).$$

A linear operator λ of the type $(\theta \rightarrow \tilde{\theta})$ is said to be *continuous* if

$$\forall \tilde{\iota}_1 \exists \iota_1 \forall \theta_1 (\mathfrak{Q}(\iota_1, \theta_1) \supset \tilde{\mathfrak{Q}}(\tilde{\iota}_1, \lambda(\theta_1))).$$

We will say that the operator λ is an *isomorphism* of the space $\mathfrak{M}$ onto the space $\tilde{\mathfrak{M}}$ if the following conditions are satisfied:

.1) λ is a one-to-one, linear and continuous mapping of $\mathfrak{M}$ onto $\tilde{\mathfrak{M}}$.

2) There exists an inverse operator $\tilde{\lambda}$ (i.e. $\tilde{\lambda}\,\underline{\text{inv}}\,\lambda$) which is a linear continuous mapping of the space $\widetilde{\mathfrak{M}}$ onto the space $\mathfrak{M}$.

We may verify that when these conditions are satisfied $\tilde{\lambda}$ is a one-to-one mapping of $\widetilde{\mathfrak{M}}$ onto $\mathfrak{M}$ and $(\tilde{\lambda}\,\underline{\text{inv}}\,\lambda)$, so that $\tilde{\lambda}$ is an isomorphism of $\widetilde{\mathfrak{M}}$ onto $\mathfrak{M}$.

We shall say that the space $\widetilde{\mathfrak{M}}$ is *isomorphic* to the space $\mathfrak{M}$ if an isomorphism of $\mathfrak{M}$ onto $\widetilde{\mathfrak{M}}$ is potentially realizable. Obviously, if $\widetilde{\mathfrak{M}}$ is isomorphic to $\mathfrak{M}$, then $\mathfrak{M}$ is isomorphic to $\widetilde{\mathfrak{M}}$, so that we also will say that the spaces $\mathfrak{M}$ and $\widetilde{\mathfrak{M}}$ are isomorphic to one another.

§2. Separation. Open and closed sets

2.1. Suppose that $\mathfrak{M}$ is a constructive locally convex space, given by the list

$$\mathrm{A}_{\mathfrak{m}},\ \mathfrak{P},\ \mathfrak{E},\ \underset{\theta}{+},\ \underset{\theta}{\cdot},\ \mathfrak{D},\ \mathrm{A}_{\mathfrak{n}},\ \mathfrak{I},\ \mathfrak{Q}. \tag{1}$$

We will say that the space $\mathfrak{M}$ is *T-separated* if

$$\forall\theta_1 (\neg(\theta_1 = \mathfrak{D}) \supset \exists\iota_1 \neg\mathfrak{Q}(\iota_1, \theta_1)),$$

and that it is *T′-separated* if

$$\forall\theta_1 (\forall\iota_1\mathfrak{Q}(\iota_1, \theta_1) \supset \theta_1 = \mathfrak{D}).$$

It is easy to see that if $\mathfrak{E}$ is a normal formula, T-separation implies T'-separation.

If the space $\mathfrak{M}$ is T-separated (T'-separated), then we will also say that its topology is T-separated (T'-separated).

It is possible to prove that if two equivalent locally convex topologies are defined on a constructive linear space, and if one of them is T-separated (T'-separated), then the other is also T-separated (T'-separated). Moreover, if the spaces $\mathfrak{M}$ and $\widetilde{\mathfrak{M}}$ are isomorphic to one another and if one of them is T-separated (T'-separated), then the other is also T-separated (T'-separated).

In classical mathematics, i.e. with the classical interpretation of terms and inferences, the analogues of the concepts of T-separation and T'-separation are obviously equivalent to one another. However, in the constructive theory the concepts of T-separation and T'-separation are not equivalent to one another. In fact, we have the following.

2.1.1. *It is possible to determine a constructive locally convex space which is T'-separated but not T-separated.*

Proof. Consider the space $\mathscr{S}$ constructed in Example 2 of §1.4. We shall show that $\mathscr{S}$ is T'-separated, i.e. that

$$\forall x (\forall i \mathfrak{Q}_S(i, x) \supset x = 0). \tag{2}$$

Suppose the duplex x is such that $\forall i \mathfrak{Q}_S(i, x)$. Then

$$\forall i \exists j \forall kl (k, l \geqslant j \supset |x| \cdot |S(k) - S(l)| < 2^{-i}). \tag{3}$$

Suppose that $x \neq 0$. Then we can realize a natural number n such that

$|x| > 2^{-n}$ (see [**27**], Lemma {18}). Hence $1 : |x| < 2^n$. Therefore we find that

$$\forall i \exists j \forall kl\,(k,\, l \geqslant j \supset |S(k) - S(l)| < 2^{n-i}).$$

This formula is equivalent to the following formula:

$$\forall i \exists j \mathrm{A} kl\,(k,\, l \geqslant j \supset |S(k) - S(l)| < 2^{-i}).$$

In view of the property of the sequence S (see [**26**], 8.3.1), this is impossible. Therefore $\neg\,(x \neq 0)$, so that $x = 0$. Formula (2) is proved.

Now we shall prove that $\mathcal{S}$ is not T-separated. As was proved in [**26**], 8.3.2, we have

$$\forall i \neg\neg \exists j \forall kl\,(k,\, l \geqslant j \supset |S(k) - S(l)| < 2^{-i}).$$

Hence it follows that $\neg\,\exists i \neg\,\mathfrak{Q}_S(i, 1)$. Thus it is possible to indicate a point $1 \neq 0$ such that $\exists i \neg\,\mathfrak{Q}_S(i, 1)$ is not true. The space $\mathcal{S}$ is not T-separated.

Theorem 2.1.1 shows that the concept of separation of a locally convex space splits into essentially distinct concepts on being carried over to constructive mathematics.

2.2. Suppose that $\mathfrak{A}$ is a set of points of the space $\mathfrak{M}$. We will say that the set $\mathfrak{A}$ is $\mathrm{З}$*-closed* in the space $\mathfrak{M}$ if

$$\forall \theta_1\,(\neg\,(\theta_1 \in \mathfrak{A}) \supset \exists \iota_1 \forall \theta_2\,(\mathfrak{Q}\,(\iota_1,\, \theta_2 - \theta_1) \supset \neg\,(\theta_2 \in \mathfrak{A}))).$$

We will say that it is $\mathrm{З}'$*-closed* in $\mathfrak{M}$ if

$$\forall \theta_1\,(\forall \iota_1 \exists \theta_2\,(\theta_2 \in \mathfrak{A}\,\&\,\mathfrak{Q}\,(\iota_1,\, \theta_2 - \theta_1)) \supset \theta_1 \in \mathfrak{A}).$$

The set $\mathfrak{A}$ is said to be *O-open* in $\mathfrak{M}$ if

$$\forall \theta_1\,(\theta_1 \in \mathfrak{A} \supset \exists \iota_1 \forall \theta_2\,(\mathfrak{Q}\,(\iota_1,\, \theta_2 - \theta_1) \supset \theta_2 \in \mathfrak{A})).$$

It is said to be *O′-open* in $\mathfrak{M}$ if

$$\forall \theta_1\,(\forall \iota \exists \theta_2\,(\neg\,(\theta_2 \in \mathfrak{A})\,\&\,\mathfrak{Q}\,(\iota_1,\, \theta_2 - \theta_1)) \supset \neg\,(\theta_1 \in \mathfrak{A})).$$

In [**36**] the concepts of З-closed, З′-closed, *O*-open and *O*′-open were defined for constructive topological spaces as well.

Suppose that $\mathfrak{A}$ is a set of points of the space $\mathfrak{M}$. The set defined by the one-parameter formula

$$\forall \iota_1 \exists \theta_2\,(\theta_2 \in \mathfrak{A}\,\&\,\mathfrak{Q}(\iota_1,\, \theta_2 - \theta_1))^{2)}$$

is said to be the З′-closure of the set $\mathfrak{A}$ in $\mathfrak{M}$. The З′-closure of any set $\mathfrak{A}$ is a З′-closed set in $\mathfrak{M}$. Every З′-closed set in $\mathfrak{M}$ coincides with its З′-closure.

Suppose that $\mathfrak{A}$ is a set of points of $\mathfrak{M}$. The set defined by the formula

$$(\alpha \in \mathfrak{P})\,\&\,\neg\,(\alpha \in \mathfrak{A}),$$

[2)] In this formula θ_1 plays the role of a parameter.

where α is a variable of genus $t_{\mathfrak{m}}$, is said to be the *complement* of the set $\mathfrak{A}$ in the space $\mathfrak{M}$.

The following assertions are obvious.

2.2.1. *Every O-open set in $\mathfrak{M}$ is O′-open in $\mathfrak{M}$.*

2.2.2. *If the set $\mathfrak{A}$ is З-closed in $\mathfrak{M}$, then its complement is O-open in $\mathfrak{M}$, and conversely.*

2.2.3. *If the set $\mathfrak{A}$ is O′-open in $\mathfrak{M}$, then its complement is З′-closed in $\mathfrak{M}$, and conversely.*

Now we shall prove the following theorems.

2.2.6. *In the space R one can realize a set which is З-closed but not З′-closed.*

Here R is the space of all real duplexes with the usual topology.

Proof. Consider the following set $\mathfrak{A}$:

$$(x \in \mathfrak{A}) \rightleftharpoons (x \leqslant 0 \vee x \geqslant 0).$$

The set $\mathfrak{A}$ is З-closed in R because $\forall x \neg\neg (x \in \mathfrak{A})$ (see [**26**], 4.5.1). Suppose that $\mathfrak{A}$ is З′-closed in R. Then

(4) $$\forall x (\forall i \exists y ((y \leqslant 0 \vee y \geqslant 0) \,\&\, |y - x| < 2^{-i}) \supset (x \leqslant 0 \vee x \geqslant 0)).$$

We shall show that for any duplex x the following holds:

(5) $$\forall i \exists y ((y \leqslant 0 \vee y \geqslant 0) \,\&\, |y - x| < 2^{-i}).$$

Suppose i is given. Put $y \leftrightharpoons \underline{x}(\overline{x}(i))$. y is a rational number, so that $(y \leqq 0 \vee y \geqq 0)$. Moreover, we have $|y - x| < 2^{-i}$ (see [**26**], 4.9.3). Thus, formula (5) is proved. Hence it follows from (4) that $(x \leqq 0 \vee x \geqq 0)$. Thus we get

$$\forall x (x \leqslant 0 \vee x \geqslant 0).$$

But this is impossible ([**28**], §2, Lemma 2). Therefore the set $\mathfrak{A}$ cannot be З′-closed, as was required to be proved.

2.2.7. *In the space $\mathscr{S}$ a set is realizable which is З′-closed but not З-closed.*

Here $\mathscr{S}$ is the space defined in Example 2, §1.4.

Proof. Consider a set $\mathfrak{A}$ defined as follows:

$$(x \in \mathfrak{A}) \rightleftharpoons (|x| \leqslant 1).$$

We shall show that the set $\mathfrak{A}$ is З′-closed in $\mathscr{S}$, i.e. that

(6) $$\forall x (\forall i \exists y (|y| \leqslant 1 \,\&\, \mathfrak{Q}_S (i, y - x)) \supset |x| \leqslant 1).$$

Suppose given a duplex x such that

(7) $$\forall i \exists y (|y| \leqslant 1 \,\&\, \mathfrak{Q}_S (i, y - x)).$$

Suppose that $|x| > 1$. Put $x_0 \leftrightharpoons |x| - 1$. Then $x_0 > 0$ and for any duplex y with $|y| \leqq 1$ we have

$$|x_0| \leqslant |x| - |y| \leqslant |y - x|.$$

Suppose given an integer i. In view of (7) we may find a duplex y_0 such that

$$|y_0| \leqslant 1 \,\&\, \mathfrak{Q}_S(i, y_0 - x).$$

Obviously $|x_0| \leqq |y_0 - x|$. Accordingly $\mathfrak{Q}_S(i, y_0 - x)$ implies that $\mathfrak{Q}_S(i, x_0)$ (see 1.3.4). So we find that

$$\forall i \mathfrak{Q}_S(i, x_0).$$

Since the space $\mathscr{S}$ is T'-separated, it follows that $x_0 = 0$. We have obtained a contradiction. Therefore $\neg(|x| > 1)$, i.e. $|x| \leqq 1$. Formula (6) is proved.

Now we shall show that the set $\mathfrak{A}$ is not $\mathfrak{Z}$-closed in $\mathscr{S}$. To this end we suppose that $\mathfrak{A}$ is $\mathfrak{Z}$-closed in $\mathscr{S}$. Then

$$\forall x (|x| > 1 \supset \exists i \forall y (\mathfrak{Q}_S(i, y - x) \supset |y| > 1)).$$

For $x = 2$ we can find a natural number i such that

$$\forall y (\mathfrak{Q}_S(i, y - 2) \supset |y| > 1). \tag{8}$$

Now we choose $y = 1$. Then from $\neg(|1| > 1)$, in view of (8), we find that $\neg\mathfrak{Q}_S(i, 1 - 2)$, so that $\neg\mathfrak{Q}_S(i, 1)$. Thus we have found that $\exists i \neg\mathfrak{Q}_S(i, 1)$. But this is impossible. Therefore the set $\mathfrak{A}$ cannot be $\mathfrak{Z}$-closed in $\mathscr{S}$. The theorem is proved.

2.2.8. *In the space $\mathscr{S}$ a set is realizable which is O'-open but not O-open.*

Proof. Suppose that $\mathfrak{B}$ is the set of duplexes such that $|x| > 1$. Then $\mathfrak{B}$ is the complement of the set $\mathfrak{A}$ constructed in the proof of 2.2.7, and, conversely, $\mathfrak{A}$ is the complement of $\mathfrak{B}$. From the fact that $\mathfrak{A}$ is $\mathfrak{Z}'$-closed in $\mathscr{S}$ it follows that $\mathfrak{B}$ is O'-open in $\mathscr{S}$. From the fact that $\mathfrak{A}$ is not $\mathfrak{Z}$-closed in $\mathscr{S}$ it follows that $\mathfrak{B}$ is not O-open in $\mathscr{S}$. The theorem is proved.

Remark. It is possible to construct in the space R an example of a set which is $\mathfrak{Z}'$-closed and not $\mathfrak{Z}$-closed, and an example of a set which is O'-open but not O-open. Such examples are given in [**36**].

2.3. Suppose that $\mathfrak{M}$ is a constructive locally convex space given by the list (1). Suppose further that $\mathfrak{P}^*$ is the set of points of the space $\mathfrak{M}$ satisfying the following conditions:

a) $\mathfrak{O} \in \mathfrak{P}^*$;

b) $\forall \theta_1\theta_2 (\theta_1 \in \mathfrak{P}^* \,\&\, \theta_2 \in \mathfrak{P}^* \supset \theta_1 + \theta_2 \in \mathfrak{P}^*)$;

c) $\forall x\theta_1 (\theta_1 \in \mathfrak{P}^* \supset x \cdot \theta_1 \in \mathfrak{P}^*)$.

Then the list

$$\mathrm{A}_{\mathfrak{m}}, \mathfrak{P}^*, \mathfrak{E}, \underset{\theta}{+}, \underset{\theta}{\cdot}, \mathfrak{O}, \mathrm{A}_{\mathfrak{n}}, \mathfrak{J}, \mathfrak{Q} \tag{9}$$

defines a subspace of $\mathfrak{M}$.

We suppose that, besides conditions a)–c), the set $\mathfrak{P}^*$ satisfies the following further condition:

d) $\forall\theta_1\theta_2\,(\theta_1 = \theta_2 \,\&\, \theta_1 \in \mathfrak{P}^* \supset \theta_2 \in \mathfrak{P}^*)$.

Then we will say that the space (9) is a *regular* subspace of $\mathfrak{M}$.

In this subsection, when we use the word subspace, we shall understand regular subspace.

We denote the subspace (9) by $\mathfrak{M}^*$.

It is possible to define a two-parameter formula $\mathfrak{E}_\Phi$ such that for any words θ_1 and θ_2 of type θ

$$\mathfrak{E}_\Phi(\theta_1, \theta_2) \equiv (\theta_1 - \theta_2 \in \mathfrak{P}^*).$$

In place of the notation $\mathfrak{E}_\Phi(\theta_1,\theta_2)$ we shall use the short notation $\theta_1 \underset{\Phi}{=} \theta_2$. Using properties a) and d), we prove easily that

$$\forall\theta_1\theta_2\,(\theta_1 = \theta_2 \supset \theta_1 \underset{\Phi}{=} \theta_2).$$

Now we define a formula $\mathfrak{Q}_\Phi$ such that for any words ι_1 of type ι and θ_1 of type θ,

$$\mathfrak{Q}_\Phi(\iota_1, \theta_1) \equiv \exists\theta_2\,(\theta_2 \underset{\Phi}{=} \theta_1 \,\&\, \mathfrak{Q}(\iota_1, \theta_2)).$$

Obviously

$$\forall\iota_1\theta_1\,(\mathfrak{Q}(\iota_1, \theta_1) \supset \mathfrak{Q}_\Phi(\iota_1, \theta_1)).$$

Now we consider the following list:

$$(10) \qquad A_m,\ \mathfrak{P},\ \mathfrak{E}_\Phi,\ \underset{\theta}{+},\ \underset{\theta}{\cdot},\ \mathfrak{D},\ A_n,\ \mathfrak{J},\ \mathfrak{Q}_\Phi.$$

It is easy to verify that this list is a constructive locally convex space. We will call this space the *quotient space of the space* $\mathfrak{M}$ *relative to the subspace* $\mathfrak{M}^*$, and denote it by $\mathfrak{M}/\mathfrak{M}^*$.

We shall say that $\mathfrak{M}^*$ is a $\mathfrak{Z}$-closed ($\mathfrak{Z}'$-closed) subspace of the space $\mathfrak{M}$ if $\mathfrak{M}^*$ is a $\mathfrak{Z}$-closed ($\mathfrak{Z}'$-closed) set in $\mathfrak{M}$.

The following assertions are valid.

2.3.1. *If $\mathfrak{M}^*$ is a $\mathfrak{Z}$-closed subspace of $\mathfrak{M}$, then the quotient space $\mathfrak{M}/\mathfrak{M}^*$ is T-separated.*

2.3.2. *If $\mathfrak{M}^*$ is a $\mathfrak{Z}'$-closed subspace of $\mathfrak{M}$, then the quotient space $\mathfrak{M}/\mathfrak{M}^*$ is T′-separated.*

Now we shall prove the following theorem.

2.3.3. *It is possible to construct a locally convex space $\mathfrak{M}$ and a $\mathfrak{Z}'$-closed subspace $\mathfrak{M}^*$ of it such that the quotient space $\mathfrak{M}/\mathfrak{M}^*$ is not T-separated.*

Proof. Suppose that τ is a letter not entering into the alphabet $Ч_3$.

Denote by $\mathfrak{m}$ the index of the alphabet $Ч_3 \cup \{\tau\}$, i.e. $A_{\mathfrak{m}} \rightleftharpoons Ч_3 \cup \{\tau\}$. Suppose that $\mathfrak{P}$ is a formula defining the set of all possible words in $A_{\mathfrak{m}}$ of the form $\alpha_1 \tau \alpha_2$, where α_1 and α_2 are real duplexes. We denote by θ a subordinate generic letter assigned to the formula $\mathfrak{P}$. We construct a formula $\mathfrak{E}$ such that for any words θ_1 and θ_2 of type θ, if $\theta_1 \overline{\underline{\circ}}\, \alpha_1 \tau \alpha_2$ and $\theta_2 \overline{\underline{\circ}}\, \beta_1 \tau \beta_2$, then

$$\mathfrak{E}(\theta_1, \theta_2) \equiv (\alpha_1 = \beta_1) \,\&\, (\alpha_2 = \beta_2).$$

It is possible to construct algorithms $\underset{\theta}{+}$ and $\underset{\theta}{\cdot}$ such that for any words θ_1 and θ_2 of type θ and any real duplex x, if $\theta_1 \overline{\underline{\circ}}\, \alpha_1 \tau \alpha_2$ and $\theta_2 \overline{\underline{\circ}}\, \beta_1 \tau \beta_2$, then

$$\underset{\theta}{+}(\theta_1 \square \theta_2) \simeq (\alpha_1 + \beta_1)\, \tau\, (\alpha_2 + \beta_2);$$

$$\underset{\theta}{\cdot}(x \square \theta_1) \simeq (x \cdot \alpha_1)\, \tau\, (x \cdot \alpha_2).$$

Write $\mathfrak{O} \leftrightharpoons 0 \tau 0$. We denote by $\mathfrak{Z}$ the set of all natural numbers. Finally, we construct a formula $\mathfrak{Q}$ such that for any natural number i and any word θ_1 of type θ, if $\theta_1 \overline{\underline{\circ}}\, \alpha_1 \tau \alpha_2$, then

$$\mathfrak{Q}(i, \theta_1) \equiv (|\alpha_1| \leqslant 2^{-i}) \,\&\, (\mathfrak{Q}_S(i, \alpha_2),$$

where $\mathfrak{Q}_S$ is the formula defined in Example 2, §1.4.

It is not hard to verify that the list

$$A_{\mathfrak{m}},\ \mathfrak{P},\ \mathfrak{E},\ \underset{\theta}{+},\ \underset{\theta}{\cdot},\ \mathfrak{O},\ Ч_0,\ \mathfrak{Z},\ \mathfrak{Q}$$

defines a locally convex space. This space will be denoted by $\mathfrak{M}$. We construct a set $\mathfrak{P}^*$ as follows:

$$(\theta_1 \in \mathfrak{P}^*) \rightleftharpoons ([\theta_1]_2^\tau = 0).$$

Here $[\theta_1]_i^\tau$, $i = 1, 2$, denotes the ith τ-term of the word θ_1. Obviously $\mathfrak{P}^*$ satisfies conditions a) – d). Therefore the list

$$A_{\mathfrak{m}},\ \mathfrak{P}^*,\ \mathfrak{E},\ \underset{\theta}{+},\ \underset{\theta}{\cdot},\ \mathfrak{O},\ Ч_0,\ \mathfrak{Z},\ \mathfrak{Q}$$

forms a subspace of the space $\mathfrak{M}$, which we denote by $\mathfrak{M}^*$.

We shall show that $\mathfrak{M}^*$ is a $\mathrm{З}'$-closed subspace of $\mathfrak{M}$. To this end it suffices to show that

(11) $$\forall \theta_1 (\forall i \exists \theta_2 ((\theta_2 \in \mathfrak{P}^*) \,\&\, \mathfrak{Q}(i, \theta_2 - \theta_1)) \supset (\theta_1 \in \mathfrak{P}^*)).$$

Suppose that θ_1 is a word of type θ such that

(12) $$\forall i \exists \theta_2 (\theta_2 \in \mathfrak{P}^* \,\&\, \mathfrak{Q}(i, \theta_2 - \theta_1)).$$

Suppose that i is any natural number. In view of (12) we can realize a word θ_2 of type θ such that $[\theta_2]_2^\tau = 0$ and $\mathfrak{Q}(i, \theta_2 - \theta_2)$. Hence it follows that $\mathfrak{Q}_S(i, [\theta_1]_2^\tau)$. Thus we find that

$$\forall i \mathfrak{Q}_S(i, [\theta_1]_2^\tau).$$

Therefore $[\theta_2]_2^\tau = 0$, i.e. $\theta_1 \in \mathfrak{P}^*$. Formula (11) is proved.

Now we show that the quotient space $\mathfrak{M}/\mathfrak{M}^*$ is not T-separated. To this end we suppose that it is T-separated, i.e. that

$$\forall\theta_1(\neg(\theta_1\in\mathfrak{P}^*)\supset\exists i\neg\exists\theta_2((\theta_2-\theta_1\in\mathfrak{P}^*)\,\&\,\mathfrak{Q}(i,\theta_2))).$$

Suppose that $X\leftrightharpoons 0_T 1$. Then X is a word of type θ and $\neg(X\in\mathfrak{P}^*)$. Accordingly it is possible to find a natural number i such that

$$\neg\exists\theta_2((\theta_2-X\in\mathfrak{P}^*)\,\&\,\mathfrak{Q}(i,\theta_2)),$$

i.e. such that

(13) $$\forall\theta_2\neg((\theta_2-X\in\mathfrak{P}^*)\,\&\,\mathfrak{Q}(i,\theta_2)).$$

Obviously $X-X\in\mathfrak{P}^*$. Therefore from (13) we get $\mathfrak{Q}(i,X)$. This means that

$$\neg(|0|\leqslant 2^{-i}\,\&\,\mathfrak{Q}_S(i,1)).$$

Hence it follows that $\neg\mathfrak{Q}_S(i,1)$. Thus we have proved that $\exists i\neg\mathfrak{Q}_S(i,1)$. This is impossible. Therefore the quotient space $\mathfrak{M}/\mathfrak{M}^*$ cannot be T-separated. The theorem is proved.

§3. Completeness of constructive locally convex spaces

3.1. Suppose that A_q is some basic alphabet of the language $\mathfrak{R}_\Sigma$ and that $\mathscr{A}$ is some one-parameter formula of the list Σ whose parameter is a variable of genus t_q. In the language $\mathfrak{R}_\Sigma$ we assign to the formula $\mathscr{A}$ a subordinate generic letter which we denote by ν. We suppose that the set $\mathscr{A}$ contains at least one element. Suppose further that $\mathfrak{X}$ is a two-parameter formula, whose parameters are variables ν_1 and ν_2 of genus ν. We write

$$(P\underset{\nu}{\prec}Q)\rightleftharpoons F^{\nu_1,\,\nu_2}_{P,\,Q}\llcorner\mathfrak{X}\lrcorner,$$

where P and Q are arbitrary object terms. If P and Q are words of type ν and $(P\underset{\nu}{\prec}Q)$, then we will say that P is majorized by the word Q, or that Q majorizes P. The list $(\mathscr{A},\mathfrak{X})$ is called a *set with order relation*. Words of the type ν are regarded as elements of this set.

A set $(\mathscr{A},\mathfrak{X})$ with order relation will be called a *directed set* if the following conditions are satisfied:

(1) $$\forall\nu_1(\nu_1\prec\nu_2);$$

(2) $$\forall\nu_1\nu_2\nu_3(\nu_1\prec\nu_2\,\&\,\nu_2\prec\nu_3\supset\nu_1\prec\nu_3);$$

(3) $$\forall\nu_1\nu_2\exists\nu_3(\nu_1\prec\nu_3\,\&\,\nu_2\prec\nu_3).$$

Here and in what follows the letter ν below the sign $\prec$ will be dropped in those cases when it does not give rise to misunderstandings.

The set of all natural numbers with the usual order relation $\leqq$ is a directed set, which we shall denote by N.

Suppose that $\mathfrak{M}$ is a locally convex space given by the list

(4) $$\mathrm{A}_{\mathfrak{m}},\ \mathfrak{P},\ \mathfrak{C},\ \underset{\theta}{+},\ \underset{\theta}{\cdot},\ \mathfrak{D},\ \mathrm{A}_{\mathfrak{n}},\ \mathfrak{J},\ \mathfrak{Q}.$$

We introduce the formula

$$\mathfrak{X}^{\iota} \rightleftharpoons V\theta_1 (\mathfrak{Q}(\iota_2, \theta_1) \supset \mathfrak{Q}(\iota_1, \theta_1)).$$

It is easy to verify that the list $(\mathfrak{J}, \mathfrak{X}^{\iota})$ is a directed set. We shall call it *the directed set of indices of neighborhoods of the origin in the space* $\mathfrak{M}$.

Suppose that $(\mathscr{A}, \mathfrak{X})$ is a directed set. We denote by p and s the indices of the alphabets $A_{\text{m}} \cup A_q$ and $A_{\text{n}} \cup A_q$, i.e. $A_p = A_{\text{m}} \cup A_q$ and $A_s = A_{\text{n}} \cup A_q$.

Every algorithm $\lambda_{p,1}$ of the type $(\nu \rightarrow \theta)$ in the alphabet A_p^{sa} will be called an $(\mathscr{A}, \mathfrak{X})$-*sequence* of points of the space $\mathfrak{M}$.

We will say that the point X of the space $\mathfrak{M}$ is a limit in $\mathfrak{M}$ of the $(\mathscr{A}, \mathfrak{X})$-sequence $\lambda_{p,1}$, and write

$$(X \,\underline{\lim}_{\mathfrak{M}} \lambda_{p,1}),$$

if

(5) $$\forall \iota_1 \exists \nu_1 \forall \nu_2 (\nu_1 \prec \nu_2 \supset \mathfrak{Q}(\iota_1, \lambda_{p,1}(\nu_2) - X)).$$

Since by hypothesis $\mathfrak{J}$ is a normal formula, by the principles of constructive interpretation of mathematical inferences, condition (5) is satisfied if and only if we can realize an algorithm $\lambda_{s,1}$ of the type $(\iota \rightarrow \nu)$ such that

(6) $$\forall \iota_1 \nu_2 (\lambda_{s,1}(\iota_1) \prec \nu_2 \supset \mathfrak{Q}(\iota_1, \lambda_{p,1}(\nu_2) - X)).$$

We will employ the following notation:

$$(U_1, \ldots, U_i \prec V_1, \ldots, V_j) \rightleftharpoons (U_1 \prec V_1) \& \ldots \& (U_1 \prec V_j) \& \ldots$$
$$\ldots \& (U_i \prec V_1) \& \ldots \& (U_i \prec V_j),$$

where $U_1, \cdots, U_i, V_1, \cdots, V_j$ are arbitrary object terms.

An $(\mathscr{A}, \mathfrak{X})$-sequence $\lambda_{p,1}$ of points of $\mathfrak{M}$ will be called *fundamental*, and also will be said to *converge in itself in the space* $\mathfrak{M}$, if

(7) $$\forall \iota_1 \exists \nu_1 \forall \nu_2 \nu_3 (\nu_1 \prec \nu_2, \nu_3 \supset \mathfrak{Q}(\iota_1, \lambda_{p,1}(\nu_2) - \lambda_{p,1}(\nu_3))).$$

In that case we will write

$$(\lambda_{p,1} \in \underline{(\mathscr{A}, \mathfrak{X}) - \text{fund}}_{\mathfrak{M}}),$$

Condition (7) is satisfied if and only if we can realize an algorithm $\lambda_{s,2}$ of the type $(\iota \rightarrow \nu)$ such that

(8) $$\forall \iota_1 \nu_2 \nu_3 (\lambda_{s,2}(\iota_1) \prec \nu_2, \nu_3 \supset \mathfrak{Q}(\iota_1, \lambda_{p,1}(\nu_2) - \lambda_{p,1}(\nu_3))).$$

Concerning the algorithm $\lambda_{s,2}$, we will say that it is a *regulator of convergence into itself* of the $(\mathscr{A}, \mathfrak{X})$-sequence $\lambda_{p,1}$, and write

$$(\lambda_{s,2} \,\underline{\text{reg conv}}_{\mathfrak{M}} \lambda_{p,1}),$$

if $\lambda_{s,2}$ is an algorithm of the type $(\iota \rightarrow \nu)$ and satisfies condition (8).

The following assertions are valid.

3.1.1. *Suppose that* $\mathfrak{M}$ *is a* T'*-separated locally convex space. Then, for any*

$(\mathscr{A}, \mathfrak{X})$-sequence $\lambda_{p,1}$ of points of the space $\mathfrak{M}$ and any points X and Y,

$$(X \underline{\lim}_{\mathfrak{M}} \lambda_{p,1}) \& (Y \underline{\lim}_{\mathfrak{M}} \lambda_{p,1}) \supset (X = Y).$$

3.1.2. *If $\lambda_{p,1}$ is an $(\mathscr{A}, \mathfrak{X})$-sequence of points of the space $\mathfrak{M}$, then*

$$\exists \theta_1 (\theta_1 \underline{\lim}_{\mathfrak{M}} \lambda_{p,1}) \supset (\lambda_{p,1} \in \underline{(\mathcal{A}, \mathfrak{X}) - \mathbf{fund}}_{\mathfrak{M}}).$$

The space $\mathfrak{M}$ is said to be *$(\mathscr{A}, \mathfrak{X})$-almost complete,* if

$$\forall \lambda_{p,1} (\lambda_{p,1} \in \underline{(\mathcal{A}, \mathfrak{X}) - \mathbf{fund}}_{\mathfrak{M}}) \supset \exists \theta_1 (\theta_1 \underline{\lim}_{\mathfrak{M}} \lambda_{p,1}).$$

Turning to the definition of the concept of $(\mathscr{A}, \mathfrak{X})$-completeness, we first introduce certain concepts and notation.

Suppose that $\odot$ is a letter not entering into the alphabet A_{m}. We adjoin the alphabet $\mathrm{A}_{\mathrm{m}} \cup \mathrm{Ч}_0 \cup \{\odot\}$ to the list of basic alphabets of the language $Я_\Sigma$ and denote this alphabet by ψ, i.e. $\mathrm{A}_\psi = \mathrm{A}_{\mathrm{m}} \cup \mathrm{Ч}_0 \cup \{\odot\}$.

Suppose that $t_{\psi,1}$ is a word in A_ψ. We will say that $t_{\psi,1}$ is a *meaningful word in* A_ψ, and write

$$(t_{\psi,1} \in \underline{\mathbf{mean}_\psi}),$$

where $t_{\psi,1}$ is a word in A_{m} or a word of the type $P \odot Q$, where P and Q are words in $\mathrm{Ч}_0$. We will say that $t_{\psi,1}$ is a *one-sided meaningful word* in A_ψ, and write

$$(t_{\psi,1} \in \underline{\mathbf{one\ mean}_\psi}),$$

if $t_{\psi,1}$ is a word in A_{m} or a word of the type $P\odot$, where P is a word in $\mathrm{Ч}_0$. Obviously every one-sided meaningful word in A_ψ is a meaningful word in A_ψ.

We construct and fix an algorithm χ in the alphabet $\mathrm{A}_{\mathrm{m}}^{\mathrm{sa}} \cup \mathrm{Ч}_0$ satisfying the following condition:

$$\forall t_{\mathrm{m},1} \mathrm{v}_1 (\langle \chi (t_{\mathrm{m},1}) \rangle_p (\mathrm{v}_1) = t_{\mathrm{m},1}).$$

Fix on some word of the type ν and denote it by Y. Then we construct an algorithm in $\mathrm{A}_s^{\mathrm{sa}}$, transforming any word in A_s into the word Y. The transcription of this algorithm is denoted by $\overline{\text{Ъ}}$.

We now construct in $\mathrm{A}_\psi^{\mathrm{sa}}$ algorithms $\overline{\mathrm{lef}}_\psi$ and $\overline{\mathrm{rig}}_\psi$ with the following properties:

1) They are applicable to any meaningful word of A_ψ.

2) If $t_{\psi,1}$ is a word in A_{m}, then

$$\overline{\mathbf{lef}}_\psi (t_{\psi,1}) \simeq \chi (t_{\psi,1}); \qquad \overline{\mathbf{rig}}_\psi (t_{\psi,1}) \simeq \text{Ъ}.$$

3) If $t_{\psi,1} \overline{\underline{\circ}} P \odot Q$, where P and Q are words in $\mathrm{Ч}_0$, then

$$\overline{\mathbf{lef}}_\psi (t_{\psi,1}) = P; \qquad \overline{\mathbf{rig}}_\psi (t_{\psi,1}) = Q$$

it being possible that Q is the empty word.

Suppose that θ is an object term. We introduce the notation

$$\underline{\Theta}_\psi \rightleftharpoons \langle \overline{\mathbf{lef}}_\psi (\Theta) \rangle_p; \qquad \overrightarrow{\Theta}^\psi \rightleftharpoons \langle \overline{\mathbf{rig}}_\psi (\Theta) \rangle_s.$$

We will say that the word $t_{\psi,1}$ is an *F-$(\mathscr{A}, \mathfrak{X})$-construct* in the space $\mathfrak{M}$, and write

$$(t_{\psi,1} \in \underline{F\text{-}(\mathscr{A}, \mathfrak{X})\ \mathbf{constr}}_{\mathfrak{M}}),$$

if the following condition is satisfied:

$$(9) \qquad (t_{\psi,1} \in \underline{\mathbf{one\ mean}}_{\psi}) \,\&\, (t_{\underbrace{\psi,1}\psi}) \in (\nu \mapsto \theta)) \,\&\, \exists \lambda_{s,1} (\lambda_{s,1}\ \underline{\mathbf{reg.\,conv.}}_{\mathfrak{M}} t_{\underbrace{\psi,1\psi}}).$$

We will say that the word $t_{\psi,1}$ is an *FR-*$(\mathscr{A}, \mathfrak{X})$*-construct* in the space $\mathfrak{M}$, and write

$$(t_{\psi,1} \in \underline{FR\text{-}(\mathscr{A}, \mathfrak{X})\ \mathbf{constr}}_{\mathfrak{M}}),$$

if the following condition holds:

$$(10) \qquad (t_{\psi,1} \in \underline{\mathbf{mean}}_{\psi}) \,\&\, (\underbrace{t_{\psi,1}}_{\psi} \in (\nu \to \theta)) \,\&\, (\overbrace{t_{\psi,1}}^{\psi}\ \underline{\mathbf{reg\ conv}}_{\mathfrak{M}} \underbrace{t_{\psi,1}}_{\psi})$$

Expressions (9) and (10) are the transcriptions of certain one-parameter formulas of the language $\mathfrak{A}$, whose parameter is the variable $t_{\psi,1}$. We adjoin these formulas to the list Σ of formulas of the language $\mathfrak{A}_{\Sigma}$ and denote by V a subordinate generic letter assigned to the formula (9), and by U a subordinate generic letter assigned to formula (10).

Analogously we may define the concept of "F-$(\mathfrak{J}, \mathfrak{X}')$-construct" and "$FR$-$(\mathfrak{J}, \mathfrak{X}')$-construct" in the space $\mathfrak{M}$. We note that in this case the role of the alphabet A_p is played by the alphabet A_η, where η denotes the index of the alphabet $A_{\mathrm{m}} \cup A_{\mathrm{n}}$, and the role of the alphabet A_s is played by the alphabet A_{n}. We adjoin the formulas

$$(11) \qquad (t_{\psi,1} \in \underline{F\text{-}(\mathfrak{J}, \mathfrak{X}')\text{-}\mathbf{constr}}_{\mathfrak{M}})$$

and

$$(12) \qquad (t_{\psi,1} \in \underline{FR\text{-}(\mathfrak{J}, \mathfrak{X}')\text{-}\mathbf{constr}}_{\mathfrak{M}})$$

to the list Σ of formulas of the language $\mathfrak{A}_{\Sigma}$. We denote by $\mathfrak{v}$ a subordinate generic letter assigned to formula (11), and by $\mathfrak{w}$ a subordinate generic letter assigned to the formula (12).

We note that if $\mathfrak{P}$ is a normal formula, then formula (12) is also normal.

Concerning the algorithm $\lambda_{\psi,1}$ in the alphabet A_{ψ}^{sa}, we shall say that it is an $(\mathscr{A}, \mathfrak{X})$*-algorithm of passage to the limit in* $\mathfrak{M}$ if

$$\forall U_1 ((\lambda_{\psi,1}(U_1) \overset{+}{\in} \theta) \,\&\, (\lambda_{\psi,1}(U_1)\ \underline{\mathbf{lim}}_{\mathfrak{M}} \underbrace{U_1}_{\psi})),$$

where U_1 is a variable of genus U, i.e. is a variable for FR-$(\mathscr{A}, \mathfrak{X})$-construct in $\mathfrak{M}$.

A locally convex space $\mathfrak{M}$ is said to be $(\mathscr{A}, \mathfrak{X})$*-complete* if an $(\mathscr{A}, \mathfrak{X})$-algorithm of passage to the limit in $\mathfrak{M}$ is realizable.

In what follows the following theorem will play a significant role.

3.1.3. *Suppose that* $(\mathscr{A}, \mathfrak{X})$ *is any directed set. If the space* $\mathfrak{M}$ *is* $(\mathfrak{J}, \mathfrak{X}')$*-complete, then it is* $(\mathscr{A}, \mathfrak{X})$*-complete.*

Proof. Suppose that the space $\mathfrak{M}$ is $(\mathfrak{J}, \mathfrak{X}')$-complete. Then a $(\mathfrak{J}, \mathfrak{X}')$-algorithm of passage to the limit in $\mathfrak{M}$ is realizable. We denote this algorithm by $\mathfrak{I}$ lim. We have to prove that an $(\mathscr{A}, \mathfrak{X})$-algorithm of passage to the limit in $\mathfrak{M}$ is also realizable.

It follows from Axiom III5 that we can realize in A^{sa}_{π} an algorithm ξ of the type $(\iota \to \iota)$ such that

$$\forall \iota \theta_1 \theta_2 (\mathfrak{Q}(\xi(\iota_1), \theta_1) \,\&\, \mathfrak{Q}(\xi(\iota_1), \theta_2) \supset \mathfrak{Q}(\iota_1, \theta_1 + \theta_2)).$$

Here we recall that by hypothesis $\mathfrak{I}$ is a normal formula.

We construct in $A^{sa}_{\eta} \cup A_{\psi}$ an algorithm γ of the type $(U\iota \to \theta)$ such that

$$\forall U_1 \iota_1 (\gamma(U_1 \square \iota_2) \simeq \underbrace{U_1}_{\psi} (\overline{U_1}^{\psi}(\iota))).$$

Recalling from the Introduction, subsection 7, the notation $(\gamma \uparrow U_1 \square)$, we then construct in A^{ca}_{ψ} an algorithm ζ such that

$$\forall U_1 (\zeta(U_1) \simeq \mathfrak{E}(\gamma \uparrow U_1 \square), \ A^{sa}_{\eta}\, 3 \odot \mathfrak{E}\xi, \ A^{sa}_{\pi}\, 3).$$

We shall prove that

$$\forall U_1 (\zeta(U_1) \overset{+}{\in} \mathfrak{w}). \tag{13}$$

To this end it suffices to show that for any word U_1 of type U the following holds:

$$\forall \iota_1 \iota_2 \iota_3 (\xi(\iota_1) \underset{\iota}{\prec} \iota_2, \iota_3 \supset \mathfrak{Q}(\iota_1, \underbrace{U_1}_{\psi}(\overline{U_1}^{\psi}(\iota_2)) - \underbrace{U_1}_{\psi}(\overline{U_1}^{\psi}(\iota_3)))). \tag{14}$$

Suppose that $\iota_1, \iota_2, \iota_3$ are words of type ι such that $\xi(\iota_1) \prec \iota_2, \iota_3$. This means that

$$\begin{gathered} \forall \theta_1 (\mathfrak{Q}(\iota_2, \theta_1) \supset \mathfrak{Q}(\xi(\iota_1), \theta_1)); \\ \forall \theta_1 (\mathfrak{Q}(\iota_3, \theta_1) \supset \mathfrak{Q}(\xi(\iota_1), \theta_1)). \end{gathered} \tag{15}$$

Since U_1 is an FR-$(\mathscr{A}, \mathfrak{X})$-construct in $\mathfrak{M}$, we have

$$\begin{gathered} \forall \nu_1 (\overline{U_1}^{\psi}(\iota_2) \underset{\nu}{\prec} \nu_1 \supset \mathfrak{Q}(\iota_2, \underbrace{U_1}_{\psi}(\nu_1) - \underbrace{U_1}_{\psi}(\overline{U_1}^{\psi}(\iota_2)))); \\ \forall \nu_1 (\overline{U_1}^{\psi}(\iota_3) \underset{\nu}{\prec} \nu_1 \supset \mathfrak{Q}(\iota_3, \underbrace{U_1}_{\psi}(\nu_1) - \underbrace{U_1}_{\psi}(\overline{U_1}^{\psi}(\iota_3)))). \end{gathered} \tag{16}$$

We find a word Y of type ν such that $\overline{U_1}^{\psi}(\iota_2), \overline{U_1}^{\psi}(\iota_3) \prec Y$. In view of (15) and (16), for that word Y we have

$$\mathfrak{Q}(\xi(\iota_1), \underbrace{U_1}_{\psi}(Y) - \underbrace{U_1}_{\psi}(\overline{U_1}^{\psi}(\iota_2))) \,\&\, \mathfrak{Q}(\xi(\iota_1), \underbrace{U_1}_{\psi}(Y) - \underbrace{U_1}_{\psi}(\overline{U_1}^{\psi}(\iota_3))),$$

so that

$$\mathfrak{Q}(\iota_1, \underbrace{U_1}_{\psi}(\overline{U_1}^{\psi}(\iota_2)) - \underbrace{U_1}_{\psi}(\overline{U_1}^{\psi}(\iota_3))).$$

Formula (14), and hence (13), are proved.

Now we denote the composition of the algorithms $\mathfrak{I}$ lim and ζ by $\mathscr{A}$lim, i.e. $\mathscr{A}\mathrm{lim} \leftrightharpoons (\mathfrak{I}\,\mathrm{lim} \bigcirc \zeta)$ (see [11]). It follows from (13) that $\mathscr{A}$lim is an algorithm of the type $(U \to \theta)$. We shall show that it is an $(\mathscr{A}, \mathfrak{X})$-algorithm of passage to the limit in $\mathfrak{M}$. Suppose that U_1 is a word of the type U and that $X \leftrightarrows \mathscr{A}\lim(U_1)$. Then $X \doteq \mathfrak{Z}\lim(\zeta(U_1))$. By definition we have

$$(X \ \lim_{\mathfrak{M}} (\gamma \uparrow U_1 \square)). \tag{17}$$

We need to prove that

$$(X \ \underline{\lim}_{\mathfrak{M}} \underline{U_1}_{\psi}). \tag{18}$$

It follows from (17) that an algorithm χ of the type $(\iota \to \iota)$ is realizable such that

$$\forall \iota_1 \iota_2 (\chi(\iota_1) \prec_{\iota} \iota_2 \supset \mathfrak{Q}(\iota_1, \underline{U_1}_{\psi} (\overline{U_1}^{\psi} (\iota_2)) - X)).$$

Denote by δ an algorithm of the type $(\iota\iota \to \iota)$ having the property

$$\forall \iota_1 \iota_2 (\iota_1 ; \iota_2 \prec_{\iota} \delta(\iota_1 \square \iota_2)).$$

That such an algorithm may be constructed follows from Axiom III2.

Now we construct algorithms μ and φ such that

$$\forall \iota_1 (\mu(\iota_1) \simeq \delta(\chi(\xi(\iota_1)) \square \xi(\iota_1));$$

$$\forall \iota_1 (\varphi(\iota_1) \simeq \overline{U_1}^{\psi} (\mu(\iota_1))).$$

For the proof of (18) it suffices for us to show that

$$\forall \iota_1 \nu_1 (\varphi(\iota_1) \prec_{\nu} \nu_1 \supset \mathfrak{Q}(\iota_1, \underline{U_1}_{\varphi} (\nu_1) - X)). \tag{19}$$

Since $\chi(\xi(\iota_1)) \prec \mu(\iota_1)$, we have

$$\forall \iota_1 \mathfrak{Q}(\xi(\iota_1), \underline{U_1}_{\psi} (\overline{U_1}^{\psi} (\mu(\iota_1))) - X). \tag{20}$$

On the other hand, it is obvious that

$$\forall \iota_1 \nu_1 (\varphi(\iota_1) \prec_{\nu} \nu_1 \supset \mathfrak{Q}(\mu(\iota_1), \underline{U_1}_{\psi} (\overline{U_1}^{\psi} (\mu(\iota_1))) - \underline{U_1}_{\psi} (\nu_1))),$$

so that

$$\forall \iota_1 \nu_1 (\varphi(\iota_1) \prec_{\nu} \nu_1 \supset \mathfrak{Q}(\xi(\iota_1), \underline{U_1}_{\psi} (\overline{U_1}^{\psi} (\mu(\iota_1))) - \underline{U_1}_{\psi} (\nu_1))). \tag{21}$$

(19) follows from (20) and (21). Hence formula (19) is proved. Thus $\mathscr{A}$ is an $(\mathscr{A}, \mathfrak{X})$-algorithm of passage to the limit in $\mathfrak{M}$, and that space is $(\mathscr{A}, \mathfrak{X})$-complete. The theorem is completely proved.

One proves the following theorem in a quite analogous way.

3.1.4. *Suppose that $(\mathscr{A}, \mathfrak{X})$ is a directed set. If the space $\mathfrak{M}$ is $(\mathfrak{Z}, \mathfrak{X}')$-almost complete, then it is also $(\mathscr{A}, \mathfrak{X})$-almost complete.*

On the basis of Theorems 3.1.3 and 3.1.4 we introduce the following definitions.

A locally convex space $\mathfrak{M}$ is said to be *complete* if it is $(\mathfrak{Z}, \mathfrak{X}')$-complete. It is said to be *almost* complete if it is $(\mathfrak{Z}, \mathfrak{X}')$-almost complete.

It is not difficult to verify the following assertion.

3.1.5. *Suppose that* $\mathfrak{M}$ *and* $\widetilde{\mathfrak{M}}$ *are isomorphic locally convex spaces. Then if one of them is complete* (*almost complete*), *then the other is also complete* (*almost complete*).

In what follows we will say that the space $\mathfrak{M}$ is enumerably complete (enumerably almost complete), if $\mathfrak{M}$ is N-complete (N-almost complete).

3.2. Now we turn to the definition of F-completion and FR-completion of a locally convex space.

Suppose that $\mathfrak{M}$ is a locally convex space. Concerning the locally convex space $\widetilde{\mathfrak{M}}$, we shall say that it is the *F-completion* of the space $\mathfrak{M}$, if $\widetilde{\mathfrak{M}}$ is almost complete and $\mathfrak{M}$ is an almost dense subspace of $\widetilde{\mathfrak{M}}$. We say that $\widetilde{\mathfrak{M}}$ is the *FR-completion* of $\mathfrak{M}$ if $\widetilde{\mathfrak{M}}$ is complete and $\mathfrak{M}$ is a dense subspace of $\widetilde{\mathfrak{M}}$.

First of all we note that for the topology $(\mathrm{A}_{\mathrm{n}}, \mathfrak{J}, \mathfrak{Q})$ of the space $\mathfrak{M}$ it is always possible to construct a topology $(\mathrm{A}_{\mathrm{n}}, \mathfrak{J}, \widetilde{\mathfrak{Q}})$ equivalent to it such that each neighborhood of the origin of the topology $(\mathrm{A}_{\mathrm{n}}, \mathfrak{J}, \widetilde{\mathfrak{Q}})$ is a $\mathfrak{Z}'$-closed set. Indeed, we shall construct a formula $\widetilde{\mathfrak{Q}}$ such that

$$\widetilde{\mathfrak{Q}}(\iota_1, \theta_1) \equiv \forall \iota_2 \exists \theta_2 (\mathfrak{Q}(\iota_1, \theta_2) \,\&\, \mathfrak{Q}(\iota_2, \theta_2 - \theta_1)).$$

Here we recall that for each word U of type ι the set $\widetilde{\mathfrak{Q}}(U,\theta_1)$ is the $\mathfrak{Z}'$-closure of the set $\mathfrak{Q}(U,\theta_1)$ in the space $\mathfrak{M}$ (see §2.2). It is easy to verify that the list $(\mathrm{A}_{\mathrm{n}}, \mathfrak{J}, \widetilde{\mathfrak{Q}})$ defines a locally convex topology equivalent to the topology $(\mathrm{A}_{\mathrm{n}}, \mathfrak{J}, \mathfrak{Q})$, and in the space

$$\mathrm{A}_{\mathrm{m}}, \mathfrak{P}, \mathfrak{E}, \underset{\theta}{+}, \underset{\theta}{\cdot}, \mathfrak{O}, \mathrm{A}_{\mathrm{n}}, \mathfrak{J}, \widetilde{\mathfrak{Q}}$$

each neighborhood of the origin, i.e. each set of the type $\widetilde{\mathfrak{Q}}(U,\theta_1)$, where U is a word of type ι, is a $\mathfrak{Z}'$-closed set.

In connection with this we may always assume that the topology of a locally convex space is given by a fundamental system of neighborhoods of the origin, each of which is a $\mathfrak{Z}'$-closed set.

Now we shall construct the F-completion and the FR-completion of a locally convex set. Suppose that $\mathfrak{M}$ is a locally convex set. We will suppose that $\mathfrak{M}$ *is a* T'-*separated space and that each neighborhood of the origin from the topology* $(\mathrm{A}_{\mathrm{n}}, \mathfrak{X}, \mathfrak{Q})$ *is a* $\mathfrak{Z}'$-*closed set in* $\mathfrak{M}$. These assumptions are expressed by the following formulas:

(22) $$\forall \theta_1 ((\theta_1 = \mathfrak{O}) \equiv \forall \iota_1 \mathfrak{Q}(\iota, \theta_1));$$

(23) $$\forall \iota \theta_1 (\forall \iota_2 \exists \theta_2 (\mathfrak{Q}(\iota_1, \theta_2) \,\&\, \mathfrak{Q}(\iota_2, \theta_2 - \theta_1)) \supset \mathfrak{Q}(\iota_1, \theta)).$$

We denote formula (11) by $\mathfrak{P}^{\mathfrak{v}}$ and formula (12) by $\mathfrak{P}^{\mathfrak{w}}$. We define the formula $\mathfrak{E}^{\mathrm{II}}$ as follows:

$$\mathfrak{E}^{\mathrm{II}} \rightleftharpoons \forall \iota_1 \exists \iota_2 \forall \iota_3 (\iota_2 \prec \iota_3 \supset \mathfrak{Q}(\iota_1, \underset{\psi}{\underline{t_{\psi,1}}}(\iota_3) - \underset{\psi}{\underline{t_{\psi,2}}}(\iota_3))).$$

Suppose that P and Q are arbitrary object terms. We introduce the notation

$$(P \underset{\mathrm{II}}{=} Q) \rightleftharpoons F_{P,\,Q}^{t_{\psi,1},\, t_{\psi,2}} \llcorner \mathfrak{E}^{\mathrm{II}} \lrcorner.$$

In view of Axioms III5 and III2, we can realize an algorithm ξ of the type $(\iota\to\iota)$, and an algorithm δ of the type $(\iota\iota\to\iota)$, satisfying the following formulas:

(24) $$\forall\iota_1\theta_1\theta_2(\mathfrak{Q}(\xi(\iota_1), \theta_1)\&\mathfrak{Q}(\xi(\iota_1), \theta_2)\supset\mathfrak{Q}(\iota_1, \theta_1+\theta_2));$$

(25) $$\forall\iota_1\iota_2\theta_1(\mathfrak{Q}(\delta(\iota_1\Box\iota_2), \theta_1)\supset\mathfrak{Q}(\iota_1, \theta_1)\&\mathfrak{Q}(\iota_2, \theta_1)).$$

We construct in A_ψ^{sa} algorithms μ and ν such that for any words P and Q in the alphabet A_ψ

$$\mu(P\Box Q)\overset{+}{\in}c_0;\qquad \nu(P\Box Q)\overset{+}{\in}c_0;$$

$$\forall\iota_1(\langle\mu(P\Box Q)\rangle_\eta(\iota_1)\simeq\underset{\longrightarrow\psi}{P}(\iota_1)\underset{\theta}{+}\underset{\longrightarrow\psi}{Q}(\iota_1));$$

$$\forall\iota_1(\langle\nu(P\Box Q)\rangle_n(\iota_1)\simeq\delta(\overline{P}^{\longrightarrow\psi}(\xi(\iota_1))\Box\overline{Q}^{\longrightarrow\psi}(\xi(\iota_1)))).$$

Then we construct algorithms $\underset{\mathfrak{v}}{+}$ and $\underset{\mathfrak{w}}{+}$ such that for any words $\mathfrak{v}_1$ and $\mathfrak{v}_2$ of type $\mathfrak{v}$ and words $\mathfrak{w}_1$ and $\mathfrak{w}_2$ of type $\mathfrak{w}$ the following holds:

$$\underset{\mathfrak{v}}{+}(\mathfrak{v}_1\Box\mathfrak{v}_2)\simeq\begin{cases}\underset{\theta}{+}(\mathfrak{v}_1\Box\mathfrak{v}_2), & \text{if } \odot \text{ does not enter into the word } \mathfrak{v}_1\Box\mathfrak{v}_2;\\ \mu(\mathfrak{v}_1\Box\mathfrak{v}_2)\odot, & \text{if } \odot \text{ enters into the word } \mathfrak{v}_1\Box\mathfrak{v}_2.\end{cases}$$

$$\underset{\mathfrak{w}}{+}(\mathfrak{w}_1\Box\mathfrak{w}_2)\simeq\begin{cases}\underset{\theta}{+}(\mathfrak{w}_1\Box\mathfrak{w}_2), & \text{if } \odot \text{ does not enter into the word } \mathfrak{w}_1\Box\mathfrak{w}_2;\\ \mu(\mathfrak{w}_1\Box\mathfrak{w}_2)\odot\nu(\mathfrak{w}_1\Box\mathfrak{w}_2), & \text{if } \odot \text{ enters into the word } \mathfrak{w}_1\Box\mathfrak{w}_2.\end{cases}$$

It is easy to see that

1) $\underset{\mathfrak{v}}{+}$ is an algorithm of the type $(\mathfrak{v}\mathfrak{v}\to\mathfrak{v})$;

2) $\underset{\mathfrak{w}}{+}$ is an algorithm of the type $(\mathfrak{w}\mathfrak{w}\to\mathfrak{w})$.

Now we turn to the construction of the multiplication algorithm. In view of property 1.3.7 we can realize an algorithm λ of the type $(\text{д}\iota\to\iota)$ such that

$$\forall x\iota_1\theta_1(\mathfrak{Q}(\lambda(x\Box\iota_1), \theta_1)\supset\mathfrak{Q}(\iota_1, x\cdot\theta_1)).$$

We construct algorithms $\bar{\mu}$ and $\bar{\nu}$ in the alphabet $A_\psi^{sa}\cup Ч_3$ such that, for any real duplex x and any word P in A_ψ,

$$\bar{\mu}(x\Box P)\overset{+}{\in}c_0;\qquad \bar{\nu}(x\Box P)\overset{+}{\in}c_0;$$

$$\forall\iota_1(\langle\bar{\mu}(x\Box P)\rangle_\eta(\iota_1)\simeq x\theta\underset{\longrightarrow\psi}{P}(\iota_1));$$

$$\forall\iota_1(\langle\bar{\nu}(x\Box P)\rangle_n(\iota_1)\simeq\overline{P}^{\longrightarrow\psi}(\lambda(x\Box\iota_1))).$$

Then we construct algorithms $\underset{\mathfrak{v}}{\cdot}$ and $\underset{\mathfrak{w}}{\cdot}$ in the alphabet $A_\psi^{sa}\cup Ч_3$ such that for any real duplex x, any word $\mathfrak{v}_1$ of type $\mathfrak{v}$ and any word $\mathfrak{w}_1$ of type $\mathfrak{w}$

$$\underset{\mathfrak{v}}{\cdot}(x\Box\mathfrak{v}_1)\simeq\begin{cases}\underset{\theta}{\cdot}(x\Box\mathfrak{v}_1), & \text{if } \odot \text{ does not enter into the word } \mathfrak{v}_1;\\ \underset{\mu}{-}(x\Box\mathfrak{v}_1)\odot, & \text{if } \odot \text{ enters into the word } \mathfrak{v}_1.\end{cases}$$

$$\underset{\mathfrak{w}}{\cdot}(x\square\mathfrak{w}_1) \simeq \begin{cases} \underset{\theta}{\cdot}(x\square\mathfrak{w}_1), & \text{if } \odot \text{ does not enter into the word } \mathfrak{w}_1;\\ \underset{\mu}{-}(x\square\mathfrak{w}_1)\odot\bar{\mathfrak{v}}(x\square\mathfrak{w}_1), & \text{if } \odot \text{ enters into the word } \mathfrak{w}_1.\end{cases}$$

It is easy to see that

1) $\underset{\mathfrak{v}}{\cdot}$ is an algorithm of the type $(\text{д}\,\mathfrak{v}\to\mathfrak{v})$;

2) $\underset{\mathfrak{w}}{\cdot}$ is an algorithm of the type $(\text{д}\,\mathfrak{w}\to\mathfrak{w})$.

In what follows we shall employ the following notation:

$$(P\underset{\mathfrak{v}}{+}Q)\rightleftharpoons\underset{\mathfrak{v}}{+}(P\square Q); \qquad (P\underset{\mathfrak{w}}{+}Q)\rightleftharpoons\underset{\mathfrak{w}}{+}(P\square Q);$$

$$(X\underset{\mathfrak{v}}{\cdot}P)\rightleftharpoons\underset{\mathfrak{v}}{\cdot}(X\square P); \qquad (X\underset{\mathfrak{w}}{\cdot}P)\rightleftharpoons\underset{\mathfrak{w}}{\cdot}(X\square P),$$

where P, Q, and X are arbitrary object terms.

Finally, we construct a formula $\mathfrak{Q}^{\text{п}}$ as follows:

$$\mathfrak{Q}^{\text{п}}\rightleftharpoons\exists\mathfrak{v}_2((\mathfrak{v}_2\underset{\text{п}}{=}t_{\psi,1})\,\&\,\exists\iota_2\forall\iota_3(\iota_2\prec\iota_3\supset\mathfrak{Q}(t_{\text{п},1},\underbrace{\mathfrak{v}_2}_{\psi}(\iota_3)))).$$

This is a two-parameter formula, whose parameters are the variables $t_{\text{п},1}$ and $t_{\psi,1}$. We write

$$\mathfrak{Q}^{\text{п}}(u,P)\rightleftharpoons F^{t_{\text{п},1},t_{\psi,1}}_{u,P}\llcorner\mathfrak{Q}^{\text{п}}\lrcorner.$$

Obviously

$$\mathfrak{Q}^{\text{п}}(\iota_1,\mathfrak{v}_1)\equiv\exists\mathfrak{v}_2((\mathfrak{v}_2\underset{\text{п}}{=}\mathfrak{v}_1)\,\&\,\exists\iota_2\forall\iota_3(\iota_2\prec\iota_3\supset\mathfrak{Q}(\iota_1,\underbrace{\mathfrak{v}_2}_{\psi}(\iota_3))));$$

$$\mathfrak{Q}^{\text{п}}(\iota_1,\mathfrak{w}_1)\equiv\exists\mathfrak{w}_2((\mathfrak{w}_2\underset{\text{п}}{=}\mathfrak{w}_1)\,\&\,\exists\iota_2\forall\iota_3(\iota_2\prec\iota_3\supset\mathfrak{Q}(\iota_1,\underbrace{\mathfrak{w}_2}_{\psi}(\iota_3)))),$$

where ι_1 is a word of type ι, $\mathfrak{v}_1$ a word of type $\mathfrak{v}$ and $\mathfrak{w}_1$ a word of type $\mathfrak{w}$.

The following assertions are valid:

(26) $$\forall\theta_1\theta_2((\theta_1\underset{\text{п}}{=}\theta_2)\equiv(\theta_1\underset{\theta}{=}\theta_2));$$

(27) $$\forall\theta_1\theta_2((\theta_1\underset{\mathfrak{v}}{+}\theta_2)\doteq(\theta_1\underset{\theta}{+}\theta_2)); \qquad \forall\theta_1\theta_2((\theta_1\underset{\mathfrak{w}}{+}\theta_2)\doteq(\theta_1\underset{\theta}{+}\theta_2));$$

(28) $$\forall x\theta_1((x\underset{\mathfrak{v}}{\cdot}\theta_1)\doteq(x\underset{\theta}{\cdot}\theta_1)); \qquad \forall x\theta_1((x\underset{\mathfrak{w}}{\cdot}\theta_1)\doteq(x\underset{\theta}{\cdot}\theta_1));$$

(29) $$\forall\iota_1\theta_1(\mathfrak{Q}^{\text{п}}(\iota_1,\theta_1)\equiv\mathfrak{Q}(\iota_1,\theta_1)).$$

The proofs of these assertions do not present any difficulties. We need only note that in the proofs of (26) and (29) assumptions (22) and (23) are used.

Now we consider the following lists:

(30) $$A_\psi,\ \mathfrak{P}^{\mathfrak{v}},\ \mathfrak{E}^{\text{п}},\ \underset{\mathfrak{v}}{+},\ \underset{\mathfrak{v}}{\cdot},\ \mathfrak{D},\ A_{\text{п}},\ \mathfrak{J},\ \mathfrak{Q}^{\text{п}}$$

and

(31) $$A_\psi,\ \mathfrak{P}^{\mathfrak{w}},\ \mathfrak{E}^{\text{п}},\ \underset{\mathfrak{w}}{+},\ \underset{\mathfrak{w}}{\cdot},\ \mathfrak{D},\ A_{\text{п}},\ \mathfrak{J},\ \mathfrak{Q}^{\text{п}}.$$

It is not hard to verify that these lists represent locally convex spaces. We shall denote the space (30) by $\mathfrak{M}_1$ and the space (31) by $\mathfrak{M}_2$.

3.2.1. *The space* $\mathfrak{M}_1$ *is the F-completion of* $\mathfrak{M}$, *and* $\mathfrak{M}_1$ *is* T'*-separated.*

3.2.2. *The space* $\mathfrak{M}_2$ *is the FR-completion of* $\mathfrak{M}$, *and* $\mathfrak{M}_2$ *is* T'*-separated.*

We shall call $\mathfrak{M}_1$ the *standard F-completion* of the space $\mathfrak{M}$, and $\mathfrak{M}_2$ the *standard FR-completion* of the space $\mathfrak{M}$.

The proofs of 3.2.1 and 3.2.2 are quite analogous. We present the proof of 3.2.2.

First of all we prove the following formulas:

(32) $$\forall \iota_1 \mathfrak{w}_1 (\mathfrak{Q}^{\text{п}}(\xi(\iota_1), \mathfrak{w}_1) \supset \exists \iota_2 \forall \iota_3 (\iota_2 \prec \iota_3 \supset \mathfrak{Q}(\iota_1, \underbrace{\mathfrak{w}_1}_{\psi}(\iota_3))));$$

(33) $$\forall \iota_1 \mathfrak{w}_1 (\exists \iota_2 \forall \iota_3 (\iota_2 \prec \iota_3 \supset \mathfrak{Q}(\iota_1, \underbrace{\mathfrak{w}_1}_{\psi}(\iota_3))) \supset \mathfrak{Q}^{\text{п}}(\iota_1, \mathfrak{w}_1)).$$

where ξ is an algorithm of type $(\iota \to \iota)$, having property (24). Formula (33) follows directly from the definition of the formula $\mathfrak{Q}^{\text{п}}$. We shall prove formula (32). Suppose that we are given words U of the type ι and $\mathfrak{w}_1$ of type $\mathfrak{w}$ such that $\mathfrak{Q}^{\text{п}}(\xi(U), \mathfrak{w}_1)$. Then we can realize a word $\mathfrak{w}_2$ of type $\mathfrak{w}$ satisfying the condition

$$\forall \iota_1 \exists \iota_2 \forall \iota_3 (\iota_2 \prec \iota_3 \supset \mathfrak{Q}(\iota_1, \underbrace{\mathfrak{w}_2}_{\psi}(\iota_3) - \underbrace{\mathfrak{w}_1}_{\psi}(\iota_3))) \,\&\, \exists \iota_2 \forall \iota_3 (\iota_2 \prec \iota_3 \supset \mathfrak{Q}(\xi(U), \underbrace{\mathfrak{w}_2}_{\psi}(\iota_3))).$$

Consequently it is possible to find a word V of type ι such that

$$\forall \iota_3 (V \prec \iota_3 \supset \mathfrak{Q}(\xi(U), \underbrace{\mathfrak{w}_2}_{\psi}(\iota_3) - \underbrace{\mathfrak{w}_1}_{\psi}(\iota_3)) \,\&\, \mathfrak{Q}(\xi(U), \underbrace{\mathfrak{w}_2}_{\psi}(\iota_3))).$$

Hence

$$\forall \iota_3 (V \prec \iota_3 \supset \mathfrak{Q}(U, \underbrace{\mathfrak{w}_1}_{\psi}(\iota_3))).$$

Formula (32) is proved. From (32) it easily follows that

$$\forall \mathfrak{w}_1 (\forall \iota_1 \mathfrak{Q}^{\text{п}}(\iota_1, \mathfrak{w}_1) \supset (\mathfrak{w}_1 \underset{\text{п}}{=} \mathfrak{O})),$$

i.e. the space $\mathfrak{M}_2$ is T'-separated.

Now we shall prove that the space $\mathfrak{M}_2$ is complete. Suppose that the letter $\odot$ does not enter into the alphabet A_ψ. We adjoin the alphabet $A_\psi \cup \{\odot\}$ to the list of basic alphabets of the language $\mathcal{A}_\Sigma$ and denote its index by γ, i.e. $A_\gamma = A_\psi \cup \{\odot\}$. This is the alphabet intended for the construction of FR-$(\mathfrak{J}, \mathfrak{X}^{\iota})$-constructs in the space $\mathfrak{M}_2$.[3)]

In view of property 1.3.6 it is possible to construct an algorithm ζ of type $(\iota \to \iota)$ such that

(34) $$\forall \iota_1 \theta_1 \theta_2 \theta_3 (\mathfrak{Q}(\zeta(\iota_1), \theta_1) \,\&\, \mathfrak{Q}(\zeta(\iota_1), \theta_2) \,\&\, \mathfrak{Q}(\zeta(\iota_1), \theta_3) \supset \mathfrak{Q}(\iota_1, \theta_1 + \theta_2 + \theta_3)).$$

Suppose that $t_{\gamma,1}$ is any FR-$(\mathfrak{J}, \mathfrak{X}^{\iota})$-construct in $\mathfrak{M}_2$. Put

---

[3)] It may be verified that

$$\forall \iota_1 \iota_2 (\forall \theta_1 (\mathfrak{Q}(\iota_1, \theta_1) \supset \mathfrak{Q}(\iota_2, \theta_2)) \equiv \forall \mathfrak{w}_1 (\mathfrak{Q}^{\text{п}}(\iota_1, \mathfrak{w}_1) \supset \mathfrak{Q}^{\text{п}}(\iota_2, \mathfrak{w}_1))).$$

We can therefore assume that the directed sets of indices of the neighborhoods of the origin in $\mathfrak{M}$ and $\mathfrak{M}_2$ coincide.

$$B \rightleftharpoons \underline{t_{\Upsilon,1}}_{\Upsilon}; \qquad D \rightleftharpoons \overline{t_{\Upsilon,1}}^{\Upsilon}.$$

We construct in the alphabet $\mathrm{A}_{\Upsilon}^{\mathrm{sa}}$ algorithms λ and χ such that, for any FR-$(\mathfrak{J}, \mathfrak{X}')$-construct $t_{\Upsilon,1}$ in the space $\mathfrak{M}_2$,

$$\lambda(t_{\Upsilon,1}) \overset{+}{\in} c_0; \ \chi(t_{\Upsilon,1}) \overset{+}{\in} c_0;$$

$$\forall \iota_1 (\langle \lambda(t_{\Upsilon,1}) \rangle_n (\iota_1) \simeq \underline{B(\iota_1)}_{\psi} \ \overline{(B(\iota_1)}^{\psi} (\iota_1)));$$

$$\forall \iota_1 (\langle \chi(t_{\Upsilon,1}) \rangle_n (\iota_1) \simeq \delta(\zeta(\iota_1) \square D(\xi(\zeta(\iota_1))))),$$

where ξ, δ and ζ are algorithms having respectively properties (24), (25) and (34).

Then we construct an algorithm $\mathscr{G}$ as follows:

$$\forall t_{\Upsilon,1} (\mathscr{G}(t_{\Upsilon,1}) \simeq \lambda(t_{\Upsilon,1}) \odot \chi(t_{\Upsilon,1})).$$

We shall show that $\mathscr{G}$ is a $(\mathfrak{J}, \mathfrak{X}')$-algorithm of passage to the limit in $\mathfrak{M}_2$. First of all we show that for any FR-$(\mathfrak{J}, \mathfrak{X}')$-construct in $\mathfrak{M}_2$

$$\mathscr{G}(t_{\Upsilon,1}) \overset{+}{\in} \mathfrak{w},$$

i.e.

$$(35) \qquad \forall \iota_1 \iota_2 \iota_3 (\langle \chi(t_{\Upsilon,1}) \rangle_n (\iota_1) \prec \iota_2, \iota_3 \supset \mathfrak{Q}(\iota_1, \langle \lambda(t_{\Upsilon,1}) \rangle (\iota_2) - \langle \lambda(t_{\Upsilon,1}) \rangle (\iota_3))).$$

By hypothesis we have

$$\forall \iota_1 \iota_2 \iota_3 (D(\iota_1) \prec \iota_2, \iota_3 \supset \mathfrak{Q}^{\Pi}(\iota_1, B(\iota_2) \underset{\mathfrak{w}}{-} B(\iota_3))).$$

Suppose that ι_1, ι_2 and ι_3 are words of type ι such that $\langle \chi(t_{\Upsilon,1}) \rangle_n (\iota_1) \prec \iota_2, \iota_3$. Then

$$\zeta(\iota_1) \prec \iota_2, \iota_3 \quad \text{and} \quad D(\xi(\zeta(\iota_1))) \prec \iota_2, \iota_3.$$

From the last formula it follows that $\mathfrak{Q}^{\Pi}(\xi(\zeta(\iota_1)), B(\iota_2) \underset{\mathfrak{w}}{-} B(\iota_3))$. Hence, in view of (32), we can realize a word Z of type ι with the property that

$$(36) \qquad \forall \theta_4 (Z \prec \iota_4 \supset \mathfrak{Q}(\zeta(\iota_1), \underline{B(\iota_2)}_{\psi} (\iota_4) - \underline{B(\iota_3)}_{\psi} (\iota_4))).$$

Put

$$S \rightleftharpoons \delta(Z \square \delta(\overline{B(\iota_2)}^{\psi} (\iota_2) \square \overline{B(\iota_3)}^{\psi} (\iota_3))).$$

Obviously

$$\mathfrak{Q}(\iota_2, \underline{B(\iota_2)}_{\psi} (\overline{B(\iota_2)}^{\psi} (\iota_2)) - \underline{B(\iota_2)}_{\psi} (S))$$

and

$$\mathfrak{Q}(\iota_3, \underline{B(\iota_3)}_{\psi} (\overline{B(\iota_3)}^{\psi} (\iota_3)) - \underline{B(\iota_3)}_{\psi} (S)).$$

Therefore

$$\mathfrak{Q}\,(\zeta\,(\theta_1),\, \underset{\psi}{\underline{B\,(\iota_2)}}\,(\overline{B\,(\iota_2)}^{\psi}\,(\iota_2) - \underset{\psi}{\underline{B\,(\iota_2)}}\,(S))$$

and

$$\mathfrak{Q}\,(\zeta\,(\iota_1),\, \underset{\psi}{\underline{B\,(\iota_3)}}\,(\overline{B\,(\iota_3)}^{\psi}\cdot(\iota_3) - \underset{\psi}{\underline{B\,(\iota_3)}}\,(S)).$$

On the other hand, it follows from (36) that

$$\mathfrak{Q}\,(\zeta\,(\iota_1),\, \underset{\psi}{\underline{B\,(\iota_2)}}\,(S) - \underset{\psi}{\underline{B\,(\iota_3)}}\,(S)).$$

From these three last formulas and from (34) we find that

$$\mathfrak{Q}\,(\iota_1,\, \underset{\psi}{\underline{B\,(\iota_2)}}\,(\overline{B\,(\iota_2)}^{\psi}\,(\iota_2)) - \underset{\psi}{\underline{B\,(\iota_3)}}\,(\overline{B\,(\iota_3)}^{\psi}\,(\iota_2))).$$

Thus formula (35) is proved. Now we need to show that for any FR-$(\mathfrak{J}, \mathfrak{X}')$-construct $t_{\gamma,1}$ in $\mathfrak{M}_2$

$$\mathcal{F}\,(t_{\gamma,1})\ \underset{\mathfrak{w}_2}{\underline{\lim}}\ \underset{\gamma}{\underline{t_{\gamma,1}}}\,,$$

i.e. that

(37) $$\forall\iota_1 \exists\iota_2 \forall\iota_3\,(\iota_2 \prec \iota_3 \supset \mathfrak{Q}^{\text{II}}\,(\iota_1,\, B\,(\iota_3) \underset{\mathfrak{w}}{-} T)),$$

where $T \leftrightharpoons \lambda(t_{\gamma,1}) \odot \chi(t_{\gamma,1})$. Suppose given a word ι_1 of type ι. Put $\iota_2 \leftrightharpoons \delta(\zeta(\iota_1) \square D(\xi(\zeta(\iota_1))))$. We shall show that

(38) $$\forall\iota_3\,(\iota_2 \cdot\!\prec \iota_3 \supset \mathfrak{Q}^{\text{II}}\,(\iota_1,\, B\,(\iota_3) \underset{\mathfrak{w}}{-} T)).$$

Suppose that ι_3 is a word of type ι such that $\iota_2 \prec \iota_3$. Put

$$Y \rightleftharpoons \delta\,(\overline{B\,(\iota_3)}^{\psi}\,(\zeta\,(\iota_1)) \square\, \iota_2).$$

We shall show that

(39) $$\forall\iota_4\,(Y \prec \iota_4 \supset \mathfrak{Q}\,(\iota_1,\, \underset{\psi}{\underline{B\,(\iota_3)}}\,(\iota_4) - \underset{\psi}{\underline{T}}\,(\iota_4))).$$

Suppose that $Y \prec \iota_4$. Then $\iota_2 \prec \iota_4$. From $\iota_2 \prec \iota_3$ and $\iota_2 \prec \iota_4$ it follows that $D(\xi(\zeta(\iota_1))) \prec \iota_3, \iota_4$. Hence $\mathfrak{Q}^{\text{II}}(\xi(\zeta(\iota_1)),\ B(\iota_3) \underset{\mathfrak{w}}{-} B(\iota_4))$. In view of (32) a word Z of type ι is realizable such that

(40) $$\forall\iota_5\,(Z \prec \iota_5 \supset \mathfrak{Q}\,(\zeta\,(\iota_1),\, \underset{\psi}{\underline{B\,(\iota_3)}}\,(\iota_5) - \underset{\psi}{\underline{B\,(\iota_4)}}\,(\iota_5))).$$

Put

$$S \rightleftharpoons \delta\,(Z \square\, \delta\,(\overline{B\,(\iota_3)}^{\psi}\,(\zeta\,(\iota_1)) \square\, \overline{B\,(\iota_4)}^{\psi}\,(\iota_4))).$$

Obviously

$$\mathfrak{Q}\,(\iota_4,\, \underset{\psi}{\underline{B\,(\iota_4)}}\,(S) - \underset{\psi}{\underline{B\,(\iota_4)}}\,(\overline{B\,(\iota_4)}^{\psi}\,(\iota_4))).$$

From $\iota_2 \prec \iota_4$ it follows that $\zeta(\iota_1) \prec \iota_4$. Hence

(41) $$\mathfrak{Q}(\zeta(\iota_1), \underbrace{B(\iota_4)}_{\psi}(S) - \underbrace{B(\iota_4)}_{\psi}(\overbrace{B(\iota_4)}^{\psi}(\iota_4))).$$

From $Y \prec \iota_4$ it follows that $\overbrace{B(\iota_3)}^{\psi}(\zeta(\iota_1)) \prec \iota_4$. Hence

(42) $$\mathfrak{Q}(\zeta(\iota_1), \underbrace{B(\iota_3)}_{\psi}(\iota_4) - \underbrace{B(\iota_3)}_{\psi}(S)).$$

Moreover, it follows from (40) that

(43) $$\mathfrak{Q}(\zeta(\iota_1), \underbrace{B(\iota_3)}_{\psi}(S) - \underbrace{B(\iota_4)}_{\psi}(S)).$$

From formulas (41) – (43) we get

$$\mathfrak{Q}(\iota_1, \underbrace{B(\iota_3)}_{\psi}(\iota_4) - \underbrace{B(\iota_4)}_{\psi}(\overbrace{B(\iota_4)}^{\psi}(\iota_4))).$$

Thus formula (39) is proved. In view of formula (33) we therefore find that $\mathfrak{Q}^{\text{п}}(\iota_1, B(\iota_3) \underset{\mathfrak{w}}{-} T)$. Formula (38), and therefore (37), are proved.

Thus we have proved that the space $\mathfrak{M}_2$ is complete.

One may conclude from formulas (26) – (29) that $\mathfrak{M}$ is a subspace of $\mathfrak{M}_2$. We shall construct an algorithm $\mathfrak{A}$ such that

$$\forall \mathfrak{w}_1 \iota_1 (\mathfrak{A}(\mathfrak{w}_1 \square \iota_1) \simeq \underbrace{\mathfrak{w}_1}_{\psi}(\overbrace{\mathfrak{w}_1}^{\psi}(\iota_1))).$$

$\mathfrak{A}$ is an algorithm of the type $(\mathfrak{w} \to \theta)$. One easily verifies that

$$\forall \mathfrak{w}_1 \iota_1 (\forall \iota_3 (\overbrace{\mathfrak{w}_1}^{\psi}(\iota_1) \prec \iota_3 \supset \iota_3 \supset \mathfrak{Q}(\iota_1, \underbrace{\mathfrak{w}_1 \underset{\mathfrak{w}}{-} \mathfrak{A}(\mathfrak{w}_1 \square \iota_1)}_{\psi}(\iota_3)))).$$

Therefore, and from (33), it follows that

$$\forall \mathfrak{w}_1 \iota_1 \mathfrak{Q}^{\text{п}}(\iota_1, \mathfrak{w}_1 \underset{\mathfrak{w}}{-} \mathfrak{A}(\mathfrak{w}_1 \square \iota_1)).$$

Therefore the space $\mathfrak{M}$ is dense in $\mathfrak{M}_2$. Theorem 3.2.2 is completely proved.

Now we formulate a theorem on the extension of continuous linear operators. Suppose that $\mathfrak{M}$ is a T'-separated locally convex space and that $\mathfrak{M}_2$ is its *FR*-standard completion. Suppose further that $\widetilde{\mathfrak{M}}$ is a locally convex space given by the list

$$A_{\widetilde{\mathfrak{m}}}, \widetilde{\mathfrak{P}}, \widetilde{\mathfrak{E}}, \widetilde{+}, \widetilde{\cdot}, \widetilde{\mathfrak{H}}, A_{\widetilde{\mathfrak{n}}}, \widetilde{\mathfrak{J}}, \widetilde{\mathfrak{Q}}.$$

Suppose that λ is a linear operator of the type $(\theta \to \tilde{\theta})$. We shall say that the algorithm μ of type $(\tau \to \iota)$ is a *regulator of continuity* of the operator λ if it satisfies the following condition:

$$\forall \tilde{\iota}_1 \theta_1 (\mathfrak{Q}(\mu(\tilde{\iota}_1), \theta_1) \supset \widetilde{\mathfrak{Q}}(\tilde{\iota}_1, \lambda(\theta_1))).$$

Analogously one defines the concept of regulator of continuity of a linear operator of the type $(\mathfrak{w} \to \tilde{\theta})$.

Suppose that λ_1 is an operator of type $(\theta \to \tilde{\theta})$ and λ_2 an operator of type $(\mathfrak{w} \to \tilde{\theta})$. We will say that λ_2 is an *extension* of λ_1 to the space $\mathfrak{M}_2$ if

$$\forall \theta_1 (\lambda_1(\theta_1) \stackrel{\sim}{=} \lambda_2(\theta_1)).$$

Here the sign $\cong$ denotes equality in $\tilde{\mathfrak{M}}$.

Suppose that $\otimes$ denotes a letter not appearing in the alphabet $A_\psi \cup A_{\tilde{\mathfrak{m}}}$. The *complete cipher of a continuous linear operator of the type* $(\theta \to \tilde{\theta})$ *(type* $(\mathfrak{w} \to \tilde{\theta})$*)* is a word of the form $P \otimes Q$, where P is the transcription of a linear operator of the type $(\theta \to \tilde{\theta})$ (of the type $(\mathfrak{w} \to \tilde{\theta})$) and Q is the transcription of the regulator of continuity of that operator.

Suppose that $P_1 \otimes Q_1$ is the complete cipher of a continuous linear operator of the type $(\theta \to \tilde{\theta})$ and that $P_2 \otimes Q_2$ is the complete cipher of a continuous linear operator of the type $(\mathfrak{w} \to \tilde{\theta})$. We will say that $P_2 \otimes Q_2$ is an *extension of the complete cipher* $P_1 \otimes Q_1$ to the space $\mathfrak{M}_2$ if the operator $\langle P_2 \rangle$ is an extension of the operator $\langle P_1 \rangle$ to $\mathfrak{M}_2$.

The following theorem holds.

3.2.3. *Suppose that $\mathfrak{M}$ is a T-separated locally convex space and that $\mathfrak{M}_2$ is its standard FR-completion. Suppose that $\tilde{\mathfrak{M}}$ is a complete, T'-separated, locally convex space. Then one can realize an algorithm ξ which for any complete cipher X of a continuous linear operator of the type $(\theta \to \tilde{\theta})$ yields an extension $\xi(X)$ of the complete cipher X to the space $\mathfrak{M}_2$.*

The proof of this theorem is analogous to the proof of Theorem 10.4.9 in [**26**].

§4. Constructive multinormed spaces

4.1. Suppose given a locally convex space $\mathfrak{M}$, defined by the list

(1) $$A_{\mathfrak{m}}, \mathfrak{P}, \mathfrak{E}, +, \cdot, \mathfrak{D}, A_{\mathfrak{n}}, \mathfrak{J}, \mathfrak{Q}.$$

The space $\mathfrak{M}$ will be said to be *metrizable* (*semimetrizable*)[4)] if one can realize a metric (semimetric) function ρ in the space $\mathfrak{P}$ (see [**26**], §9) such that the following conditions are satisfied:

a) $\forall \theta_1 \theta_2 ((\theta_1 \underset{\theta}{=} \theta_2) \equiv (\rho(\theta_1 \square \theta_2) = 0))$;

b) $\forall \iota_1 \exists n \forall \theta_1 \theta_2 (\rho(\theta_1 \square \theta_2) \leqslant 2^{-n} \supset \mathfrak{Q}(\iota_1, \theta_1 - \theta_2))$;

c) $\forall n \exists \iota_1 \forall \theta_1 \theta_2 (\mathfrak{Q}(\iota_1, \theta_1 - \theta_2) \supset \rho(\theta_1 \square \theta_2) \leqslant 2^{-n})$.

We will say that the topology of the space $\mathfrak{M}$ has an *enumerable fundamental system of neighborhoods of the origin* if the set $\mathfrak{J}$ is enumerable. The space $\mathfrak{M}$ is said to be *separated* if it is both T-separated and T'-separated.

The following assertion is obvious:

---

4) In [**30**] the concept of metrizability was introduced not only for constructive locally convex spaces, but also for constructive uniform spaces.

4.1.1. *If the locally convex space* $\mathfrak{M}$ *is metrizable* (*semimetrizable*), *then* $\mathfrak{M}$ *is separated and its topology is equivalent to the topology of the enumerable fundamental system of neighborhoods of the origin.*

Now we shall prove the following theorem.

4.1.2. *It is possible to construct a locally convex space* $\mathfrak{M}$ *having the following properties*: 1) $\mathfrak{M}$ *is separated*; 2) *the topology of* $\mathfrak{M}$ *has an enumerable fundamental system of neighborhoods of the origin*; 3) $\mathfrak{M}$ *is not metrizable* (*in fact, not even semimetrizable*).

PROOF. Denote by m the index of the alphabet $Ч_3 \cup \{\tau\}$, and by $\mathfrak{P}$ the formula defining a set of words in $А_m$ of the form $P\tau\alpha$, where P is the transcription of the sequence of real duplexes and α is a real duplex, such that the condition $\forall n(|\langle P\rangle_3(n)| \leqq \alpha)$ is satisfied. The elements of the set $\mathfrak{P}$ will be called words of type θ. Suppose that θ_1 is a word of type θ and that $\theta_1 \overset{\circ}{=} P\tau\alpha$. Write

$$\{\theta_1\} \rightleftharpoons \langle P_3\rangle; \qquad [\theta_1]_2^\tau \rightleftharpoons \alpha.$$

The formula $\mathfrak{E}$ defining equality of words of type θ is constructed so that for any words θ_1 and θ_2 of type θ

$$\mathfrak{E}(\theta_1, \theta_2) \equiv \forall n\{\theta_1\}(n) \underset{\text{B}}{=} \{\theta_2\}(n)).$$

We construct algorithms $\underset{\theta}{+}$ of type $(\theta\theta \to \theta)$ and $\underset{\theta}{\cdot}$ of type $(д\theta \to \theta)$ such that, for any words θ_1 and θ_2 of type θ and any real duplex α,

$$\forall n(\{\underset{\theta}{+}(\theta_1 \square \theta_2)\}(n) \simeq \{\theta_1\}(n) + \{\theta_2\}(n));$$

$$[\underset{\theta}{+}(\theta_1 \square \theta_2)]_2^\tau \doteqdot [\theta_1]_2^\tau + [\theta_2]_2^\tau;$$

$$\forall n(\{\underset{\theta}{\cdot}(\alpha \square \theta_1)\}(n) \simeq \alpha \underset{\theta}{\cdot} \{\theta_1\}(n)); \qquad [\cdot(\alpha \square \theta_1)]_2^\tau \doteqdot |\alpha| \cdot [\theta_1]_2^\tau.$$

We write $\mathfrak{O} \leftrightharpoons Q\tau 0$, where Q is the transcription of an algorithm transforming any natural number into the number 0.

Denote by $\mathfrak{J}$ the set of all natural numbers. We shall construct a formula $\mathfrak{Q}$ such that, for any natural number i and any word θ_1 of type θ,

$$\mathfrak{Q}(\iota, \theta_1) \equiv \forall n(|\{\theta_1\}(n)| < 2^{-i}).$$

It is not hard to verify that the list

$$А_m, \mathfrak{P}, \mathfrak{E}, \underset{\theta}{+}, \underset{\theta}{\cdot}, \mathfrak{O}, А_n, \mathfrak{J}, \mathfrak{Q} \tag{2}$$

defines a locally convex space. We will denote this space by $\mathfrak{M}$. Obviously the topology of $\mathfrak{M}$ has an enumerable fundamental system of neighborhoods of the origin. From the definition of the formulas $\mathfrak{E}$ and $\mathfrak{Q}$ it follows that the space $\mathfrak{M}$ is T'-separated. The T-separation of $\mathfrak{M}$ is proved as follows. Let θ_1 be a point of $\mathfrak{M}$ such that $\neg(\theta_1 \underset{\theta}{=} \mathfrak{O})$. Then $\neg \forall n(\{\theta_1\}(n) = 0)$. Hence we can realize a natural number n_0 such that $\{\theta_1\}(n_0) \neq 0$ (see the Appendix, Lemma 1). In view of Lemma $\{18\}$ of [27] we can realize a natural number i such that $|\{\theta_1\}(n_0)| > 2^{-i}$. Hence we find $\neg\mathfrak{Q}(i, \theta_1)$, as was required to be proved.

Now we prove that $\mathfrak{M}$ is not semimetrizable, so that it is also not metrizable. To this end we suppose that $\mathfrak{M}$ is semimetrizable. Then we can realize a semimetric function ρ in the set $\mathfrak{P}$ such that

$$\forall i \exists k \forall \theta_1 \theta_2 (\rho(\theta_1 \square \theta_2) \leqslant 2^{-k} \supset \mathfrak{Q}(i, \theta_1 - \theta_2)).$$

Here we recall that ρ is an algorithm of the type $(\theta \to \text{в})$. Hence a natural number R_0 is realizable such that

$$\forall \theta_1 (\rho(\theta_1 \square \mathfrak{O}) \leqslant 2^{-k_0} \supset \mathfrak{Q}(0, \theta_1)),$$

so that

(3) $$\forall \theta_1 (\neg \forall n (|\{\theta_1\}(n)| \leqslant 1) \supset \rho(\theta_1 \square \mathfrak{O}) > 2^{-k_0}).$$

Suppose that Ω is an algorithm of the type $(\text{нн} \to \text{н})$ such that it is not possible to have an algorithm φ of the type $(\text{н} \to \text{н})$ satisfying the condition

(4) $$\forall k ((\varphi(k) = 0) \equiv \exists n (\Omega(k \square n) = 0))$$

(see [12] or [26], §2). Construct an algorithm $\mathfrak{B}$ such that

$$\mathfrak{B}(k \square n) \doteqdot \begin{cases} 0, & \text{if } \Omega(k \square n) \neq 0; \\ 2, & \text{if } \Omega(k \square n) = 0. \end{cases}$$

Then we construct an algorithm $\mathfrak{C}$ such that

$$\forall k (\mathfrak{C}(k) \simeq \mathcal{E}(\mathfrak{B} \uparrow k \square), Ч_3^{sa} З \tau 2).$$

Obviously $\mathfrak{C}(k)$ is a point of $\mathfrak{M}$ for any k. We shall construct an algorithm $\mathfrak{D}$ of type $(\text{н} \to \text{в})$ such that

$$\forall k (\mathfrak{D}(k) \simeq \rho(\mathfrak{C}(k) \square \mathfrak{O})).$$

From $\forall k (\mathfrak{D}(k) \in \text{в})$ it follows that $\forall k \exists x (\mathfrak{D}(k) \underset{\text{в}}{=} x)$, where x is a variable of genus д. Thus we can realize an algorithm $\mathfrak{F}$ of type $(\text{н} \to \text{д})$ with the property

$$\forall k (\mathfrak{F}(k) \underset{\text{в}}{=} \mathfrak{D}(k)).$$

From the definitions of the algorithms $\mathfrak{B}$, $\mathfrak{C}$, and $\mathfrak{F}$, and from (3), it follows that

(5) $$\exists n (\Omega(k \square n) = 0) \supset \mathfrak{F}(k) > 2^{-k_0},$$

(6) $$\forall n (\Omega(k \square n) \neq 0) \supset \mathfrak{F}(k) = 0.$$

From these formulas and from the formula

$$\neg \forall n (\Omega(k \square n) \neq 0) \supset \exists n (\Omega(k \square n) = 0)$$

it follows that for any natural number k

(7) $$\mathfrak{F}(k) > 0 \equiv \exists n(\Omega(k \square n) = 0).$$

In view of Theorem 1.3 of [4] we have

$$\forall k (\mathfrak{F}(k) > 0 \vee \mathfrak{F}(k) < 2^{-k_0}).$$

This means that we can realize an algorithm φ of type $(\text{н} \to \text{н})$ such that

(8) $$\forall k ((\varphi(k) = 0 \supset \mathfrak{F}(k) > 0) \& (\varphi(k) \neq 0 \supset \mathfrak{F}(k) < 2^{-k_0})).$$

Suppose that k is a number such that $\mathfrak{F}(k) > 0$. Then it follows from (7) and (5) that $\mathfrak{F}(k) > 2^{-k_0}$. Hence from (8) we find that $\neg(\varphi(k) \neq 0)$, i.e. $\varphi(k) = 0$. Thus, if $\mathfrak{F}(k) > 0$, then $\varphi(k) = 0$. Hence, using the first implication of formula (8), we find that

(9) $$\forall k ((\varphi(k) = 0) \equiv (\mathfrak{F}(k) > 0)).$$

(4) follows from (7) and (9). Thus we have obtained a contradiction. Therefore the space $\mathfrak{M}$ cannot be semimetrizable. The theorem is proved.

4.2. Suppose given a constructive linear space

(10) $$\text{A}_{\mathfrak{m}}, \mathfrak{P}, \mathfrak{E}, +, \cdot, \mathfrak{O}.$$

Concerning the algorithm $\mathcal{N}$ in the alphabet $\text{A}_{\mathfrak{m}}^{\text{sa}} \cup \text{Ч}_3$, we will say that it is a *norm* (*seminorm*) in the space (10) if it is an algorithm of the type $(\theta \to \text{д})$ (of the type $(\theta \to \text{в})$) and satisfies the following conditions:

IV 1. $\forall \theta_1 ((\theta_1 \underset{\theta}{=} \mathfrak{O}) \equiv (\mathcal{N}(\theta_1) = 0))$;

IV 2. $\forall x \theta_1 (\mathcal{N}(x \cdot \theta_1) = |x| \cdot \mathcal{N}(\theta_1))$;

IV 3. $\forall \theta_1 \theta_2 (\mathcal{N}(\theta_1 + \theta_2) \leqslant \mathcal{N}(\theta_1) + \mathcal{N}(\theta_2))$.

The list

(11) $$\text{A}_{\mathfrak{m}}, \mathfrak{P}, \mathfrak{E}, +, \cdot, \mathfrak{O}, \mathcal{N}$$

will be called a *constructive normed* (*seminormed*) *space* if $(\text{A}_{\mathfrak{m}}, \mathfrak{P}, \mathfrak{E}, +, \cdot, \mathfrak{O})$ is a linear space and $\mathcal{N}$ is a norm (seminorm) in it.

This definition is equivalent to the definition introduced by N. A. Šanin in [26].

Suppose that $\mathcal{N}$ is a norm (seminorm) in the linear space (10). The list $(\text{Ч}_0, \mathfrak{J}, \mathfrak{Q})$, where $\mathfrak{J}$ is the set of all natural numbers and the formula $\mathfrak{Q}$ is defined in such a way that $\mathfrak{Q}(i, \theta_1) \equiv \mathcal{N}(\theta_1) \leqq 2^{-i}$, defines a locally convex topology on the space (10). This topology will be called *the topology corresponding to the norm* (*seminorm*) $\mathcal{N}$.

Equipping each normed (seminormed) space with a locally convex topology corresponding to its norm (seminorm), we may regard it as a locally convex space.

From this definition and from IV 1 it follows that every normed (seminormed) space is a separated locally convex space.

We will say that the locally convex space

(12) $$\mathrm{A_m}, \mathfrak{P}, \mathfrak{E}, +, \cdot, \mathfrak{D}, \mathrm{A_n}, \mathfrak{J}, \mathfrak{Q}$$

(or its topology) is *normable* (*seminormable*) if in the linear space ($\mathrm{A_m}, \mathfrak{P}, \mathfrak{E}, +, \cdot, \mathfrak{D}$) a norm (seminorm) $\mathscr{N}$ is realizable, with the property that the topology corresponding to $\mathscr{N}$ is equivalent to the topology ($\mathrm{A_n}, \mathfrak{J}, \mathfrak{Q}$) of the space (12).

We note that it is possible to realize locally convex spaces which are seminormable but not normable. This follows, for example, from Theorem 4.1.2 of [**16**].

Suppose that A_q is an alphabet, $\mathscr{A}$ a normal formula of the list Σ, defining some set of words in A_q, and $\mathscr{N}$ an algorithm in the alphabet $\mathrm{A_m^{sa}} \cup \mathrm{A}_q \cup \text{Ч}_3$. We suppose that the set $\mathscr{A}$ contains at least one element. We denote by ν a subordinate generic letter put into correspondence with the formula $\mathscr{A}$. Concerning the list ($\mathrm{A}_q, \mathscr{A}, \mathscr{N}$), we will say that it is a *multinorm* (*multiseminorm*) in the linear space (10) if $\mathscr{N}$ is an algorithm of the type ($\nu\theta \to$ д) (the type ($\nu\theta \to$ в)), and if the following conditions are satisfied:

V 1. $\forall \theta_1 ((\theta_1 = \mathfrak{D}) \equiv \forall \nu_1 (\mathscr{N}(\nu_1 \square \theta_1) = 0))$;

V 2. $\forall \nu_1 x \theta_1 (\mathscr{N}(\nu_1 \square x \cdot \theta_1) = |x| \cdot \mathscr{N}(\nu_1 \square \theta_1))$;

V 3. $\forall \nu_1 \theta_1 \theta_2 (\mathscr{N}(\nu_1 \square \theta_1 + \theta_2) \leqslant \mathscr{N}(\nu_1 \square \theta_1) + \mathscr{N}(\nu_1 \square \theta_2))$.

The list

(13) $$\mathrm{A_m}, \mathfrak{P}, \mathfrak{E}, +, \cdot, \mathfrak{D}, \mathrm{A}_q, \mathscr{A}, \mathscr{N}$$

will be called a *constructive multinormed* (*multiseminormed*) space if ($\mathrm{A_m}, \mathfrak{P}, \mathfrak{E}, +, \cdot, \mathfrak{D}$) is a linear space and if ($\mathrm{A}_q, \mathscr{A}, \mathscr{N}$) is a multinorm (multiseminorm) in it.

Suppose that ($\mathrm{A}_q, \mathscr{A}, \mathscr{N}$) is a multinorm (multiseminorm) in the linear space (10). We denote by $\mathfrak{n}$ the index of the alphabet $\mathrm{A}_q \cup \text{Ч}_0 \cup \{\tau\}$, where τ is a letter not entering into $\mathrm{A}_q \cup \text{Ч}_0$. It is possible to construct a normal formula $\mathfrak{J}$ defining the set of all possible words of the form $n \tau \nu_1 \tau \cdots \tau \nu_k$, where n and k are natural numbers ($k \geqq 1$) and $\nu_1, \cdots, \nu_k$ are words of type ν. We adjoin the formula $\mathfrak{J}$ to the list Σ of formulas of the language Я_Σ and denote by ι the corresponding subordinate generic letter. Then we construct a formula $\mathfrak{Q}$ such that for any words ι_1 of type ι and θ_1 of type θ, if $\iota_1 \overline{\underline{\circ}}\, n \tau \nu_1 \tau \cdots \tau \nu_k$, then

$$\mathfrak{Q}(\nu_1, \theta_1) \equiv (\mathscr{N}(\nu_1 \square \theta_1) \leqslant 2^{-n}) \& \ldots \& (\mathscr{N}(\nu_l \square \theta_1) \leqslant 2^{-n}).$$

The list ($\mathrm{A_n}, \mathfrak{J}, \mathfrak{Q}$) defined in this way is a locally convex topology on the linear space (10). This topology will be called *the topology corresponding to the multinorm* (*multiseminorm*) ($\mathrm{A}_q, \mathscr{A}, \mathscr{N}$).

Equipping each multinormed (multiseminormed) space with the locally convex topology corresponding to its multinorm (multiseminorm), we may regard it as a locally convex space.

From the definition of the topology corresponding to the multinorm (multiseminorm), and from V 1, it follows that every multinormed (multiseminormed) space is a T'-separated locally convex space.

REMARK 1. If the multinorm (multiseminorm) $(\mathrm{A}_q, \mathscr{A}, \mathscr{N})$ satisfies further the condition

V 4. $\forall \mathrm{v}_1 \mathrm{v}_2 \exists \mathrm{v}_3 \forall \theta_1 (\mathscr{N}(\mathrm{v}_3 \square \theta_1) \geqslant \max(\mathscr{N}(\mathrm{v}_1 \square \theta_1) \square \mathscr{N}(\mathrm{v}_2 \square \theta_1)))$,

then the topology corresponding to this multinorm (multiseminorm) may be defined by the list $(\mathrm{A}_{\mathrm{II}}, \tilde{\mathfrak{J}}, \tilde{\mathfrak{Q}})$; here $\tilde{\mathfrak{J}}$ is the set of all possible words of the form $n \tau \nu_1$, where n is a natural number and ν_1 is a word of type ν. In addition, for any element $n \tau \nu_1$ of the set $\tilde{\mathfrak{J}}$ and any word θ_1 of type θ, $\tilde{\mathfrak{Q}}(n \tau \nu_1, \theta_1) \equiv \mathscr{N}(\nu_1 \square \theta_1) \leqq 2^{-n}$. This topology is equivalent to the topology $(\mathrm{A}_{\mathrm{II}}, \mathfrak{J}, \mathfrak{Q})$ defined above.

REMARK 2. If, in addition to conditions V 1 – V 3, the multinorm (multiseminorm) $(\mathrm{A}_q, \mathscr{A}, \mathscr{N})$ satisfies also the condition

V 4′. $\forall \mathrm{v}_1 \mathrm{v}_2 \exists \mathrm{v}_3 \forall \theta_1 (\mathscr{N}(\mathrm{v}_3 \square \theta_1) \geqslant \max(\mathscr{N}(\mathrm{v}_1 \square 2 \cdot \theta_1) \square \mathscr{N}(\mathrm{v}_2 \square 2 \cdot \theta_1)))$,

then the topology corresponding to it may be defined by the list $(\mathrm{A}_q, \mathscr{A}, \tilde{\mathfrak{Q}})$, where $\tilde{\mathfrak{Q}}(\nu_1, \theta_1) \equiv (\mathscr{N}(\nu_1 \square \theta_1) \leqq 1)$ for any words ν_1 of type ν and θ_1 of type θ. This topology is equivalent to the topology $(\mathrm{A}_{\mathrm{II}}, \mathfrak{J}, \mathfrak{Q})$ defined above.

We will say that the locally convex space (12) (or its topology) is *multinormed* (*multiseminormed*) if a multinorm (multiseminorm) $(\mathrm{A}_q, \mathscr{A}, \mathscr{N})$ is realizable in the linear space $(\mathrm{A}_{\mathrm{III}}, \mathfrak{P}, \mathfrak{E}, +, \cdot, \mathfrak{O})$ and is such that the topology corresponding to $(\mathrm{A}_q, \mathscr{A}, \mathscr{N})$ is equivalent to the topology $(\mathrm{A}_{\mathrm{II}}, \mathfrak{J}, \mathfrak{Q})$ of the space (12).

The multinorm $(\mathrm{A}_q, \mathscr{A}, \mathscr{N})$ is said to be an *enumerable* norm if $\mathscr{A}$ is an enumerable set. A multinormed space is said to be *enumerably multinormed*, if its multinorm is enumerable. A locally convex space is said to be *enumerably normable* if it is normable by means of an enumerable multinorm.

We note that if $(\mathrm{A}_q, \mathscr{A}, \mathscr{N})$ is an enumerable norm, then for any word θ_1 of type θ we have

$$\neg \forall \mathrm{v}_1 (\mathscr{N}(\mathrm{v}_1 \square \theta_1) = 0) \supset \exists \mathrm{v}_1 (\mathscr{N}(\mathrm{v}_1 \square \theta_1) > 0)$$

(see Lemma 1 of the Appendix). Therefore it is easily proved that every enumerably normed space is separated, i.e. both T-separated and T'-separated.

4.2.1. *If a locally convex space* $\mathfrak{M}$ *is enumerably normable, then it is metrizable.*

The proof of this theorem is completely analogous to the proof of the corresponding theorem in classical functional analysis (see, for example, [7], Chapter XI).

Now we consider the space $\mathfrak{M}$ constructed in the proof of Theorem 4.1.2.

It is easy to see that for that space we have

$$\forall ij\exists x\,(x > 0\ \&\ \forall\theta_1\,(\mathfrak{Q}\,(i, \theta_1) \supset \mathfrak{Q}\,(j, x\cdot\theta_1))).$$

Thus, for any integer i, the neighborhood of the origin with the index i is a bounded set in $\mathfrak{M}$. On the other hand, it follows from the nonmetrizability of $\mathfrak{M}$ that it is not normable. Thus we have obtained the following:

4.2.2. *It is possible to construct a locally convex space which is separated and has a bounded neighborhood of the origin, but which is not normable.*

Now we shall prove the following theorem.

4.2.3. *Suppose that $\mathfrak{M}$ is a locally convex space, T'-separated and having a bounded neighborhood of the origin. Then, if $\mathfrak{M}$ is multinormable, it is also normable.*

PROOF. Suppose given a space $\mathfrak{M}$ defined by the list (1). We denote by Y the index of a bounded neighborhood of the origin in $\mathfrak{M}$.

Then

(14) $$\forall\iota_1\exists x\,(x > 0\ \&\ \forall\theta_1\,(\mathfrak{Q}\,(Y, \theta_1) \supset \mathfrak{Q}\,(\iota_1, x\cdot\theta_1))).$$

Suppose that $\mathfrak{M}$ is multinormable using the multinorm $(A_q, \mathscr{A}, \mathscr{N})$. Then the topology corresponding to the multinorm $(A_q, \mathscr{A}, \mathscr{N})$ and the topology $(A_n, \mathfrak{J}, \mathfrak{Q})$ are equivalent to one another. Therefore we can find a natural number n and words $U_1, \cdots, U_k$, with k a positive natural number, such that

(15) $$\forall\theta_1\,(\mathscr{N}\,(U_1\square\theta_1) \leqslant 2^{-n}\ \&\dots\&\ \mathscr{N}\,(U_k\square\theta_1) \leqslant 2^{-n} \supset \mathfrak{Q}\,(Y, \theta_1)).$$

We shall construct an algorithm $\mathfrak{N}$, of the type $(\theta \rightarrow д)$, such that

$$\forall\theta_1\,(\mathfrak{N}\,(\theta_1) \simeq \max\,(\mathscr{N}\,(U_1\square\theta_1)\square\dots\square\mathscr{N}\,(U_k\square\theta_1))).$$

We shall prove the following formulas:

(16) $$\forall i\exists\iota_1\forall\theta_1\,(\mathfrak{Q}\,(\iota_1, \theta_1) \supset \mathfrak{N}\,(\theta_1) \leqslant 2^{-i}),$$

(17) $$\forall\iota_1\exists i\forall\theta_1\,(\mathfrak{N}\,(\theta_1) \leqslant 2^{-i} \subset \mathfrak{Q}\,(\iota_1, \theta_1)).$$

Consider (16). Suppose given a natural number i. For the word $i\tau U_1\tau\cdots\tau U_k$ it is possible to find a word ι_1 of type ι such that

$$\forall\theta_1\,(\mathfrak{Q}\,(\iota_1\,\theta_1) \supset \mathscr{N}\,(U_1\square\theta_1) \leqslant 2^{-i}\ \&\dots\&\ \mathscr{N}\,(U_k\square\theta_1) \leqslant 2^{-i}),$$

so that

$$\forall\theta_1\,(\mathfrak{Q}\,(i_1, \theta_1) \supset \mathfrak{N}\,(\theta_1) \leqslant 2^{-i}).$$

Now we shall prove (17). Suppose given a word ι_1 of type ι. In view of (14) we find a number $x > 0$ such that

(18) $$\forall\theta_1\,(\mathfrak{Q}\,(Y, \theta_1) \supset \mathfrak{Q}\,(\iota_1, x\cdot\theta_1)).$$

Then we find a natural number i such that $2^{-i} < x \cdot 2^{-n}$. From (15), (18), and the definition of $\mathfrak{N}$ we easily deduce that

$$\forall \theta_1 (\mathfrak{N}(\theta_1) \leqslant 2^{-i} \supset \mathfrak{Q}(\iota_1, \theta_1)),$$

as we needed to prove.

The fact that $\mathfrak{N}$ satisfies Axioms IV2 and IV3 for the norm follows directly from its definition. It remains for us to prove that

$$\forall \theta_1 (\theta_1 = \mathfrak{O} \equiv \mathfrak{N}(\theta_1) = 0).$$

Obviously $\theta_1 = \mathfrak{O} \supset \mathfrak{N}(\theta_1) = 0$. Now we suppose that $\mathfrak{N}(\theta_1) = 0$. Then it follows from (17) that $\forall \iota_1 \mathfrak{Q}(\iota_1, \theta_1)$, so that from the T'-separation of the space $\mathfrak{M}$ we find that $\theta_1 = \mathfrak{O}$.

Thus $\mathfrak{N}$ is a norm in $\mathfrak{M}$, so that $\mathfrak{M}$ is normable.

The following follows directly from 4.2.2 and 4.2.3:

4.2.4. *It is possible to realize a separable locally convex space which is not multinormable.*

REMARK 3. As we know, in classical mathematics any separated locally convex space is multinormable. Further, for a locally convex space to be normable, it is necessary and sufficient that it should be separated and that it should have a bounded neighborhood of the origin [**44**]. Thus, Theorems 4.2.2 and 4.2.4 are points of essential divergence between the classical and constructive theories of locally convex spaces.

4.3. Suppose that $\mathfrak{M}$ is a locally convex space given by the list (1). Suppose that $\mathfrak{K}$ is a normal one-parameter formula defining some subset of $\mathfrak{Z}$; we are supposing that $\mathfrak{K}$ contains at least one element. We denote by κ a subordinate generic letter put into correspondence with the formula $\mathfrak{K}$ in the language $Я_\Sigma$. Suppose further that Φ is an algorithm of the type $(\kappa\theta \to д)$.

We will call the algorithm Φ a *Minkowski $\mathfrak{K}$-algorithm relative to the topology* $(A_п, \mathfrak{Z}, \mathfrak{Q})$, if the following conditions are satisfied:

Φ 1. $\forall \varkappa_1 \theta_1 x \left(x > 0 \,\&\, \mathfrak{Q}\left(\varkappa_1, \frac{1}{x} \cdot \theta_1\right) \supset x \geqslant \Phi(\varkappa_1 \square \theta_1)\right);$

Φ 2. $\forall \varkappa_1 \theta_1 n \exists x \left(x > 0 \,\&\, \mathfrak{Q}\left(\varkappa_1, \frac{1}{x} \cdot \theta_1\right) \,\&\, x - \Phi(\varkappa_1 \square \theta_1) < 2^{-n}\right).$

4.3.1. *A Minkowski $\mathfrak{K}$-functional relative to the topology* $(A_п, \mathfrak{Z}, \mathfrak{Q})$ *has the following properties*:

a) $\forall \varkappa_1 \theta_1 (\theta_1 = \mathfrak{O} \supset \Phi(\varkappa_1 \square \theta_1) = 0);$

b) $\forall \varkappa_1 \theta_1 x (\Phi(\varkappa_1 \square x \cdot \theta_1) = |x| \cdot \Phi(\varkappa_1 \square \theta_1));$

c) $\forall \varkappa_1 \theta_1 \theta_2 (\Phi(\varkappa_1 \square \theta_1 + \theta_2) \leqslant \Phi(\varkappa_1 \square \theta_1) + \Phi(\varkappa_1 \square \theta_2));$

d) $\forall \varkappa_1 \theta_1 (\Phi(\varkappa_1 \square \theta_1) < 1 \supset \mathfrak{Q}(\varkappa_1, \theta_1));$

e) $\forall \varkappa_1 \theta_1 (\mathfrak{Q}(\varkappa_1, \theta_1) \supset \Phi(\varkappa_1 \square \theta_1) \leqslant 1)$.

Proof. These properties are easily deduced from conditions $\Phi 1$ and $\Phi 2$. We shall prove for example property c). Suppose given words κ_1 of type κ and words θ_1 and θ_2 of type θ. Suppose that n is any natural number. In view of $\Phi 2$ it is possible to find positive real duplexes x and y such that

(19) $$\mathfrak{Q}\left(\varkappa_1, \frac{1}{x} \cdot \theta_1\right), \quad \mathfrak{Q}\left(\varkappa_1, \frac{1}{y} \cdot \theta_1\right),$$

(20) $$x < \Phi(\varkappa_1 \square \theta_1) + 2^{-(n+1)}, \quad y < \Phi(\varkappa_1 \square \theta_2) + 2^{-(n+1)}.$$

Obviously

$$\frac{1}{x+y}(\theta_1 + \theta_2) = \frac{x}{x+y} \cdot \frac{1}{x}\theta_1 + \frac{y}{x+y} \cdot \frac{1}{y}\theta_2$$

and

$$\frac{x}{x+y} + \frac{y}{x+y} = 1.$$

Therefore in view of Axiom III3 and (19) we find that

$$\mathfrak{Q}\left(\varkappa_1, \frac{1}{x+y}\theta_1 + \theta_2)\right).$$

Hence in view of $\Phi 1$ we get $\Phi(\kappa_1 \square \theta_1 + \theta_2) \leqq x + y$. From this last formula and from (20) it follows that

$$\Phi(\varkappa_1 \square \theta_1 + \theta_2) < \Phi(\varkappa_1 \square \theta_1) + \Phi(\varkappa_1 \square \theta_2) + 2^{-n}.$$

This is true for any n, so that we get

$$\Phi(\varkappa_1 \square \theta_1 + \theta_2) \leqslant \Phi(\varkappa_1 \square \theta_1) + \Phi(\varkappa_1 \square \theta_2).$$

Property c) is proved.

Suppose that τ is a letter not entering into the alphabet $A_{\text{II}} \cup Ч_0$. We denote the index of the alphabet $A_{\text{II}} \cup Ч_0 \cup \{\tau\}$ by s, i.e. $A_s = A_{\text{II}} \cup Ч_0 \cup \{\tau\}$. It is possible to construct a normal alphabet $\widetilde{\mathfrak{I}}$ defining the set of all possible words of the form $n \tau P_1 \tau \cdots \tau P_k$, where n and k are natural numbers with k positive, and $P_1, \cdots, P_k$ words of type κ. We denote by $\tilde{\iota}$ a subordinate generic letter put into correspondence with the formula $\mathfrak{I}$. Obviously each element of the set $\widetilde{\mathfrak{I}}$ (i.e. word of type $\tilde{\iota}$) is a τ-system of words in $A_{\text{II}} \cup Ч_0$.

Denote by L_τ an algorithm which is applicable to any τ-system of words in $A_{\text{II}} \cup Ч_0$ and transforms each such τ-system into the number of its τ-terms. We denote by $И$ an algorithm having the following property: for any τ-system P of words in $A_{\text{II}} \cup Ч_0$ and any natural number i, if $1 \leqq i \leqq L_\tau(P)$, then $И(i \square P)$ is the ith τ-term of P. For example, if $P \,\overline{\circ}\, n \tau P_1 \tau \cdots \tau P_k$, where n is a natural number and $P_1, \cdots, P_k$ are words of A_{II}, then $И(1 \square P) \doteq n$, and $И(i+1 \square P) \doteq P_i$ $(i = 1, \cdots, k)$.

Instead of the notation $И(i \square P)$, we shall sometimes write $[P]_i^\tau$.

We will call the set $\mathfrak{K}$ a *generating subset of the set* $\mathfrak{J}$ if

(21) $\forall\iota_1\exists\tilde{\iota}_1\forall\theta_1(\forall i(1<i\leqslant L_\tau(\tilde{\iota}_1)\supset\mathfrak{L}(И(i\square\tilde{\iota}_1),2^{И(1\square\tilde{\iota}_1)}\cdot\theta_1))\supset\mathfrak{L}(\iota_1,\theta_1))$.

4.3.2. *Suppose the space* $\mathfrak{M}$ *is* T'*-separated. Suppose that one can realize a generating subset* $\mathfrak{K}$ *of the set* $\mathfrak{J}$ *with the property that a Minkowski* $\mathfrak{K}$*-functional is realizable relative to the topology* $(\mathrm{A}_{\mathrm{n}},\mathfrak{J},\mathfrak{Q})$. *Then the space* $\mathfrak{M}$ *is multinormable.*

Proof. Suppose that $\mathfrak{K}$ is a generating subset of $\mathfrak{J}$ and that Φ is a realization of a Minkowski $\mathfrak{K}$-functional relative to the topology $(\mathrm{A}_{\mathrm{n}},\mathfrak{J},\mathfrak{Q})$.

First we shall show that

$$\forall\theta_1(\forall\varkappa_1(\Phi(\varkappa_1\square\theta_1)=0)\supset\theta_1=\mathfrak{O}). \tag{22}$$

Indeed, suppose that θ_1 is a point of $\mathfrak{M}$ such that $\forall\kappa_1(\Phi(\kappa_1\square\theta_1)=0)$. Then it follows from condition $\Phi2$ that $\forall\kappa_1 n\mathfrak{Q}(\kappa_1,2^n\cdot\theta_1)$. Therefore in view of (21) we get $\forall\tilde{\iota}_1\mathfrak{Q}(\iota_1,\theta_1)$. Since the space $\mathfrak{M}$ is T'-separated, $\theta_1=\mathfrak{O}$.

From (22) and from properties a) – c) of the algorithm Φ (see 4.3.1) it follows that $(\mathrm{A}_{\mathrm{n}},\mathfrak{K},\Phi)$ is a multinorm in $\mathfrak{M}$.

We denote by $\tilde{\mathfrak{Q}}$ a two-parameter formula such that, for any words $\tilde{\iota}_1$ of type $\tilde{\iota}$ and θ_1 of type θ,

$$\tilde{\mathfrak{L}}(\tilde{\iota}_1,\theta_1)\equiv\forall i(1<i\leqslant L_\tau(\tilde{\iota}_1)\supset\Phi(И(i\square\tilde{\iota}_1)\square\theta_1)\leqslant 2^{-И(1\square\tilde{\iota}_1)}).$$

Obviously the list $(\mathrm{A}_s,\tilde{\mathfrak{J}},\tilde{\mathfrak{Q}})$ defines a locally convex topology corresponding to the multinorm $(\mathrm{A}_{\mathrm{n}},\mathfrak{K},\Phi)$. We shall prove that this topology is equivalent to the topology $(\mathrm{A}_{\mathrm{n}},\mathfrak{J},\mathfrak{Q})$, i.e. that

$$\forall\iota_1\exists\tilde{\iota}_1\forall\theta_1(\tilde{\mathfrak{L}}(\tilde{\iota}_1,\theta_1)\supset\mathfrak{L}(\iota_1,\theta_1)); \tag{23}$$

$$\forall\tilde{\iota}_1\exists\iota_1\forall\theta_1(\mathfrak{L}(\iota_1,\theta_1)\supset\tilde{\mathfrak{L}}(\tilde{\iota}_1,\theta_1)). \tag{24}$$

Let us prove (23). Suppose given a word ι_1 of type ι. In view of (21) we may find a word Z of type $\tilde{\iota}$ such that

$$\forall\theta_1(\forall i(1<i\leqslant L_\tau(Z)\supset\mathfrak{L}(И(i\square Z),2^{И(1\square Z)}\cdot\theta_1))\supset\mathfrak{L}(\iota_1,\theta_1)). \tag{25}$$

Suppose that $Z\,\underline{\underline{\circ}}\,n\tau P_1\tau\cdots\tau P_k$, where n is a natural number and $P_1,\cdots,P_k$ are words of type κ. Put $\tilde{\iota}_1\leftrightharpoons n+1\tau P_1\tau\cdots\tau P_k$. Suppose that θ_1 is a point of the space $\mathfrak{M}$ such that $\tilde{\mathfrak{Q}}(\tilde{\iota}_1,\theta_1)$. Then

$$\forall i(1<i\leqslant L_\tau(\tilde{\iota}_1)\supset\Phi(И(i\square\tilde{\iota}_1)\square\theta_1)<2^{-n}),$$

so that

$$\forall i(1<i\leqslant L_\tau(\tilde{\iota}_1)\supset\Phi(И(i\square\tilde{\iota}_1)\square 2^n\cdot\theta_1)<1).$$

In view of property d) of 4.3.1 it follows that

$$\forall i(1<i\leqslant L_\tau(Z)\supset\mathfrak{L}(И(i\square Z),2^{И(1\square Z)}\cdot\theta_1)).$$

Therefore in view of (25) we get $\mathfrak{Q}(\iota_1,\theta_1)$. Formula (23) is proved.

Now we prove (24). Suppose given a word $\tilde{\iota}_1$ of type $\tilde{\iota}$ and $\tilde{\iota}_1 \overline{\underline{\circ}}\, n\, \tau P_1 \tau \cdots \tau P_k$, where n is a natural number and $P_1, \cdots, P_k$ are words of type κ. For any i with $1 \leqq i \leqq k$ it is possible to find a word Y_i of type ι such that

$$(26) \qquad \forall \theta_1 (\mathfrak{L}(Y_i, \theta_1) \supset \mathfrak{L}(P_i, 2^n \cdot \theta_1))$$

(see 1.3.7). Thereupon, in view of Axiom III2, one can find a word ι_1 of type ι such that

$$(27) \qquad \forall \theta_1 (\mathfrak{L}(\iota_1, \theta_1) \supset \mathfrak{L}(Y_1, \theta_1) \& \ldots \& \mathfrak{L}(Y_k, \theta_1)).$$

From (26), (27), and property e) of 4.3.1 we easily deduce that

$$\forall \theta_1 (\mathfrak{L}(\iota_1, \theta_1) \supset \tilde{\mathfrak{L}}(\tilde{\iota}_1, \theta_1)).$$

Formula (24), and therefore Theorem 4.3.2 are proved.

Theorem 4.3.2 shows that the existence of a generating subset $\mathfrak{K}$ of the set $\mathfrak{J}$ such that we can realize a Minkowski $\mathfrak{K}$-functional relative to the topology $(\text{A}_{\text{п}}, \mathfrak{J}, \mathfrak{Q})$ is a sufficient condition for the locally convex space $\mathfrak{M}$ to be multinormable. However, this condition is not necessary. Indeed, we have the theorem stated below.

By a nonempty subset of a set we mean a subset containing at least one element.

4.3.3. *It is possible to construct a multinormable locally convex space* $\mathfrak{M}$ *which has the following property: no nonempty subset* $\mathfrak{K}$ *of* $\mathfrak{J}$ *can have a Minkowski* $\mathfrak{K}$*-functional relative to the topology* $(\text{A}_{\text{п}}, \mathfrak{J}, \mathfrak{Q})$ *of the space* $\mathfrak{M}$.

PROOF. Suppose that S is a Specker sequence of rational numbers which has the following properties:

1) $\forall k (0 < S(k) \leqq S(k+1) < 1)$;

2) it is not possible to have a real duplex which is a limit of the sequence S.

We denote by $\mathfrak{J}$ the set of natural numbers and by $\mathfrak{Q}$ a two-parameter formula such that for any natural number i and real duplex x

$$\mathfrak{L}(i, x) \equiv \exists k (|x| \leqslant 2^{-i} \cdot S(k)).$$

It is not hard to verify that the list $(\text{Ч}_0, \mathfrak{J}, \mathfrak{Q})$ defines a locally convex topology on the linear space of all real duplexes. In other words, the list

$$(28) \qquad \text{Ч}_3, \mathfrak{P}_{\text{Д}}, \mathfrak{C}_{\text{Д}}, \underset{\text{в}}{+}, \underset{\text{в}}{\cdot}, 0, \text{Ч}_0, \mathfrak{J}, \mathfrak{L}$$

defines a locally convex space. We denote it by $\mathfrak{M}$.

It is easy to prove that

$$\forall i \exists j \forall x (\mathfrak{L}(j, x) \supset |x| \leqslant 2^{-i});$$

$$\forall i \exists j \forall x (|x| \leqslant 2^{-j} \supset \mathfrak{L}'(i, x)).$$

Therefore the topology $(\text{Ч}_0, \mathfrak{J}, \mathfrak{Q})$ is equivalent to the usual topology of the space of real duplexes, i.e. the space of Example 1, §1.4. Hence $\mathfrak{M}$ is multinormable.

Now suppose that $\mathfrak{K}$ is some nonempty subset of the set $\mathfrak{I}$, i.e. some nonempty set of natural numbers. We shall fix a natural number i_0 lying in $\mathfrak{K}$.

We suppose that we can realize a Minkowski $\mathfrak{K}$-functional relative to the topology $(\mathrm{Ч}_0, \mathfrak{I}, \mathfrak{Q})$. We denote this functional by Φ. Then we write $X \leftrightarrows \Phi(i_0 \square 1)$. It follows from properties $\Phi 1$ and $\Phi 2$ of the Minkowski $\mathfrak{K}$-functional that

$$\forall x \left(x > 0 \,\&\, \exists k\left(\frac{1}{x} \leqslant 2^{-i_0} \cdot S(k)\right) \supset x \geqslant X\right); \tag{29}$$

$$\forall n \exists x \left(x > 0 \,\&\, \exists k\left(\frac{1}{x} \leqslant 2^{-i_0} \cdot S(k)\right) \& x - X < 2^{-n}\right). \tag{30}$$

Since $1:(2^{i_0}:S(m)) \leqq 2^{-i_0} \cdot S(m)$ for any m, it follows from (29) that

$$\forall m\,(2^{i_0}:S(m) \geqslant X). \tag{31}$$

Let n be any natural number. By (30) we find a real duplex $x > 0$ and a natural number k such that

$$\frac{1}{x} \leqslant 2^{-i_0} \cdot S(k) \quad \text{and} \quad x - X < 2^{-n}.$$

Hence $x \geqq 2^{i_0}:S(k)$. From $x \geqq 2^{i_0}:S(k) \geqq X$ (see (31)) and from $x - X < 2^{-n}$ it follows that

$$0 \leqslant 2^{i_0}:S(k) - X < 2^{-n}.$$

Suppose that l is a natural number $\geqq k$. Then $S(k) \leqq S(l)$, so that $2^{i_0}:S(l) \leqq 2^{i_0}:S(k)$. On the other hand, (31) implies $2^{i_0}:S(l) \geqq X$. Thus we have proved that

$$\forall n \exists k \forall l\,(l \geqslant k \supset |2^{i_0}:S(l) - X| < 2^{-n}). \tag{32}$$

We shall show that $X > 0$. Indeed, in view of (30) it is possible to find a real duplex x such that

$$x > 0, \frac{1}{x} \leqslant 2^{-i_0} \quad \text{and} \quad X > x - 1.$$

Hence it follows that $X > 2^{i_0} - 1$, so that $X > 0$. It is not hard to see from (32) that

$$\forall n \exists k \forall l\,(l \geqslant k \supset |S(l) - 2^{i_0}:X| < 2^{-n}).$$

Thus the sequence S has a limit. But this is impossible. Therefore it is impossible to have such a Minkowski $\mathfrak{K}$-functional relative to the topology $(\mathrm{Ч}_0, \mathfrak{I}, \mathfrak{Q})$. Theorem 4.3.3 is proved.

Remark. In a quite analogous way one may prove a more general proposition. Let $\mathfrak{M}$ be a multinormable space. Suppose that its multinorm $(\mathrm{A}_q, \mathscr{A}, \mathscr{N})$ satisfies the condition

$$\forall \nu_1 \exists \theta_1 (\mathscr{N}(\nu_1 \square \theta_1) > 0)$$

where ν_1 is a variable for elements of the space $\mathscr{A}$. Then we can construct in $\mathfrak{M}$ a locally convex topology $(A_\text{п}, \mathfrak{J}, \mathfrak{Q})$ which has the following properties:

1) $(A_\text{п}, \mathfrak{J}, \mathfrak{Q})$ is multinormable, and equivalent to the topology of $\mathfrak{M}$;

2) for any nonempty subset $\mathfrak{K}$ of $\mathfrak{J}$ it is not possible to have a Minkowski $\mathfrak{K}$-functional relative to the topology $(A_\text{п}, \mathfrak{J}, \mathfrak{Q})$.

4.4. Now we shall say some words on constructive enumerably normable spaces. By definition, a multinormable space $\mathfrak{M}$ is said to be enumerably normable if its multinorm $(A_q, \mathscr{A}, \mathscr{N})$ is an enumerable norm, i.e. if the set $\mathscr{A}$ is enumerable. Without loss of generality we will suppose in what follows that in an enumerably normed space the set $\mathscr{A}$ is the set of all natural numbers, or, in some cases, the set of positive natural numbers. Under this hypothesis $\mathscr{N}$ becomes an algorithm of the type $(\text{н}\theta \to \text{д})$. In place of the notation $(A_q, \mathscr{A}, \mathscr{N})$ for an enumerable norm we shall use for short only the notation of the algorithm $\mathscr{N}$.

Suppose that $\mathfrak{M}$ is an enumerably normed space defined by the list

(33) $$A_\text{т}, \mathfrak{P}, \mathfrak{C}, +, \cdot, \mathfrak{D}, \mathscr{N}.$$

The locally convex topology corresponding to $\mathscr{N}$ will be called the topology of the space $\mathfrak{M}$. It follows from 4.2.1 and 4.1.1 that the topology of $\mathfrak{M}$ is equivalent to a topology having an enumerable fundamental system of neighborhoods of the origin, while the directed set of neighborhoods of the origin in this last topology coincides with a set of natural numbers with the usual order relation.

It is possible to construct a concrete topology equivalent to the topology of the space $\mathfrak{M}$ and having an enumerable fundamental system of neighborhoods of the origin, as follows. Denote by $\mathfrak{J}$ the set of natural numbers and by $\mathfrak{Q}$ a two-parameter formula such that for any natural number i and word θ_1 of type θ

$$\mathfrak{Q}(i, \theta_1) \equiv \max(\mathscr{N}(0 \square \theta_1) \square \ldots \square \mathscr{N}(i \square \theta_1)) \leqslant 2^{-i}.$$

It is not hard to see that the list $(\text{Ч}_0, \mathfrak{J}, \mathfrak{Q})$ is a locally convex topology equivalent to the topology of the space $\mathfrak{M}$. This topology has an enumerable fundamental system of neighborhoods of the origin, and the directed set of neighborhoods of the origin in this topology corresponds to the directed set of natural numbers. The topology $(\text{Ч}_0, \mathfrak{J}, \mathfrak{Q})$, defined in this way, will also be called the topology of the enumerably normed space $\mathfrak{M}$, and the locally convex space

(34) $$A_\text{т}, \mathfrak{P}, \mathfrak{C}, +, \cdot, \mathfrak{D}, \text{Ч}_0, \mathfrak{J}, \mathfrak{Q}$$

will be called a *locally convex space associated with the enumerably normed space* $\mathfrak{M}$.

REMARK. Suppose that φ is a sequence of points of the space $\mathfrak{M}$. The sequence φ converges in itself (converges to the point X) in $\mathfrak{M}$ if there exists an algorithm γ of the type $(\text{H} \to \text{H})$ such that

$$\forall imn\,(m, n \geqslant \gamma(i) \supset \mathfrak{Q}\,(i, \varphi(m) - \varphi(n))) \tag{35}$$

(respectively,

$$\forall in\,(n \geqslant \gamma(i) \supset \mathfrak{Q}\,(i, X - \varphi(n)))). \tag{35'}$$

Suppose that δ is an algorithm of the type $(\text{H}\,\text{H} \to \text{H})$, defined as follows:

$$\forall ki\,(\delta(k \,\square\, i) \simeq \gamma(\max(k \,\square\, i))).$$

Then δ satisfies the condition

$$\forall kimn\,(m, n \geqslant \delta(k \,\square\, i) \supset \mathcal{N}\,(k \,\square\, \varphi(m) - \varphi(n)) \leqslant 2^{-i}) \tag{36}$$

(respectively,

$$\forall kin\,(n \geqslant \delta(k \,\square\, i) \supset \mathcal{N}\,(k \,\square\, X - \varphi(n)) \leqslant 2^{-i})). \tag{36'}$$

Conversely, if the algorithm δ satisfies condition (36) (respectively (36′)), then the algorithm γ defined by the formula

$$\forall i\,(\gamma(i) \simeq \max_{j \leqslant i} \delta(j \,\square\, i)),$$

satisfies condition (35) ((35′)).

Suppose that the standard F-completion and the standard FR-completion of the space (34) are defined by the following lists respectively:

$$\text{A}_\psi,\ \mathfrak{P}^{\mathfrak{v}},\ \mathfrak{C}^{\Pi},\ \underset{\mathfrak{v}}{+},\ \underset{\mathfrak{v}}{\cdot},\ \mathfrak{D},\ \text{Ч}_0,\ \mathfrak{J},\ \mathfrak{Q}^{\Pi}, \tag{37}$$

$$\text{A}_\psi,\ \mathfrak{P}^{\mathfrak{w}},\ \mathfrak{C}^{\Pi},\ \underset{\mathfrak{w}}{+},\ \underset{\mathfrak{w}}{\cdot},\ \mathfrak{D},\ \text{Ч}_0,\ \mathfrak{J},\ \mathfrak{Q}^{\Pi} \tag{38}$$

(see §3.2). Here we use the notation introduced in §3. We note that in this case words of the type $\mathfrak{v}$ are F-N-constructs, and words of type $\mathfrak{w}$ are FR-N-constructs in the space (34).[5] In what follows, instead of "F-N-construct" and "FR-N-construct", we will write for short "F-construct" and "FR-construct".

By the definition of F-construct and FR-construct we have, for any words $\mathfrak{v}_1$ of type $\mathfrak{v}$ and $\mathfrak{w}_1$ of type $\mathfrak{w}$,

$$\forall i \exists l \forall mn\,(m, n \geqslant l \supset \mathfrak{Q}\,(i, \underline{\mathfrak{v}_{1_\psi}}(m) - (\underline{\mathfrak{v}_{1_\psi}}(n)))); \tag{39}$$

$$\forall imn\,(m, n \geqslant \overline{\mathfrak{w}}_1^{\psi}(i) \supset \mathfrak{Q}\,(i, \mathfrak{w}_{1_\psi}(m) - \underline{\mathfrak{w}_{1_\psi}}(n))). \tag{40}$$

Now we shall construct an algorithm φ such that, for any word P in A_ψ and any natural numbers k and n, if P is a word of type $\mathfrak{v}$ or $\mathfrak{w}$, then

[5] N is the set of natural numbers with the usual order relation.

$$\varphi(k\square P\square n)\simeq\mathcal{N}(k\square \underline{P_{\psi}}(n)).$$

From (39), (40) and the definition of the formula $\mathfrak{Q}$ it follows that

(41) $\forall \mathfrak{v}_1 k i \exists l \forall m n (m, n \geqslant l \supset |\varphi(k\square \mathfrak{v}_1 \square m) - \varphi(k\square \mathfrak{v}_1\square n)| < 2^{-i})$;

$$\forall \mathfrak{w}_1 k i m n (m, n \geqslant \overline{\mathfrak{w}_1}^{\psi}(\max(k\square i)) \supset |\varphi(k\square \mathfrak{w}_1\square m_1) - \varphi(k\square \mathfrak{w}_1\square n)| \leqslant 2^{-i}).$$

(42)

We construct an algorithm ψ:

$$\forall k\mathfrak{w}_1 i(\psi(k\square\mathfrak{w}_1\square i)\simeq \overline{\mathfrak{w}_1^{\psi}}(\max(k\square i))).$$

It is clear from (41) and (42) that for any word $\mathfrak{v}_1$ of type $\mathfrak{v}$, any word $\mathfrak{w}_1$ of type $\mathfrak{w}$ and any natural number k the sequences $(\varphi \uparrow k \square \mathfrak{v}_1 \square)$ and $(\varphi \uparrow k \square \mathfrak{w}_1 \square)$ are fundamental sequences of real duplexes, and $(\psi \uparrow k \square \mathfrak{w}_1 \square)$ is a regulator of convergence in itself of the sequence $(\varphi \uparrow k \square \mathfrak{w}_1 \square)$. Starting from the algorithms φ and ψ, we may construct algorithms $\mathcal{N}^{\mathfrak{v}}$ of the type (н $\mathfrak{v}\rightarrow$ в) and $\mathcal{N}^{\mathfrak{w}}$ of type (н $\mathfrak{w}\rightarrow$ д), having the following properties: 1) for any word $\mathfrak{v}_1$ of type $\mathfrak{v}$ and any natural number k, $\mathcal{N}^{\mathfrak{v}}(k\square\mathfrak{v}_1)$ is an F-letter, which is the limit in the space of F-numbers of a sequence χ defined as follows:

$$\forall n(\chi(n)\simeq \overline{\text{prin}}(\varphi(k\square\mathfrak{v}_1\square n)));$$

2) for any word $\mathfrak{w}_1$ of type $\mathfrak{w}$ and any natural number k, $\mathcal{N}^{\mathfrak{w}}(k\square\mathfrak{w}_1)$ is an FR-number which is the limit in the space of FR-numbers of the sequence $(\varphi \uparrow k \square \mathfrak{w}_1 \square)$ (with the regulator of convergence $(\psi \uparrow k \square \mathfrak{w}_1 \square)$. (See [26], Theorem 10.4.6 and Remark 2 to it.)

It is possible to prove that $\mathcal{N}^{\mathfrak{v}}$ defines an enumerable norm in the space (37), and that $\mathcal{N}^{\mathfrak{w}}$ defines an enumerable norm in the space (38). Moreover, the space (37) is enumerably seminormable by means of the enumerable seminorm $\mathcal{N}^{\mathfrak{v}}$, and the space (38) is enumerably normable by means of the enumerable norm $\mathcal{N}^{\mathfrak{w}}$.

The enumerably seminormed space

$$\mathrm{A}_{\psi},\ \mathfrak{P}^{\mathfrak{v}},\ \mathfrak{E}^{\Pi},\ \underset{\mathfrak{v}}{+},\ \underset{\mathfrak{v}}{\cdot},\ \mathfrak{O},\ \mathcal{N}^{\mathfrak{v}}$$

will be called *the standard F-completion* of the enumerably normed space $\mathfrak{M}$, and the enumerably normed space

$$\mathrm{A}_{\psi},\mathfrak{P}^{\mathfrak{w}},\ \mathfrak{E}^{\Pi},\ \underset{\mathfrak{w}}{+},\ \underset{\mathfrak{w}}{\cdot},\ \mathfrak{O},\ \mathcal{N}^{\mathfrak{w}}$$

will be called *the standard FR-completion* of the enumerably normed space $\mathfrak{M}$.

CHAPTER II

CERTAIN BASIC STRUCTURES OF THE THEORY OF CONSTRUCTIVE LOCALLY CONVEX SPACES

§1. Product of a family of constructive locally convex spaces

1.1. In this subsection we present a method of representing a family of constructive locally convex spaces. For simplicity of exposition we will use the terminology and notation of the extended language whose description was given in the paper [**35**].

We suppose that we are given a logico-mathematical language $\mathfrak{A}_\Sigma^\Delta$ of the type described in [**35**], described on the basis of a list of basic alphabets

$$\text{(1)} \qquad A_0, A_1, \cdots, A_h,$$

a list Σ of one-parameter formulas of the language $\mathfrak{A}$ and a list Δ of two-parameter formulas of the language $\mathfrak{A}_\Sigma$ (see [**35**]).

Suppose that A_q is an alphabet of the list (1) and $\mathscr{A}$ some normal formula of the list Σ, whose parameter is a variable of the genus t_q. The subordinate generic letter assigned to the formula $\mathscr{A}$ will be denoted by ν. We suppose that the set $\mathscr{A}$ contains at least one element. Suppose further that: 1) A_{m} and A_{n} are alphabets of the list (1); 2) $\mathfrak{P}$ is a two-parameter formula of the list Δ, whose parameters are a variable u of genus ν and a parameter α of genus t_{m}; 3) $\mathfrak{J}$ is a two-parameter formula of the list Δ, whose parameters are a variable u of genus ν and a variable β of genus t_{n}; 4) $\mathfrak{E}$ is a three-parameter formula, whose parameters are a variable u of genus ν and variables γ and δ of genus t_{m}; 5) $\mathfrak{Q}$ is a three-parameter formula, whose variables are a variable u of genus ν, a variable ζ of genus t_{n}, and a variable η of genus t_{m}.[1)]

We suppose that $\mathfrak{J}$ is a normal formula. More precisely speaking, the equivalent of $\mathfrak{J}$ in the language $\mathfrak{A}$ is a normal formula. We denote the pregeneric letter assigned to the formula $\mathfrak{P}$ by θ, and the pregeneric letter assigned to the formula $\mathfrak{J}$ by ι (see the definition of pregeneric letter in [**35**]). We introduce the following notation:

$$\mathfrak{P}^U \leftrightharpoons (U \overset{+}{\in} \mathrm{v}) \,\&\, F^u_{U}\llcorner \mathfrak{P} \lrcorner; \qquad \mathfrak{J}^U \leftrightharpoons (U \overset{+}{\in} \mathrm{v}) \,\&\, F^u_{U}\llcorner \mathfrak{J} \lrcorner;$$

$$\mathfrak{E}^U \leftrightharpoons F^u_{U}\llcorner \mathfrak{E} \lrcorner; \qquad \mathfrak{Q}^U \leftrightharpoons F^u_{U}\llcorner \mathfrak{Q} \lrcorner;$$

---

1) We are supposing that $\mathfrak{P}$, $\mathfrak{J}$, $\mathfrak{E}$ and $\mathfrak{Q}$ are formulas of the language $\mathfrak{A}_\Sigma$. The parameters of genus ν of these formulas may fail to be the same. However, for economy of symbols (and without loss of generality) we denote all of these parameters of genus ν by the common letter U. In what follows the variables of genus ν will frequently be denoted by u, v and w.

$$(P \underset{U}{=} Q) \quad [\text{or} \quad \mathfrak{C}(U, P, Q)] \leftrightharpoons F^{u,\,\gamma,\,\delta}_{U,\,P,\,Q}\llcorner \mathfrak{C} \lrcorner;$$

$$\mathfrak{Q}(U, R, P) \; [\text{or} \quad \mathfrak{Q}^U (R, P)] \rightleftharpoons F^{u,\,\zeta,\,\eta}_{U,\,R,\,P}\llcorner \mathfrak{Q} \lrcorner,$$

where U, P, Q and R are arbitrary object terms.

We note that, for any word U of type ν, words of type θ^U are elements of the set $\mathfrak{P}^U$, and words of the type ι^U are elements of the set $\mathfrak{I}^U$ (see [**35**]).

Suppose given further the following objects: 1) an algorithm $+$ such that, for any word U of type ν, $(+ \uparrow U\square)$ is an algorithm of the type $(\theta^U\theta^U \to \theta^U)$; 2) an algorithm $\cdot$ such that, for any word U of the type ν, $(\cdot \uparrow U\square)$ is an algorithm of the type $(д\theta^U \to \theta^U)$; 3) an algorithm $\mathfrak{O}$ such that, for any word U of type ν, $\mathfrak{O}(U)$ is a word of type θ^U. The algorithms $(+ \uparrow U\square)$ and $(\cdot \uparrow U\square)$ will be denoted by $\underset{U}{+}$ and $\underset{U}{\cdot}$ respectively, and the word $\mathfrak{O}(U)$ by $\mathfrak{O}^U$.

DEFINITION. The list

$$\mathrm{A}_q,\ \mathscr{A},\ \mathrm{A}_{\mathfrak{m}},\ \mathfrak{P},\ \mathfrak{C},\ +,\ \cdot,\ \mathfrak{O},\ \mathrm{A}_{\mathfrak{n}},\ \mathfrak{I},\ \mathfrak{Q} \tag{2}$$

will be called an *$\mathscr{A}$-family of constructive locally convex spaces* if for any fixed word U of type ν the list

$$\mathrm{A}_{\mathfrak{m}},\ \mathfrak{P}^U,\ \mathfrak{C}^U,\ \underset{U}{+},\ \underset{U}{\cdot},\ \mathfrak{O}^U,\ \mathrm{A}_{\mathfrak{n}},\ \mathfrak{I}^U,\ \mathfrak{Q}^U \tag{3}$$

is a constructive locally convex space.

This definition, of course, may be formulated in the form of a system of axioms which the list (2) must satisfy. We write down only the topological axioms of this system, as follows:

III 1*. $\forall u\iota_1^u\theta_1^u\theta_2^u\,(\theta_1^u \underset{u}{=} \theta_2^u \,\&\, \mathfrak{Q}(u, \iota_1^u, \theta_1^u) \supset \mathfrak{Q}(u, \iota_1^u, \theta_2^u));$

III 2*. $\forall u\iota_1^u\iota_2^u\exists\iota_3^u\forall\theta_1^u\,(\mathfrak{Q}(u, \iota_3^u, \theta_1^u) \supset \mathfrak{Q}(u, \iota_1^u, \theta_1^u) \,\&\, \mathfrak{Q}(u, \iota_2^u, \theta_1^u));$

III 3*. $\forall u\iota_1^u\theta_1^u\theta_2^u xy\,(|x| + |y| = 1 \,\&\, \mathfrak{Q}(u, \iota_1^u, \theta_1^u) \,\&\, \mathfrak{Q}(u, \iota_1^u, \theta_2^u)$
$\supset \mathfrak{Q}(u, \iota_1^u, x_u \underset{u}{\cdot} \theta_1^u \underset{u}{+} y \underset{u}{\cdot} \theta_2^u));$

III 4*. $\forall u\iota_1^u\theta_1^u\exists x\,(x > 0 \,\&\, \mathfrak{Q}(u, \iota_1^u, x \underset{u}{\cdot} \theta_1^u));$

III 5*, $\forall u\iota_1^u\exists\iota_2^u\forall\theta_1^u\theta_2^u\,(\mathfrak{Q}(u, \iota_2^u, \theta_1^u) \,\&\, \mathfrak{Q}(u, \iota_2^u, \theta_2^u) \supset \mathfrak{Q}(u, \iota_1^u, \theta_1^u \underset{u}{+} \theta_2^u)).$

We will denote the space (3) by $\mathfrak{M}^U$.

1.2. Suppose given an $\mathscr{A}$-family of constructive locally convex spaces defined by the list (2).

We suppose that the letters τ and э do not enter into the alphabet $\mathrm{A}_{\mathfrak{n}} \cup \mathrm{A}_q$. We adjoin to the list of basic alphabets of the language Я the alphabets $\mathrm{A}_{\mathfrak{m}} \cup \mathrm{A}_q$ and $\mathrm{A}_{\mathfrak{n}} \cup \mathrm{A}_q \cup \{\tau, э\}$, and denote their indices by r and s respectively:

$$\mathrm{A}_r = \mathrm{A}_{\mathfrak{m}} \cup \mathrm{A}_q; \quad \mathrm{A}_s = \mathrm{A}_{\mathfrak{n}} \cup \mathrm{A}_q \cup \{\tau, э\}.$$

We shall construct in A_s^{sa} an algorithm $\mathfrak{A}$ such that, for any word P in A_s, $\mathfrak{A}(P) \doteq \Lambda$ if and only if P has the form

(4) $$P_1 \ni Q_1 \tau \cdots \tau P_k \ni Q_k,$$

where $P_1, \cdots, P_k$ are words in A_q and $Q_1, \cdots, Q_k$ are words in A_n $(k \geqq 1)$.

We denote by L_τ an algorithm which is applicable to any word of the form (4) and which transforms each such word into the number of its τ-terms. Then we construct algorithms $И_1$ and $И_2$ having the following properties: for any word P in A_s and any natural number i, if $\mathfrak{A}(P) = \Lambda$ and $1 \leqq i \leqq L_\tau(P)$, then $И_1(i \,\Box\, P)$ is the first $\ni$-term of the ith τ-term of the word P, and $И_2(i \,\Box\, P)$ is the second $\ni$-term of the ith τ-term of the word P. For example, if P is the word (4) and $1 \leqq i \leqq k$, then $И_1(i \,\Box\, P) \doteq P_i$ and $И_2(i \,\Box\, P) \doteq Q_i$.

Now we introduce the following formulas:

$$\mathfrak{P}^* \leftrightharpoons \forall u (\langle c_{0,1}\rangle_r (u) \overset{+}{\in} \mathfrak{P}^u);$$

$$\mathfrak{C}^* \leftrightharpoons \forall u (\langle c_{0,1}\rangle_r (u) \underset{u}{==} \langle c_{0,2}\rangle_r (u)),$$

where u is a variable of genus ν, $c_{0,1}$ and $c_{0,2}$ are variables of genus c_0 (i.e. variables for words in $Ч_0$), and $t_{s,1}$ is a variable of genus t_s (i.e. a variable for words in A_s).

The word P in $Ч_0$ is an element of the set $\mathfrak{P}^*$ if and only if P is the transcription of an algorithm which transforms each word U of type ν into a word of type θ^U. The word P in A_s lies in the set $\mathfrak{J}^*$ if and only if P has the form (4), while P_i is a word of type ν and Q_i a word of type ι^{P_i} $(i = 1, \cdots, k)$. Since $\mathscr{A}$ and $\mathfrak{J}$ are normal formulas, $\mathfrak{J}^*$ (more precisely, the equivalent to $\mathfrak{J}^*$ in the language Я) is also a normal formula.

We adjoin the formulas $\mathfrak{P}^*$ and $\mathfrak{J}^*$ to the list Σ of formulas of the language $Я_{\Sigma'}$ and denote the subordinate generic letters assigned to those formulas by θ and ι. We note that the signs θ and ι have already been used to denote the pregeneric letters assigned to the formulas $\mathfrak{P}$ and $\mathfrak{J}$. Thus we choose one and the same sign θ to denote the pregeneric letter assigned to the formula $\mathfrak{P}$ and for the generic letter assigned to the formula $\mathfrak{P}^*$, and analogously for ι. Such a use of the signs in question will not lead to confusion in the further exposition. We shall proceed in this way below in other analogous situations as well.

We construct in $Ч_0^{sa}$ an algorithm $\underset{\theta}{+}$ of the type $(\theta\theta \to \theta)$, and in $Ч_3^{sa}$ an algorithm $\underset{\theta}{\cdot}$ of the type $(д\theta \to \theta)$, such that

$$\forall \theta_1 \theta_2 u (\langle \underset{\theta}{+} (\theta_1 \Box \theta_2)\rangle_r (u) \simeq \langle\theta_1\rangle_r (u) \underset{u}{+} \langle\theta_2\rangle_r (u));$$

$$\forall x \theta_1 u (\langle \underset{\theta}{\cdot} (x \Box \theta_1)\rangle_r (u) \simeq x \underset{u}{\cdot} \langle\theta_1\rangle_r (u)).$$

Finally we denote by $\mathfrak{D}^*$ the transcription of the algorithm $\mathfrak{D}$ relative to the alphabet A_r^{sa}. Obviously $\mathfrak{D}^*$ is a word of type θ.

We agree to write

$$(P \underset{\theta}{=} Q) \leftrightharpoons F_{P,Q}^{c_{0,1}, c_{0,2}} \llcorner \mathfrak{E}^* \lrcorner; \qquad \mathfrak{Q}^*(R, P) \leftrightharpoons F_{R,P}^{t_{s,1}, c_{0,1}} \llcorner \mathfrak{Q}^* \lrcorner;$$

$$(P \underset{\theta}{+} Q) \leftrightharpoons \underset{\theta}{+} (P \square Q); \qquad (P \underset{\theta}{\cdot} Q) \leftrightharpoons \underset{\theta}{\cdot} (P \square Q),$$

where P, Q, and R are arbitrary object terms.

Choosing the objects just constructed, we obtain the list

(5) $$\mathfrak{U}_0, \mathfrak{P}^*, \mathfrak{E}^*, \underset{\theta}{+}, \underset{\theta}{\cdot}, \mathfrak{D}^*, \mathrm{A}_s, \mathfrak{J}^*, \mathfrak{Q}^*.$$

It is easily verified that this list defines a constructive locally convex space, which we shall call the *product* of the $\mathscr{A}$-family (2) of locally convex spaces and denote by $\mathfrak{M}$.

Now we shall prove some theorems on the separation of a product of an $\mathscr{A}$-family of locally convex spaces.

1.2.1. *The product of an $\mathscr{A}$-family of T'-separated locally convex spaces is a T'-separated space.*

Proof. We suppose that the list (2) defines an $\mathscr{A}$-family of T'-separated spaces. This means that

(6) $$\forall u \theta_1^u (\forall \iota_1^u \mathfrak{Q}(u, \iota_1^u, \theta_1^u) \supset \theta_1^u \underset{u}{=} (\mathfrak{D}^u)).$$

We have to prove that the space $\mathfrak{M}$ is T'-separated, i.e. that

(7) $$\forall \theta_1 (\forall \iota_1 \mathfrak{Q}^*(\iota_1, \theta_1) \supset \theta_1 \underset{\theta}{=} \mathfrak{D}^*).$$

Suppose that θ_1 is a point of $\mathfrak{M}$ such that $\forall \iota_1 \mathfrak{Q}^*(\iota_1, \theta_1)$. From this property it is easily deduced that $\forall u \iota_1^U (\mathfrak{Q}(u, \iota_1^U, \langle\theta_1\rangle_r(u))$. Hence, in view of (6) we find that $\forall u (\langle\theta_1\rangle_r(u) \underset{u}{=} \mathfrak{D}^U)$, i.e. $\theta_1 \underset{\theta}{=} \mathfrak{D}^*$. Formula (7) is proved.

1.2.2. *If the product $\mathfrak{M}$ is T'-separated, then for any word U of type ν the space $\mathfrak{M}^U$ is T'-separated.*

Proof. Suppose that (7) holds. We have to prove formula (6). Suppose that U is a word of type ν and that P is a word of type θ^U such that $\forall \iota_1^U \mathfrak{Q}(U, \iota_1^U, P)$. It is possible to construct a word θ_1 of type θ such that, for any word R of type ν,

$$\langle\theta_1\rangle_r(R) = \begin{cases} P, & \text{if } R \underset{\circ}{=} U; \\ \mathfrak{D}(R), & \text{if } R \underset{\circ}{\neq} U. \end{cases}$$

Suppose that ι_1 is a word of type ι. If the word $\tau U \ni$ does not enter into the word $\tau\iota_1$, then obviously $\mathfrak{Q}^*(\iota_1, \theta_1)$. If the word $\tau U \ni$ does enter into the word $\tau\iota_1$, then from the hypothesis $\forall \iota_1^U \mathfrak{Q}(U, \iota_1^U, P)$ it also follows that $\mathfrak{Q}^*(\iota_1, \theta_1)$. Thus we have proved that $\forall \iota_1 \mathfrak{Q}^*(\iota_1, \theta_1)$. Hence from (7) we find that $\theta_1 \underset{\theta}{=} \mathfrak{D}^*$, so that $P \underset{U}{=} \mathfrak{D}(U)$. Formula (6) is proved.

1.2.3. *If the product* $\mathfrak{M}$ *is T-separated, then for any word U of type* ν *the space* $\mathfrak{M}^U$ *is T-separated.*

PROOF. Suppose that $\mathfrak{M}$ is T-separated, i.e. that

(8) $$\forall\theta_1(\urcorner(\theta_1 \underset{\theta}{=} \mathfrak{O}^*) \supset \exists\iota_1 \urcorner\mathfrak{Q}^*(\iota_1, \theta_1)).$$

Suppose that U is a word of type ν and P a word of type θ^U such that $\urcorner(P \underset{U}{=} \mathfrak{O}^U)$. We have to prove that $\exists\iota_1^U \urcorner\mathfrak{Q}(U, \iota_1^U, P)$. To this end we construct a word θ_1 of type θ the same as in the proof of 1.2.2. Obviously $\urcorner(\theta_1 \underset{\theta}{=} \mathfrak{O}^*)$. Hence from (8) it follows that a word ι_1 of type ι is realizable for which $\urcorner\mathfrak{Q}^*(\iota_1, \theta_1)$.

If the word $\tau U \ni$ does not occur in the word $\tau\iota_1$, then $\mathfrak{Q}^*(\iota_1, \theta_1)$. But this is impossible and therefore the word $\tau U \ni$ cannot not occur in the word $\tau\iota_1$, so that $\tau U \ni$ occurs j times in $\tau\iota_1$. Then we can realize words $V_1, \cdots, V_j$ of type ι^U such that the words $\tau U \ni V_1 \tau, \cdots, \tau U \ni V_j \tau$ occur in the word $\tau\iota_1\tau$. We find a word V of type ι^U such that

$$\forall\theta_1^U(\mathfrak{Q}(U, V, \theta_1^U) \supset \mathfrak{Q}(U, V_1, \theta_1^U) \& \ldots \& \mathfrak{Q}(U, V_j, \theta_1^U)).$$

From the definition of the formula $\mathfrak{Q}^*$ and from $\urcorner\mathfrak{Q}^*(\iota_1, \theta_1)$ it follows that

$$\urcorner(\mathfrak{Q}(U, V_1, P) \& \ldots \& \mathfrak{Q}(U, V_j, P)),$$

so that $\urcorner\mathfrak{Q}(U, V, P)$, as we were required to prove.

1.2.4. *If* $\mathscr{A}$ *is an algorithmically enumerable set, then the product of an* $\mathscr{A}$*-family of metrizable locally convex spaces is a T-separated space.*

PROOF. Suppose that the list (2) defines an $\mathscr{A}$-family of metrizable spaces. Taking into account the fact that $\mathscr{A}$ is a normal formula, from this assumption we deduce that an algorithm ρ having the following property is realizable:

(9) $$\forall u\theta_1^u((\rho(u\Box\theta_1^u) \overset{+}{\in} \text{Д}) \& (\theta_1^u \underset{u}{=} \mathfrak{O}^u \equiv \rho(u\Box\theta_1^u) = 0)).$$

We shall prove that the product $\mathfrak{M}$ is T-separated, i.e. that

(10) $$\forall\theta_1(\urcorner(\theta_1 \underset{\theta}{=} \mathfrak{O}^*) \supset \exists\iota_1 \urcorner\mathfrak{Q}^*(\iota_1, \theta_1)).$$

Suppose given a word θ_1 of type θ such that $\urcorner(\theta_1 \underset{\theta}{=} \mathfrak{O}^*)$. Then

$$\urcorner\forall u(\langle\theta_1\rangle_r(u) \underset{u}{=} \mathfrak{O}(u)).$$

Hence, and from (9), it follows that

(11) $$\urcorner\forall u(\rho(u\Box\langle\theta_1\rangle_r(u)) = 0).$$

Since $\mathscr{A}$ is an enumerable set, in view of Lemma 1 in the Appendix we find from (11) that

$$\exists u\,(\rho\,(u\,\square\,\langle\theta_1\rangle_r\,(u)) \neq 0).$$

Thus a word U of type ν is realizable such that $\rho(U\,\square\,\langle\theta_1\rangle_r(U)) \neq 0$. Hence, in view of (9), we find that $\neg\,(\langle\theta_1\rangle_r(U) \underset{u}{=} \mathfrak{O}(U))$. The space $\mathfrak{M}^U$ is metrizable, and so T-separated. Therefore we may find a word V of type ι^U such that $\neg\,\mathfrak{Q}(U, V, \langle\theta_1\rangle_r(U))$, from which we get $\neg\,\mathfrak{Q}^*(U \ni V, \theta_1)$. Formula (10) is thus proved.

1.2.5. *It is possible to construct an $\mathscr{A}$-family of normable spaces whose product is not T-separated.*

PROOF. Suppose that $\mathscr{A}$ is the set of words in Ч_0 constructed in Lemma 2 of the Appendix. This set $\mathscr{A}$ is defined by a normal formula and has the following property: it is not true that for any algorithm λ of the type $(\nu \to \text{Д})$ one has

$$\neg\forall \nu_1\,(\lambda\,(\nu_1) = 0) \supset \exists \nu_1\,(\lambda\,(\nu_1) \neq 0), \tag{12}$$

where ν is a subordinate generic letter assigned to the set $\mathscr{A}$.

It is not hard to construct an $\mathscr{A}$-family of locally convex spaces such that for any word U of type ν the space $\mathfrak{M}^U$ of this $\mathscr{A}$-family coincides with the space R of real duplexes (Chapter I, §1.4, Example 1). Since R is a normable space, this $\mathscr{A}$-family is an $\mathscr{A}$-family of normable spaces. We denote their product by $\mathfrak{M}$ and show that $\mathfrak{M}$ is not T-separated. To this end we suppose that $\mathfrak{M}$ is T-separated, i.e. that

$$\forall \theta_1\,(\neg\,(\theta_1 \underset{\theta}{=} \mathfrak{O}^*) \supset \exists \iota_1\,\neg\,\mathfrak{Q}^*\,(\iota_1, \theta_1)). \tag{13}$$

We note that in this case the words of type θ are the transcriptions of algorithms of the type $(\nu \to \text{Д})$, and words of type ι are words of type $P_1 \ni n_1 \tau \cdots \tau P_k \ni n_k$, where $P_1, \cdots, P_k$ are words of type ν and $n_1, \cdots, n_k$ are natural numbers ($k \geqq 1$).

Suppose that λ is some algorithm of type $(\nu \to \text{Д})$ having the property

$$\neg\,\forall \nu_1\,(\lambda\,(\nu_1) = 0).$$

Write $\theta_1 \leftrightharpoons \{\lambda, \text{Ч}_3^{\text{sa}}\}$. Then θ_1 is a word of type θ and $\neg\,(\theta_1 \underset{\theta}{=} \mathfrak{O}^*)$. Therefore it follows from (13) that we can realize a word ι_1 of type ι such that $\neg\,\mathfrak{Q}^*(\iota_1, \theta_1)$. Suppose that ι_1 is representable as follows:

$$\iota_1 \rightleftharpoons P_1 \ni n_1 \tau \ldots \tau P_k \ni n_k,$$

where $P_1, \cdots, P_k$ are words of type ν and $n_1, \cdots, n_k$ are natural numbers.

It follows from $\neg\,\mathfrak{Q}^*(\iota_1, \theta_1)$ that

$$\neg\,\forall i\,(1 \leqslant i \leqslant k \supset (|\langle\theta_1\rangle_r\,(P_i)| - 2^{-n_i}) \leqslant 0).$$

Hence, using Lemma 1 of the Appendix, we find that

$$\exists i\,(1 \leqslant i \leqslant k \,\&\, (|\langle\theta_1\rangle_r\,(P_i)| - 2^{-n_i}) > 0).$$

Thus it is possible to find a word P_i of type ν such that $|\langle\theta_1\rangle_r(P_i)|>0$, i.e. $\lambda(P_i)\neq 0$. Thus we get

$$\exists\, \nu_1(\lambda(\nu_1)\neq 0).$$

We have proved that for any algorithm λ of the type $(\nu\to \text{д})$ the relation (12) holds. But that is not true. Therefore $\mathfrak{M}$ is not T-separated.

REMARK. The set $\mathscr{A}$ used in the proof of 1.2.5 is not an algorithmically enumerable set.

1.3. In the special case of a finite family of locally convex spaces we may define the product in another way, as follows.

Suppose given locally convex spaces $\mathfrak{M}_1,\cdots,\mathfrak{M}_k$, with $\mathfrak{M}_i$ $(1\leqq i\leqq k)$ defined by the list

$$\mathrm{A}_{\mathfrak{m}_i},\ \mathfrak{P}_i,\ \mathfrak{E}_i,\ \underset{i}{+},\ \underset{i}{\cdot},\ \mathfrak{O}_i,\ \mathrm{A}_{\mathfrak{n}_i},\ \mathfrak{J}_i.\ \mathfrak{Q}_i.$$

We denote the indices of the alphabets $\mathrm{A}_{\mathfrak{m}_1}\cup\cdots\cup\mathrm{A}_{\mathfrak{m}_k}$ and $\mathrm{A}_{\mathfrak{n}_1}\cup\cdots\cup\mathrm{A}_{\mathfrak{n}_k}$ by $\mathfrak{m}$ and $\mathfrak{n}$ respectively:

$$\mathrm{A}_{\mathfrak{m}}=\mathrm{A}_{\mathfrak{m}_1}\cup\ldots\cup\mathrm{A}_{\mathfrak{m}_k},\ \mathrm{A}_{\mathfrak{n}}=\mathrm{A}_{\mathfrak{n}_1}\cup\ldots\cup\mathrm{A}_{\mathfrak{n}_k}.$$

Suppose that э is a letter not entering into $\mathrm{A}_{\mathfrak{m}}$ or into $\mathrm{A}_{\mathfrak{n}}$. We adjoin $\mathrm{A}_{\mathfrak{m}}\cup\{\text{э}\}$ and $\mathrm{A}_{\mathfrak{n}}\cup\{\text{э}\}$ to the list of basic alphabets of the language Я and denote their indices by r and s respectively:

$$\mathrm{A}_r=\mathrm{A}_{\mathfrak{m}}\cup\{\text{э}\},\qquad \mathrm{A}_s=\mathrm{A}_{\mathfrak{n}}\cup\{\text{э}\}.$$

It is possible to construct sets $\mathfrak{P}^*$ and $\mathfrak{J}^*$ such that: a) the word P in A_r belongs to the set $\mathfrak{P}^*$ if and only if P is a k-termed э-system of words in $\mathrm{A}_{\mathfrak{m}}$, while the ith э-term of the word P is a point of the space $\mathfrak{M}_i$ $(i=1,\cdots,k)$; b) a word Q in A_s belongs to the set $\mathfrak{J}^*$ if and only if Q is a k-termed э-system of words in $\mathrm{A}_{\mathfrak{n}}$, while the ith э-term of the word Q is an element of the set $\mathfrak{J}_i$. The set $\mathfrak{J}^*$ may be defined by a normal formula.

We construct algorithms $И_1$ and $И_2$ such that: a) if P is a k-termed э-system of words in $\mathrm{A}_{\mathfrak{m}}$ and $1\leqq i\leqq k$, then $И_1(i\,\square\,P)$ is the ith э-term of the word P; b) if Q is a k-termed э-system of words in $\mathrm{A}_{\mathfrak{n}}$ and $1\leqq i\leqq k$, then $И_2(i\,\square\,P)$ is the ith э-term of the word Q.

We denote the subordinate generic letters put into correspondence with the formulas $\mathfrak{P}^*$ and $\mathfrak{J}^*$ by θ and ι respectively.

Now we construct two-parameter formulas $\mathfrak{E}^*$ and $\mathfrak{Q}^*$, an algorithm $\underset{\theta}{+}$ of the type $(\theta\theta\to\theta)$ and an algorithm $\underset{\theta}{\cdot}$ of the type $(\text{д}\theta\to\theta)$ such that for any words θ_1 and θ_2 of type θ, any word ι_1 of type ι, and any real duplex x,

$$\mathfrak{E}^*(\theta_1,\theta_2)\equiv\mathfrak{E}_1(И_1(1\,\square\,\theta_1),\ И_1(1\,\square\,\theta_2))\,\&\ldots\&\,\mathfrak{E}_k(И_1(k\,\square\,\theta_1),\ И_1(k\,\square\,\theta_2));$$

$$\mathfrak{Q}^*(\iota_1,\theta_1)\equiv\mathfrak{Q}_1(И_2(1\,\square\,\iota_1),\ И_1(1\,\square\,\theta_1))\,\&\ldots\&\,\mathfrak{Q}_k(И_2(k\,\square\,\iota_1),\ И_1(k\,\square\,\theta_1));$$

$$\underset{\theta}{+}(\theta_1\square\,\theta_2)\simeq\underset{1}{+}(И_1(1\,\square\,\theta_1)\,\square\,И_1(1\,\square\,\theta_2))\,\text{э}\ldots\text{э}\,\underset{k}{+}(И_1(k\,\square\,\theta_1)\,\square\,И_1(k\,\square\,\theta_2));$$

$\underset{\theta}{\cdot}(x\square\theta_1) \simeq \underset{1}{\cdot}(x\square И_1(1\square\theta_1)) \ni \ldots \ni \underset{k}{\cdot}(x\square И_1(k\square\theta_1))$.

Finally, we write $\mathfrak{D}^* \leftrightharpoons \mathfrak{D}_1 \ni \cdots \ni \mathfrak{D}_k$. It is not hard to prove that the list

$$\mathrm{A}_r,\ \mathfrak{P}^*,\ \mathfrak{E}^*,\ \underset{\theta}{+},\ \underset{\theta}{\cdot},\ \mathfrak{D}^*,\ \mathrm{A}_s,\ \mathfrak{J}^*,\ \mathfrak{Q}^*$$

defines a locally convex space. We will call this list the product of the spaces $\mathfrak{M}_1, \cdots, \mathfrak{M}_k$ and denote it by $\mathfrak{M}_1 \times \cdots \times \mathfrak{M}_k$.

If all the spaces $\mathfrak{M}_1, \cdots, \mathfrak{M}_k$ coincide with one and the same space $\mathfrak{M}$, we shall denote this product simply by $\mathfrak{M}^{(k)}$.

§2. The projective limit of the projective spectrum of constructive locally compact spaces

2.1. Suppose given an $\mathscr{A}$-family of locally convex spaces defined by the list

$$\mathrm{A}_q,\ \mathcal{A},\ \mathrm{A}_{\mathfrak{m}},\ \mathfrak{P},\ \mathfrak{E},\ +,\ \cdot,\ \mathfrak{D},\ \mathrm{A}_{\mathfrak{n}},\ \mathfrak{J},\ \mathfrak{Q} \tag{1}$$

(see §1.1). Here we are using the notation introduced in §1.1.

Now we suppose that on $\mathscr{A}$ there is given an order relation $\mathfrak{X}$ such that the list $(\mathscr{A}, \mathfrak{X})$ becomes a directed set (see §3.1, Chapter I). Suppose that ω is an algorithm such that, for any words U and V of type ν, if $V \prec U$, then $(\omega \uparrow U\square V\square)$ is an algorithm of type $(\theta^U \to \theta^V)$, i.e. it transforms each word of the type θ^U into a word of type θ^V. We will say that the list

$$\mathrm{A}_q,\ \mathcal{A},\ \mathfrak{X},\ \mathrm{A}_{\mathfrak{m}},\ \mathfrak{P},\ \mathfrak{E},\ +,\ \cdot,\ \mathfrak{D},\ \mathrm{A}_{\mathfrak{n}},\ \mathfrak{J},\ \mathfrak{Q},\ \omega \tag{2}$$

defines a *projective spectrum of constructive locally convex spaces* if for each fixed word U of type ν the list

$$\mathrm{A}_{\mathfrak{m}},\ \mathfrak{P}^U,\ \mathfrak{E}^U,\ \underset{U}{+},\ \underset{U}{\cdot},\ \mathfrak{D}^U;\ \mathrm{A}_{\mathfrak{n}},\ \mathfrak{J}^U,\ \mathfrak{Q}^U \tag{3}$$

is a locally convex space and if the algorithm ω satisfies the following conditions:

1) $\forall u v \theta_1^u \theta_2^u (v \prec u \,\&\, \theta_1^u \underset{u}{=} \theta_2^u \supset \omega(u\square v\square\theta_1^u) \underset{v}{=} \omega(u\square v\square\theta_2^u))$;

2) $\forall u v \theta_1^u \theta_2^u (v \prec u \supset \omega(u\square v\square\theta_1^u \underset{u}{+} \theta_2^u) \underset{v}{=} \omega(u\square v\square\theta_1^u) \underset{v}{+} \omega(u\square v\square\theta_2^u))$;

3) $\forall u v x \theta_1^u (v \prec u \supset \omega(u\square v\square x \underset{u}{\cdot} \theta_1^u) \underset{v}{=} x \underset{v}{\cdot} \omega(u\square v\square\theta_1^u))$;

4) $\forall u v (v \prec u \supset \forall \iota_1^v \exists \iota_1^u \forall \theta_1^u (\mathfrak{Q}(u, \iota_1^u, \theta_1^u) \supset \mathfrak{Q}(v, \iota_1^v, \omega(u\square v\square\theta_1^u))))$;

5) $\forall u v w \theta_1^u (w \prec v \prec u \supset \omega(u\square w\square\theta_1^u) \underset{w}{=} \omega(v\square w\square\omega(u\square v\square\theta_1^u)))$.

Here we recall that x is a variable of genus д, and u, v and w are variables of genus ν.

The first four conditions show that, for any words U and V of type ν, if $V \prec U$, then $(\omega \uparrow U\square V\square)$ is a continuous linear mapping of the space $\mathfrak{M}^U$ into the space $\mathfrak{M}^V$.

Suppose that the letter τ does not enter into the alphabet $\mathrm{A}_{\mathfrak{n}} \cup \mathrm{A}_q$, and that in the list of basic alphabets of the language $\text{Я}_{\Sigma}^{\Delta}$ there are alphabets $\mathrm{A}_{\mathfrak{m}} \cup \mathrm{A}_q$ and $\mathrm{A}_{\mathfrak{n}} \cup \mathrm{A}_q \cup \{\tau\}$. We denote the indices of these alphabets by r and s correspondingly: $\mathrm{A}_r = \mathrm{A}_{\mathfrak{m}} \cup \mathrm{A}_q$ and $\mathrm{A}_s = \mathrm{A}_{\mathfrak{n}} \cup \mathrm{A}_q \cup \{\tau\}$. We introduce the following formulas:

$$\mathfrak{P}^* \leftrightharpoons \forall u (\langle c_{0,1}\rangle_r (u) \overset{+}{\in} \mathfrak{P}^u) \,\&\, \forall uv (v \prec \dot{u} \supset \langle c_{0,1}\rangle_r (v) \underset{v}{=} \omega (u \square v \square \langle c_{0,1}\rangle_r (u)));$$

$$\mathfrak{E}^* \leftrightharpoons \forall u (\langle c_{0,1}\rangle_r (u) \underset{u}{=} \langle c_{0,2}\rangle_r (u)),$$

where $c_{0,1}$ and $c_{0,2}$ are variables of genus c_0. The formula $\mathfrak{P}^*$ defines some set of words in the alphabet Ч_0. We adjoin the formula $\mathfrak{P}^*$ (more precisely speaking, its equivalent in the basic language Я) to the list of formulas Σ, and denote by θ the corresponding generic letter. We write

$$(P \underset{\theta}{=} Q) \leftrightharpoons F_{P,\,Q}^{c_{0,1},\,c_{0,2}} \llcorner \mathfrak{E}^* \lrcorner,$$

where P and Q are arbitrary object terms.

Consider the set of all possible words in the alphabet A_s of the form $U\tau R$, where U is a word of type ν and R a word of type ι^U. Since $\mathscr{A}$ and $\mathfrak{I}$ are normal formulas, this set may be defined by a normal formula, which we shall denote by $\mathfrak{I}^*$. We adjoin $\mathfrak{I}^*$ to the list of formulas Σ of the language Я_{Σ} and denote by ι the corresponding subordinate generic letter.

We construct a two-parameter formula $\mathfrak{Q}^*$ such that, for any words ι_1 of type ι and θ_1 of type θ, if $\iota_1 \overline{\underline{\circ}} U\tau R$, then

$$\mathfrak{Q}^* (\iota_1, \theta_1) \equiv \mathfrak{Q} (U, R, \langle\theta_1\rangle_r (U)).$$

Now we construct algorithms $\underset{\theta}{+}$ of type $(\theta\theta \to \theta)$ and $\underset{\theta}{\cdot}$ of type $(\text{д}\theta \to \theta)$ such that

$$\forall \theta_1\theta_2 u (\langle \underset{\theta}{+} (\theta_1 \square \theta_2)\rangle_r (u) \simeq \langle\theta_1\rangle_r (u) \underset{u}{+} \langle\theta_2\rangle_r (u);$$

$$\forall x\theta_1 u (\langle \underset{\theta}{\cdot} (x \square \theta_1)\rangle_r (u) \simeq x \underset{u}{\cdot} \langle\theta_1\rangle_r (u)).$$

Finally, we write $\mathfrak{D}^* \leftrightharpoons \mathcal{E}\mathfrak{D}, \mathrm{A}_r^{\mathrm{sa}}\mathcal{3}$.

Now we shall prove that the list

$$\text{Ч}_0,\ \mathfrak{P}^*,\ \mathfrak{E}^*,\ \underset{\theta}{+},\ \underset{\theta}{\cdot},\ \mathfrak{D}^*,\ \mathrm{A}_s,\ \mathfrak{I}^*,\ \mathfrak{Q}^* \tag{4}$$

is a locally convex space. To this end it suffices for us to verify for the list (4) the axioms of a locally convex space. This verification presents no difficulties at all. For example, the verification of Axiom III2 is carried out as follows. We need to prove that

$$\forall \iota_1\iota_2 \exists \iota_3 \forall \theta_1 (\mathfrak{Q}^* (\iota_3, \theta_1) \supset \mathfrak{Q}^* (\iota_3, \theta_1) \,\&\, \mathfrak{Q}^* (\iota_2, \theta_1)). \tag{5}$$

Suppose given words ι_1 and ι_2 of type ι, $\iota_1 \overline{\underline{\circ}} U\tau R$, $\iota_2 \overline{\underline{\circ}} V\tau S$. It is possible to find a word W of type ν such that $U \prec W$ and $V \prec W$. In view of property

4) of the algorithm ω, we can realize words X and Y of type ι^W such that

(6) $$\forall\theta_1^W(\mathfrak{Q}(W, X, \theta_1^W) \supset \mathfrak{Q}(U, R, \omega(W\square U\square\theta_1^W)));$$

(7) $$\forall\theta_1^W(\mathfrak{Q}(W, Y, \theta_1^W) \supset \mathfrak{Q}(V, S, \omega(W\square V\square\theta_1^W))).$$

Now it follows from Axiom III2* that we can realize a word Z of type ι^W such that

(8) $$\forall\theta_1^W(\mathfrak{Q}(W, Z, \theta_1^W) \supset \mathfrak{Q}(W, X, \theta_1^W) \& \mathfrak{Q}(W, Y, \theta_1^W)).$$

Put $\iota_3 \leftrightharpoons W\tau Z$. It is easy to deduce from (6) – (8) and from the definition of the set $\mathfrak{P}^*$ that

$$\forall\theta_1(\mathfrak{Q}^*(\iota_3, \theta_1) \supset \mathfrak{Q}^*(\iota, \theta_1) \& \mathfrak{Q}^*(\iota_2, \theta_1)).$$

Thus formula (5) is proved.

The locally convex space (4) will be called the *projective limit* of the projective spectrum of the locally convex spaces (2).

We denote the space (4) by $\mathfrak{M}$. We construct an algorithm Ω such that

$$\forall u\theta_1(\Omega(u\square\theta_1) \simeq \langle\theta_1\rangle_r(u)).$$

It is easy to see that for each fixed word U of type ν the algorithm $(\Omega \uparrow U\square)$ is a linear mapping of $\mathfrak{M}$ into the space $\mathfrak{M}^U$ defined by the list (3).

Suppose that in the linear space

(9) $$Ч_0, \mathfrak{P}^*, \mathfrak{E}^*, \underset{\theta}{+}, \underset{\theta}{\cdot}, \mathfrak{O}^*$$

there is defined some locally convex topology $(A_{\tilde{n}}, \tilde{\mathfrak{J}}, \tilde{\mathfrak{Q}})$. We will say that, for each word U of type ν, $(\Omega \uparrow U\square)$ is a continuous mapping of the space (9) with the topology $(A_{\tilde{n}}, \tilde{\mathfrak{J}}, \tilde{\mathfrak{Q}})$ into the space $\mathfrak{M}^U$ if

(10) $$\forall u\iota_1^u \exists\tilde{\iota}_1\forall\theta_1(\tilde{\mathfrak{Q}}(\tilde{\iota}_1, \theta_1) \supset \mathfrak{Q}(u, \iota_1^n, \Omega(u\square\theta_1))),$$

where $\tilde{\iota}_1$ is a variable for elements of the set $\tilde{\mathfrak{J}}$.

2.1.1. *For any word $\dot{U}$ of type ν, $(\Omega \uparrow U\square)$ is a continuous linear mapping of $\mathfrak{M}$ into $\mathfrak{M}^U$.*

2.1.2. *Suppose in the linear space* (9) *there is defined some locally convex topology* $(A_{\tilde{n}}, \tilde{\mathfrak{J}}, \tilde{\mathfrak{Q}})$, *satisfying condition* (10). *Then this topology majorizes the topology* $(A_s, \mathfrak{J}^*, \mathfrak{Q}^*)$ *of the projective limit* $\mathfrak{M}$.

These assertions follow directly from the definition of projective limit. Assertion 2.1.2 shows that the topology of the projective limit is the weakest among those locally convex topologies in the space (9) for which the mapping $(\Omega \uparrow U\square)$ is continuous for each word U of type ν.

2.2. In this subsection we prove certain assertions on separation and completeness of the projective limit of the projective spectrum of locally convex spaces. The following assertions are valid.

2.2.1. *The projective limit of the projective spectrum of T'-separated locally convex spaces is a T'-separated space.*

2.2.2. *If* (4) *is the projective spectrum of T-separated locally convex spaces and if, for any words U and V of type ν, $V \prec U$, $(\omega \uparrow U \square V \square)$ is a one-to-one mapping from $\mathfrak{M}^U$ into $\mathfrak{M}^V$, then the projective limit of this spectrum is a T-separated space.*

The first assertion follows simply from the definitions of T-separation and of the topology of the projective limit. We shall prove the second. By hypothesis, for any words U and V of type ν, if $V \prec U$, then $(\omega \uparrow U \square V \square)$ is a one-to-one mapping from $\mathfrak{M}^U$ into $\mathfrak{M}^V$. This means that

$$\text{(11)} \qquad \forall uv\theta_1^u\theta_2^u (v \prec u \,\&\, \omega(u \square v \square \theta_1^u) \underset{v}{=} \omega(u \square v \square \theta_2^u) \supset \theta_1^u \underset{u}{=} \theta_2^u).$$

Suppose for any word U of type ν the space $\mathfrak{M}^U$ is T-separated. We have to prove that the projective limit $\mathfrak{M}$ is T-separated, i.e. that

$$\text{(12)} \qquad \forall \theta_1 (\neg(\theta_1 \underset{\theta}{=} \mathfrak{D}^*) \supset \exists \iota_1 \neg \mathfrak{Q}^*(\iota, \theta_1)).$$

Suppose given a word θ_1 of type θ such that $\neg(\theta_1 \underset{\theta}{=} \mathfrak{D}^*)$. Then

$$\text{(13)} \qquad \neg \forall u (\langle\theta_1\rangle_r(u) \underset{u}{=} \mathfrak{D}(u)).$$

Fix on some word U of type ν. We shall prove the following formula:

$$\text{(14)} \qquad \neg(\langle\theta_1\rangle_r(U) \underset{u}{=} \mathfrak{D}(U)).$$

Suppose that $\langle\theta_1\rangle_r(U) \underset{U}{=} \mathfrak{D}(U)$. Suppose that V is any word of type ν. We find a word W of type ν such that $U \prec W$ and $V \prec W$. We have

$$\langle\theta_1\rangle_r(U) \underset{U}{=} \omega(W \square U \square \langle\theta_1\rangle_r(W)); \quad \langle\theta_1\rangle_r(V) \underset{V}{=} \omega(W \square V \square \langle\theta_1\rangle_r(W)).$$

Hence from (11) and $\langle\theta_1\rangle_r(U) \underset{U}{=} \mathfrak{D}(U)$ it follows that $\langle\theta_1\rangle_r(W) \underset{W}{=} \mathfrak{D}(W)$ (we note that $\mathfrak{D}(U) \underset{U}{=} \omega(W \square U \square \mathfrak{D}(W))$). Therefore in view of property 1) of the algorithm ω we find that $\langle\theta_1\rangle_r(V) \underset{V}{=} \mathfrak{D}(V)$. Thus we have proved that, for any word V of type ν, $\langle\theta_1\rangle_r(V) \underset{V}{=} \mathfrak{D}(V)$. This contradicts (13). Therefore formula (14) is proved. From the T-separation of the space $\mathfrak{M}^U$ it follows that there exists a word R of type ι^U such that $\neg \mathfrak{Q}(U, R, \langle\theta_1\rangle_r(U))$.

Put $\iota_1 \leftrightharpoons U \tau R$. Obviously $\neg \mathfrak{Q}^*(\iota_1, \theta_1)$. Formula (12) is proved.

2.2.3. *The projective limit of the projective spectrum of T'-separated and complete locally convex spaces is a complete locally convex space.*

Proof. Denote the indices of the alphabets $A_r \cup A_s$ and $A_r \cup Ч_0 \cup \{\odot\}$ by η and ψ respectively: $A_\eta = A_r \cup A_s$ and $A_\psi = A_r \cup Ч_0 \cup \{\odot\}$. Here we are supposing that $\odot$ does not enter into $A_r \cup Ч_0$. We shall denote the directed set of indices of the neighborhoods of zero in the projective limit $\mathfrak{M}$ by $\mathfrak{Z}^*$.

A_ψ is the alphabet intended for the construction of FR-$\mathfrak{J}$*-constructs both in the space $\mathfrak{M}$ and in the space $\mathfrak{M}^U$.

We adjoin the following formula:

$$(t_{\psi,1} \in \underline{FR\text{-}\mathfrak{J}^*\text{-constr}_{\mathfrak{M}}})$$

to the list Σ of formulas of the language $\mathcal{A}_\Sigma$, and denote by $\mathfrak{w}$ the corresponding subordinate generic letter (see §3.1, Chapter I).

We suppose that for each word U of type ν the space $\mathfrak{M}^U$ is complete and T'-separated. We need to prove that $\mathfrak{M}$ is also complete. In view of Theorem 3.1.3 of Chapter I the space $\mathfrak{M}^U$ is $\mathfrak{J}$*-complete for any word U of type ν. Since $\mathscr{A}$ is a normal formula, by the principle of constructive interpretation of mathematical inferences it follows from this supposition that we can realize an algorithm L for which, for any word U of type ν, $(L \uparrow U\square)$ is a $\mathfrak{J}$*-algorithm of passage to the limit in the space $\mathfrak{M}^U$.

Now we shall construct a $\mathfrak{J}$*-algorithm of passage to the limit in $\mathfrak{M}$. First of all we construct in A_ψ^{sa} algorithms δ, γ, and ζ satisfying the following conditions for any word U of type ν and for any word $\mathfrak{w}_1$ of type $\mathfrak{w}$:

$$\delta(U\square\mathfrak{w}_1) \overset{+}{\in} c_0\,;\quad \gamma(U\square\mathfrak{w}_1) \overset{+}{\in} c_0;$$

$$\forall\iota_1 (\langle\delta(U\square\mathfrak{w}_1)\rangle_\eta(\iota_1) \simeq \langle\underset{\longleftarrow\psi}{\mathfrak{w}_1}(\iota_1)\rangle_r(U));$$

$$\forall\iota_1^U (\langle\gamma(U\square\mathfrak{w}_1)\rangle_s(\iota_1^U) \simeq \overset{\longrightarrow\psi}{\mathfrak{w}_1}(U\tau\iota_1^U));$$

$$\zeta(U\square\mathfrak{w}_1) \simeq \delta(U\square\mathfrak{w}_1)\odot\gamma(U\square\mathfrak{w}_1).$$

Obviously, for any words U of type ν and $\mathfrak{w}_1$ of type $\mathfrak{w}$, $\langle\delta(U\square\mathfrak{w}_1)\rangle_\eta$ is a $\mathfrak{J}$*-sequence of points of the space $\mathfrak{w}^U$. Furthermore, from the property of the word $\mathfrak{w}_1$ that

$$\forall\iota_1\iota_2\iota_3(\overset{\longrightarrow\psi}{\mathfrak{w}_1}(\iota_1) \prec \iota_2.\ \iota_3 \supset \mathfrak{Q}^*(\iota_1, \underset{\longleftarrow\psi}{\mathfrak{w}_1}(\iota_2) \underset{\theta}{-} \underset{\longleftarrow\psi}{\mathfrak{w}_1}(\iota_3)))$$

it follows that

$$\forall\iota_1^U\iota_2\iota_3(\langle\gamma(U\square\mathfrak{w}_1)\rangle_s(\iota_1^U) \prec \iota_2, \iota_3 \supset \mathfrak{Q}(U, \iota_1^U, \langle\delta(U\square\mathfrak{w}_1)\rangle_\eta(\iota_2) - \langle\delta(U\square\mathfrak{w}_1)\rangle_\eta(\iota_3))).$$

Thus $\zeta(U\square\mathfrak{w}_1)$ is an FR-$\mathfrak{J}$*-construct in the space $\mathfrak{M}^U$.

Now we construct an algorithm H in the alphabet A_ψ^{sa} as follows:

$$\forall\mathfrak{w}_1 u(H(\mathfrak{w}_1\square u) \simeq L(u\square\zeta(u\square\mathfrak{w}_1))).$$

Obviously $\forall\mathfrak{w}_1 u\ (H(\mathfrak{w}_1\square u) \overset{+}{\in} \theta^U)$. We shall prove the formula

$$(15)\qquad \forall uv\ (v \underset{\nu}{\prec} u \supset H(\mathfrak{w}_1\square v) \underset{v}{=} \omega(u\square v\square H(\mathfrak{w}_1\square u))),$$

where $\mathfrak{w}_1$ is a word of type $\mathfrak{w}$, u and v are variables of genus ν and $\underset{\nu}{\prec}$ denotes the order relation in the directed set $(\mathscr{A}, \mathfrak{X})$.

Suppose given words U and V of type ν such that $V \underset{\nu}{\prec} U$. By definition $H(\mathfrak{w}_1 \square U)$ and $H(\mathfrak{w}_1 \square V)$ are limits respectively of the $\mathfrak{J}$*-sequence $\langle \delta(U \square \mathfrak{w}_1)\rangle_\eta$ in $\mathfrak{M}^U$ and the $\mathfrak{J}$*-sequence $\langle \delta(V \square \mathfrak{w}_1)\rangle_\eta$ in $\mathfrak{M}^V$. Accordingly, we have

$$\forall \iota_1^U \exists \iota_1 \forall \iota_2 (\iota_1 \prec \iota_2 \supset \mathfrak{Q}(U, \iota_1^U, H(\mathfrak{w}_1 \square U) \underset{U}{-} \langle \delta(U \square \mathfrak{w}_1)\rangle_\eta(\iota_2))), \tag{16}$$

$$\forall \iota_1^V \exists \iota_1 \forall \iota_2 (\iota_1 \prec \iota_2 \supset \mathfrak{Q}(V, \iota_1^V, H(\mathfrak{w}_1 \square V) \underset{V}{-} \langle \delta(V \square \mathfrak{w}_1)\rangle_\eta(\iota_2))). \tag{17}$$

From the definition of the algorithm δ it is obvious that

$$\forall \iota_1 (\langle \delta(V \square \mathfrak{w}_1)\rangle_\eta(\iota_1) \underset{V}{=} \omega(U \square V \square \langle \delta(U \square \mathfrak{w}_1)\rangle_\eta(\iota_1))). \tag{18}$$

Since $(\omega \uparrow U \square V \square)$ is a continuous mapping,

$$\forall \iota_1^V \exists \iota_1^U \forall \theta_1^U (\mathfrak{Q}(U, \iota_1^U, \theta_1^U) \supset \mathfrak{Q}(V, \iota_1^V, \omega(U \square V \square \theta_1^U))). \tag{19}$$

Now we prove that

$$\forall \iota_1^V \mathfrak{Q}(V, \iota_1^V, H(\mathfrak{w}_1 \square V) \underset{V}{-} \omega(U \square V \square H(\mathfrak{w}_1 \square U))). \tag{20}$$

Suppose given a word ι_1^V of type ι^V. We find a word ι_2^V of type ι^V such that

$$\forall \theta_1^V \theta_2^V (\mathfrak{Q}(V, \iota_2^V, \theta_1^V) \,\&\, \mathfrak{Q}(V, \iota_2^V, \theta_2^V) \supset \mathfrak{Q}(V, \iota_1^V, \theta_1^V + \theta_2^V)). \tag{21}$$

In view of (18) it is possible to find a word ι_1^U of type ι^U such that

$$\forall \theta_1^U (\mathfrak{Q}(U, \iota_1^U, \theta_1^U) \supset \mathfrak{Q}(V, \iota_2^V, \omega(U \square V \square \theta_1^U))). \tag{22}$$

In view of (16) and (17) it is possible to find a word Y of type ι such that

$$\mathfrak{Q}(u, \iota_1^U, H(\mathfrak{w}_1 \square U) \underset{U}{-} \langle \delta(U \square \mathfrak{w}_1)\rangle_\eta(Y)), \tag{23}$$

$$\mathfrak{Q}(V, \iota_2^V, H(\mathfrak{w}_1 \square V) \underset{V}{-} \langle \delta(V \square \mathfrak{w}_1)\rangle_\eta(Y)). \tag{24}$$

It follows from (23), (22), and (18) that

$$\mathfrak{Q}(V, \iota_2^V, \omega(U \square V \square H(\mathfrak{w}_1 \square U)) \underset{V}{-} \langle \delta(V \square \mathfrak{w}_1)\rangle_\eta(Y)). \tag{25}$$

Thereupon it follows from (24), (25) and (21) that

$$\mathfrak{Q}(V, \iota_1^V, H(\mathfrak{w}_1 \square V) \underset{V}{-} \omega(U \square V \square H(\mathfrak{w}_1 \square U))).$$

Thus formula (20) is proved. By hypothesis $\mathfrak{M}^V$ is a T'-separated space. Therefore (15) follows from (20), as was required to be proved.

Now we construct an algorithm $\mathscr{L}$ such that

$$\forall \mathfrak{w}_1 (\mathscr{L}(\mathfrak{w}_1) \simeq \mathfrak{E}(H \uparrow \mathfrak{w}_1 \square), \mathrm{A}_\Phi^{\mathrm{sa}} 3).$$

It is obvious from (15) that $\forall \mathfrak{w}_1(\mathscr{L}(\mathfrak{w}_1) \overset{+}{\in} \theta)$. We shall show that for any word $\mathfrak{w}_1$ of type $\mathfrak{w}$ we have

$$\forall \iota_1 \exists \iota_2 \forall \iota_3 (\iota_2 \prec \iota_3 \supset \mathfrak{Q}^*(\iota_1, \mathscr{L}(\mathfrak{w}_1) - \underbrace{\mathfrak{w}_1}_{\Phi}(\iota_3))). \tag{26}$$

Suppose given a word ι_1 of type ι and $\iota_1 \overline{\underline{\circ}} U_\tau R$, where U is a word of type ν and R a word of type ι^U. It follows that a word ι_2 of type ι is realizable for which

$$\forall \iota_3 (\iota_2 \prec \iota_3 \supset \mathfrak{Q}(U, R, H(\mathfrak{w}_1 \square U) \underset{U}{-} \langle \delta (U \square \mathfrak{w}_1) \rangle_\eta (\iota_3))).$$

Therefore

$$\forall \iota_3 (\iota_2 \prec \iota_3 \supset \mathfrak{Q}^* (\iota_1, \mathscr{L}(\mathfrak{w}_1) - \underset{\longrightarrow \psi}{\mathfrak{w}_1} (\iota_3))).$$

Formula (26) is proved, and $\mathscr{L}(\mathfrak{w}_1)$ is the limit of the $\mathfrak{J}^*$-sequence $\underset{\longrightarrow}{\mathfrak{w}_1}{}_\psi$ in $\mathfrak{M}$. Thus $\mathscr{L}$ is an $\mathfrak{J}^*$-algorithm for the passage to the limit in $\mathfrak{M}$. The space $\mathfrak{M}$ is complete. Theorem 2.2.3 is completely proved.

§3. Inductive limit of the inductive spectrum of constructive locally convex spaces

3.1. Suppose given an $\mathscr{A}$-family of locally convex spaces defined by the list

(1) $$\mathrm{A}_q, \mathcal{A}, \mathrm{A}_{\mathfrak{m}}, \mathfrak{P}, \mathfrak{E}, +, \cdot, \mathfrak{D}, \mathrm{A}_{\mathfrak{n}}, \mathfrak{J}, \mathfrak{Q}.$$

(See §1.1.) Here we shall use the notation introduced in §1.1.

We suppose that on the set $\mathscr{A}$ there is defined an order relation $\mathfrak{X}$ such that the list $(\mathscr{A}, \mathfrak{X})$ is a directed set. Suppose that ω is an algorithm such that, for any words U and V of type ν, if $U \prec V$, then $(\omega \uparrow U \square V \square)$ is an algorithm of type $(\theta^U \to \theta^V)$. We will say that the list

(2) $$\mathrm{A}_q, \mathcal{A}, \mathfrak{X}, \mathrm{A}_{\mathfrak{m}}, \mathfrak{P}, \mathfrak{E}, +, \cdot, \mathfrak{D}, \mathrm{A}_{\mathfrak{n}}, \mathfrak{J}, \mathfrak{Q}, \omega$$

defines an *inductive spectrum of locally convex spaces* if for each fixed word U of type ν the list

(3) $$\mathrm{A}_{\mathfrak{m}}, \mathfrak{P}^U, \mathfrak{E}^U, \underset{U}{+}, \underset{U}{\cdot}, \mathfrak{D}^U, \mathrm{A}_{\tilde{\mathfrak{m}}}, \mathfrak{J}^U, \mathfrak{Q}^U$$

is a locally convex space (denoted by $\mathfrak{M}^U$), and if the algorithm ω has the following properties:

1) $\forall u v \theta_1^u \theta_2^u (u \prec v \supset (\theta_1^u \underset{u}{=} \theta_2^u \equiv \omega (u \square v \square \theta_1^u) \underset{v}{=} \omega (u \square v \square \theta_2^u)));$

2) $\forall u v \theta_1^u \theta_2^u (u \prec v \supset \omega (u \square v \square \theta_1^u \underset{u}{+} \theta_2^u) \underset{v}{=} \omega (u \square v \square \theta_1^u) \underset{v}{+} \omega (u \square v \square \theta_2^u));$

3) $\forall u v x \theta_1^u (u \prec v \supset \omega (u \square v \square x \underset{u}{\cdot} \theta_1^u) \underset{v}{=} x \underset{v}{\cdot} \omega (u \square v \square \theta_1^u));$

4) $\forall u v (u \prec v \supset \forall \iota_1^v \exists \iota_1^u \forall \theta_1^u (\mathfrak{Q}(u, \iota_1^u, \theta_1^u) \supset \mathfrak{Q}(v, \iota_1^v, \omega (u \square v \square \theta_1^u))));$

5) $\forall u v w \theta_1^u (u \prec v \prec w \supset \omega (u \square w \square \theta_1^u) \underset{w}{=} \omega (v \square w \square \omega (u \square v \square \theta_1^u))),$

where u, v, w are variables of genus ν and x is a variable of genus д.

The first four conditions show that, for any words U and V of type ν, if $U \prec V$, then $(\omega \uparrow U \square V \square)$ is a continuous linear one-to-one mapping of the space $\mathfrak{M}^U$ into the space $\mathfrak{M}^V$.

Now we construct the inductive limit of the inductive spectrum (2). In view of the fact that

$$\forall uv \exists w \, (u \prec w \,\&\, v \prec w),$$

it is possible to construct an algorithm μ of the type $(\nu\nu \to \nu)$ such that

$$\forall uv \, (u \prec \mu\,(u \square v) \,\&\, v \prec \mu\,(u \square v)).$$

Here u, v, and w are variables of genus ν.

Suppose that the letter τ does not enter into the alphabet $A_{\mathrm{m}} \cup A_q$ and that in the list of basic alphabets of the language $Я_\Sigma^\Delta$ there are included the alphabets $A_{\mathrm{m}} \cup A_q \cup \{\tau\}$ and $A_{\mathrm{n}} \cup A_q$. We denote their indices by r and s respectively: $A_r = A_{\mathrm{m}} \cup A_q \cup \{\tau\}$ and $A_s = A_{\mathrm{n}} \cup A_q$.

It is possible to construct a formula $\mathfrak{P}^*$ defining the set of all possible words in the alphabet A_r of the form $U\tau R$, where U is a word of type ν and P is a word of type θ^U. We adjoin this formula to the list Σ of formulas of the language $Я_\Sigma$, and denote by θ the corresponding subordinate generic letter. We denote by $И_1$ and $И_2$ algorithms having the following properties: if P is a two-termed τ-system of words in $A_{\mathrm{m}} \cup A_q$ then $И_1$ and $И_2$ are applicable to P, while $И_1(P)$ is the first τ-term of the word P and $И_2(P)$ is the second.

We construct an algorithm $\tilde{\mu}$ such that

$$\forall \theta_1\theta_2 \, (\tilde{\mu}\,(\theta_1 \square \theta_2) \simeq \mu\,(И_1\,(\theta_1) \square И_1\,(\theta_2))).$$

Now we introduce a formula $\mathfrak{E}^*$ as follows:

$$\mathfrak{E}^* \leftrightharpoons \mathfrak{E}\,(\tilde{\mu}\,(\alpha \square \beta)), \quad \omega\,(И_1\,(\alpha) \square \tilde{\mu}\,(\alpha \square \beta) \square И_2\,(\alpha)),$$
$$\omega\,(И_1\,(\beta) \square \tilde{\mu}\,(\alpha \square \beta) \square И_2\,(\beta)),$$

where α and β are variables of genus t_r. We write

$$(P \underset{\theta}{=} Q) \leftrightharpoons F^{\alpha,\,\beta}_{P,\,Q} \llcorner \mathfrak{E}^* \lrcorner,$$

where P and Q are arbitrary object terms. Obviously, if θ_1 and θ_2 are words of type θ and $\theta_1 \,\overline{\underline{\circ}}\, U\tau P$, $\theta_2 \,\overline{\underline{\circ}}\, V\tau Q$, then

$$(\theta_1 \underset{\theta}{=} \theta_2) \equiv (\omega\,(U \square W \square P) \underset{W}{=} \omega\,(V \square W \square Q)),$$

where $W \leftrightharpoons \mu(U \square V)$. We note that if $\mathfrak{E}$ is a normal formula, so is $\mathfrak{E}^*$.

Suppose that θ_1 and θ_2 are words of type θ, while $\theta_1 \,\overline{\underline{\circ}}\, U\tau P$, $\theta_2 \,\overline{\underline{\circ}}\, V\tau Q$. Properties 1) and 5) of the algorithm ω imply the following properties:

(4) $$(\theta_1 \underset{\theta}{=} \theta_2) \equiv \exists u\,(U,\, V \prec u \,\&\, \omega\,(U \square u \square P) \underset{u}{=} \omega\,(V \square u \square Q));$$

(5) $$(\theta_1 \underset{\theta}{=} \theta_2) \equiv \forall u\,(U,\, V \prec u \supset \omega\,(U \square u \square P) \underset{u}{=} \omega\,(V \square u \square Q)).$$

We construct an algorithm σ such that, for any words θ_1 and θ_2 of type θ, if $\theta_1 \,\overline{\underline{\circ}}\, U\tau P$ and $\theta_2 \,\overline{\underline{\circ}}\, V\tau Q$, then

$$\sigma(\theta_1 \square \theta_2) \mathrel{\overset{\cdot}{=}} \underset{W}{+} (\omega(U \square W \square P) \square \omega(V \square W \square Q)),$$

where $W \leftrightharpoons \mu\ (U \square V)$. Now we construct algorithms $\underset{\theta}{+}$ and $\underset{\theta}{\cdot}$ such that

$$\forall \theta_1 \theta_2 (\underset{\theta}{+} (\theta_1 \square \theta_2) \simeq \widetilde{\mu} (\theta_1 \square \theta_2) \tau \sigma (\theta_1 \square \theta_2));$$

$$\forall x \theta_1 (\underset{\theta}{\cdot} (x \square \theta_1) \simeq И_1 (\theta_1) \tau (x \underset{И_1(\theta_1)}{\cdot} И_2 (\theta_1))).$$

Obviously $\underset{\theta}{+}$ is an algorithm of the type $(\theta\theta \to \theta)$ and $\underset{\theta}{\cdot}$ is an algorithm of the type $(д\theta \to \theta)$.

Fix on a word U of type ν and write $\mathfrak{O}^* \leftrightharpoons U \tau\, \mathfrak{O}(U)$. $\mathfrak{O}^*$ is a word of type θ. We note that

$$\forall u (\mathfrak{O}^* \underset{\theta}{=} u \tau\, \mathfrak{O}(u)).$$

This property is easily proved with the use of formula (4).

Suppose that $\mathrm{A}_{r'} = \mathrm{A}_r \cup Ч_0$, where r' is the index of the alphabet $\mathrm{A}_r \cup Ч_0$. We construct an algorithm S such that, for any positive natural number m, any algorithm φ of the type $(н \to д)$ and any algorithm ψ of the type $(н \to \theta)$,

$$S(m \square \mathcal{E}\varphi,\ Ч_3^{\mathrm{sa}} З \square \mathcal{E}\psi,\ \mathrm{A}_{r'}^{\mathrm{sa}} З) \mathrel{\overset{\cdot}{=}} \varphi(1) \underset{\theta}{\cdot} \psi(1) \underset{\theta}{+} \cdots \underset{\theta}{+} \varphi(m) \underset{\theta}{\cdot} \psi(m).$$

Now we introduce further the following formulas:

$$\mathfrak{J}^* \leftrightharpoons (\langle c_{0,1} \rangle_s \in (\nu \to t_{\mathrm{n}})) \,\&\, \forall u (\langle c_{0,1} \rangle_s (u) \in \mathfrak{J}^u);$$

$$\mathfrak{Q}^* \leftrightharpoons \exists m \varphi \psi ((t_{r,1} \underset{\theta}{=} S(m \square \mathcal{E}\varphi,\ Ч_3^{\mathrm{sa}} З (\square \mathcal{E}\psi,\ \mathrm{A}_{r'}^{\mathrm{sa}} З)) \,\&\, \left(\sum_{i=1}^{m} |\varphi(i)| = 1 \right)$$

$$\&\, \forall i (1 \leqslant i \leqslant m \supset \mathfrak{Q} (И_1 (\psi(i)),\ \langle c_{0,1} \rangle_s (И_1 (\psi(i))),\ И_2 (\psi(i)))).$$

where u is a variable of genus ν, φ is a functorial variable for algorithms of the type $(н \to д)$ and ψ is a functorial variable for algorithms of the type $(н \to \theta)$. The parameter of the formula $\mathfrak{J}^*$ is the variable $c_{0,1}$, and the parameters of the formula $\mathfrak{Q}^*$ are the variables $c_{0,1}$ and $t_{r,1}$. We adjoin the formula $\mathfrak{J}^*$ to the list Σ of formulas of the language $Я_\Sigma$, and denote by ι the corresponding subordinate generic letter. We note that, for any words ι_1 of type ι and θ_1 of type θ, the condition $\mathfrak{Q}^*(\iota_1, \theta_1)$ is satisfied if and only if we can realize a natural number m, words $U_1, \cdots, U_m$ of type ν, words P_1 of type $\theta^{U_1}, \cdots, P_m$ of type θ^{U_m}, and real duplexes $\alpha_1, \cdots, \alpha_m$, such that

$$\theta_1 \underset{\theta}{=} \alpha_1 \cdot U_1 \tau P_1 + \cdots + \alpha_m \cdot U_m \tau P_m, \qquad |\alpha_1| + \cdots + |\alpha_m| = 1$$

and

$$\mathfrak{Q}(U_1, R_1, P_1) \,\&\, \ldots \,\&\, \mathfrak{Q}(U_m, R_m, P_m),$$

where $R_i \leftrightharpoons \langle \iota_1 \rangle_s (U_i)$ $(i = 1, \cdots, m)$.

Consider the following list:

(6) $$A_r,\ \mathfrak{P}^*,\ \mathfrak{E}^*,\ \underset{\theta}{+},\ \underset{\theta}{\cdot},\ \mathfrak{D}^*,\ \Psi_0,\ \mathfrak{I}^*,\ \mathfrak{Q}^*.$$

One may verify that this list defines a constructive locally convex space; since this verification does not present any particular difficulties, we omit it.

The space (6) will be called *the inductive limit of the inductive spectrum of the locally convex space* (2). In what follows we shall denote the space (6) by $\mathfrak{M}$.

We construct an algorithm Ω such that

$$\forall u\theta_1^u (\Omega \{u\square\theta_1^u\} \simeq u\tau\theta_1^u).$$

Obviously for any fixed word U of type ν the algorithm $(\Omega \uparrow U\square)$ is a linear mapping of the space $\mathfrak{M}^U$ into the space $\mathfrak{M}$.

Suppose that in the linear space

(7) $$A_r,\ \mathfrak{P}^*,\ \mathfrak{E}^*,\ \underset{\theta}{+},\ \underset{\theta}{\cdot},\ \mathfrak{D}^*$$

some locally convex topology $(A_{\tilde{\pi}}, \tilde{\mathfrak{I}}, \tilde{\mathfrak{Q}})$ is defined. We will say that, for each word U of type ν, $(\Omega \uparrow U\square)$ is a continuous mapping of the space $\mathfrak{M}^U$ into the space (7) with the topology $(A_{\tilde{\pi}}, \tilde{\mathfrak{I}}, \tilde{\mathfrak{Q}})$, if

(8) $$\forall u\tilde{\iota}\exists\iota_1^u\forall\theta_1^u (\mathfrak{Q}(u, \iota_1^u, \theta_1^u) \supset \tilde{\mathfrak{Q}}(\tilde{\iota}_1, \Omega(u\square\theta_1^u))),$$

where $\tilde{\iota}_1$ is a variable for elements of the set $\tilde{\mathfrak{I}}$.

3.1.1. *For any word U of type ν, $(\Omega \uparrow U\square)$ is a continuous mapping of the space $\mathfrak{M}^U$ into the space $\mathfrak{M}$.*

3.1.2. *Suppose that in the linear space* (7) *there is defined some locally convex topology $(A_{\tilde{\pi}}, \tilde{\mathfrak{I}}, \tilde{\mathfrak{Q}})$ satisfying condition* (8). *Then this topology is majorized by the topology $(\Psi_0, \mathfrak{I}^*, \mathfrak{Q}^*)$ of the inductive limit $\mathfrak{M}$.*

Assertion 3.1.1 follows from the obvious formula

$$\forall u\iota_1\theta_1^u (\mathfrak{Q}(u, \langle\iota_1\rangle_s(u), \theta_1^u) \supset \mathfrak{Q}^*(\iota_1\, u\tau\theta_1^u)).$$

We shall prove 3.1.2. Suppose that the topology $(A_{\tilde{\pi}}, \tilde{\mathfrak{I}}, \tilde{\mathfrak{Q}})$ satisfies condition (8). We need to prove that

(9) $$\forall\tilde{\iota}_1\exists\iota_1\forall\theta_1 (\mathfrak{Q}^*(\iota_1, \theta_1) \supset (\tilde{\mathfrak{Q}}(\tilde{\iota}_1, \theta_1)).$$

Since $\mathscr{A}$ and $\tilde{\mathfrak{I}}$ are normal formulas, it follows from (8) that one can realize an algorithm α transforming each word of the form $U\square\tilde{\iota}_1$, where U is a word of type ν and $\tilde{\iota}_1$ a word of type $\tilde{\iota}$, into a word $\alpha(U\square\tilde{\iota}_1)$ of type $\mathfrak{t}^u$, and such that

(10) $$\forall u\tilde{\iota}_1\theta_1^u (\mathfrak{Q}(u, \alpha(u\square\tilde{\iota}_1), \theta_1^u) \supset \tilde{\mathfrak{Q}}(\tilde{\iota}_1, u\tau\theta_1^u)).$$

Suppose given a word $\tilde{\iota}_1$ of type $\tilde{\iota}$. We shall construct a word ι_1 of type ι such that

$$\forall u\,(\langle \iota_1\rangle_s(u) \simeq \alpha(u\Box\tilde{\iota}_1)).$$

Suppose that θ_1 is a word of type θ such that $\mathfrak{Q}^*(\iota_1,\theta_1)$. Then we can realize a natural number m, words $U_1,\cdots,U_m$ of type ν, words P_1 of type $\theta^{U_1},\cdots,P_m$ of type θ^{U_m}, and real duplexes $\alpha_1,\cdots,\alpha_m$ such that

$$(11)\qquad \theta_1 \underset{\theta}{=} \alpha_1\cdot U_1\tau P_1+\cdots+\alpha_m\cdot U_m\tau P_m,\ |\alpha_1|+\cdots+|\alpha_m|=1,$$

$$(12)\qquad \mathfrak{Q}(U_1,R_1,P_1)\&\ldots\&\mathfrak{Q}(U_m,R_m,P_m),$$

where $R_i \leftrightharpoons \langle\iota_1\rangle_s(U_i)$ $(i=1,\cdots,m)$. It follows from (10) and (12) that

$$\tilde{\mathfrak{Q}}(\tilde{\iota}_1,U_1\tau P_1)\&\ldots\&\tilde{\mathfrak{Q}}(\tilde{\iota}_1,U_m\tau P_m).$$

Hence, using (11), $\tilde{\mathfrak{Q}}(\tilde{\iota}_1,\theta_1)$ follows. Thus formula (9), and also assertion 3.1.2, are proved.

3.1.3. *Suppose that α is an algorithm of type $(\nu\to\nu)$, F an algorithm such that, for each word U of type ν, $(F \uparrow U\Box)$ is an operator of the type $\theta^U\to\theta^{\alpha(U)}$, and G an algorithm of type $(\theta\to\theta)$ defined by the formula*

$$\forall u\theta_1^u\,(G(u\tau\theta_1^u)\simeq\alpha(u)\,\tau F(u\Box\theta_1^u)).$$

If, for each word U of type ν, $(F\uparrow U\Box)$ is a continuous mapping of $\mathfrak{M}^U$ into $\mathfrak{M}^{\alpha(U)}$ and G is a linear operator in $\mathfrak{M}$, then G is a continuous mapping of $\mathfrak{M}$ into itself.

PROOF. From the continuity of the mappings $(F\uparrow U\Box)$ it follows that we can realize an algorithm γ which is applicable to any word of the type $U\Box R$, where U is a word of type ν and R is a word of the type $\iota^{\alpha(U)}$, and which transforms each such word into a word of type ι^U and satisfies the following condition:

$$(13)\qquad \forall u\iota_1^{\alpha(u)}\theta_1^u\,(\mathfrak{Q}(u,\gamma(u\Box\iota_1^{\alpha(u)}),\theta_1^u)\supset\mathfrak{Q}(\alpha(u),\iota_1^{\alpha(u)},F(u\Box\theta_1^u))).$$

In order to prove that G is continuous, it is sufficient for us to prove the formula

$$(14)\qquad \forall\iota_1\exists\iota_2\forall\theta_1\,(\mathfrak{Q}^*(\iota_2,\theta_1)\supset\mathfrak{Q}^*(\iota_1,G(\theta_1))).$$

Suppose given a word ι_1 of type ι. We shall construct a word ι_2 such that

$$\forall u\,(\langle\iota_2\rangle(u)\simeq\gamma(u\Box\langle\iota_1\rangle(\alpha(u)))).$$

Suppose that θ_1 is a word of type θ satisfying the condition $\mathfrak{Q}^*(\iota_2,\theta_1)$. Then we can realize a number m words $U_1,\cdots,U_m$ of type ν, words P_1 of type $\theta^{U_1},\cdots,P_m$ of type θ^{U_m}, and duplexes $\alpha_1,\cdots,\alpha_m$ such that

$$(15)\qquad \begin{gathered}\theta_1=\alpha_1\cdot U_1\tau P_1+\cdots+\alpha_m\cdot U_m\tau P_m,\ |\alpha_1|+\cdots+|\alpha_m|=1\\ \mathfrak{Q}(U_1,\langle\iota_2\rangle(U_1),P_1)\&\ldots\&\mathfrak{Q}(U_m,\langle\iota_2\rangle(U_m),P_m)\end{gathered}$$

We write $V_i \leftrightharpoons \alpha(U_i)$, $Q_i \leftrightharpoons F(U_i \square P_i)$ $(i = 1, \cdots, m)$.

It follows from (13) and (15) that

$$\mathfrak{Q}(V_1, \langle\iota_1\rangle(V_1), Q_1) \& \ldots \& \mathfrak{Q}(V_m, \langle\iota_1\rangle(V_m), Q_m), \ |\alpha_1| + \cdots + |\alpha_m| = 1. \quad (16)$$

Moreover, from the linearity of the operator G we get

$$G(\theta_1) = \alpha_1 \cdot G(U_1 \tau P_1) + \cdots + \alpha_m \cdot G(U_m \tau P_m) = \alpha_1 \cdot V_1 \tau Q_1 + \cdots + \alpha_m \cdot V_m \tau Q_m. \quad (17)$$

From (16) and (17) it follows that $\mathfrak{Q}^*(\iota_1, G(\theta_1))$. Thus formula (14), and with it the theorem, are proved.

3.2. We consider the special case of an inductive spectrum (2) when the notation $+$, $\cdot$, $\mathfrak{O}$ and ω is understood as follows:

a) $+$ is an algorithm of type $(t_m\ t_m \dot{\to} t_m)$, having the property

$$\forall u \theta_1^u \theta_2^u (+(\theta_1^u \square \theta_2^u) \overset{+}{\in} \theta^u);$$

b) $\cdot$ is an algorithm of the type $(Д t_m \dot{\to} t_m)$, having the property

$$\forall u x \theta_1^u (\cdot(x \square \theta_1^u) \overset{+}{\in} \theta^u);$$

c) $\mathfrak{O}$ is a word of the type t_m, having the property

$$\forall u (\mathfrak{O} \in \theta^u);$$

d) ω is an algorithm of the type $(\nu\nu t_m \dot{\to} t_m)$, having the property

$$\forall u v \theta_1^u (u \prec v \supset \omega(u \square v \square \theta_1^u) = \theta_1^u).$$

This last condition means that, for any words U and V of the type ν, $U \prec V$, $(\omega \uparrow U \square V \square)$ is the identity mapping of $\mathfrak{M}^U$ into $\mathfrak{M}^V$, and every point of $\mathfrak{M}^U$ is a point of $\mathfrak{M}^V$.

In this case the algorithms $\underset{u}{+}$ and $\underset{u}{\cdot}$ are replaced simply by the algorithms $+$ and $\cdot$, and the word $\mathfrak{O}^U$ is replaced by the word $\mathfrak{O}$. Conditions 1)–4) of the algorithm ω show that *if $U \prec V$, then $\mathfrak{M}^U$ is a linear subspace of the space $\mathfrak{M}^V$, and the identity mapping of $\mathfrak{M}^U$ into $\mathfrak{M}^V$ is one-to-one and continuous.* Condition 5), naturally, is fulfilled.

Under these hypotheses the inductive spectrum (2) will be called an *expanding inductive spectrum of locally convex spaces.*

Suppose that the list

$$\mathrm{A}_q,\ \mathcal{A},\ \mathfrak{X},\ \mathrm{A}_{\mathfrak{m}},\ \mathfrak{P},\ \mathfrak{E},\ +,\ \cdot,\ \mathfrak{O},\ \mathrm{A}_{\mathfrak{n}},\ \mathfrak{J},\ \mathfrak{Q},\ \omega \quad (18)$$

defines an expanding inductive spectrum of locally convex spaces. The inductive limit $\mathfrak{M}$ of an expanding inductive spectrum (18), constructed by the method in §3.1, will be called an *FR-inductive limit.* Along with an *FR*-inductive limit we construct now the so-called *F-inductive limit,* as follows.

We introduce the following formulas:

$$\widetilde{\mathfrak{P}} \leftrightharpoons \exists u \mathfrak{P}(u, \alpha); \qquad \widetilde{\mathfrak{E}} \leftrightharpoons \exists u (\mathfrak{P}(u, \alpha) \,\&\, \mathfrak{P}(u, \beta) \,\&\, (\alpha \underset{u}{=} \beta)),$$

where u is a variable of genus ν, and α and β are variables of genus t_m. We will write

$$(P \approx Q) \leftrightharpoons F^{\alpha, \beta}_{P, Q} \llcorner \widetilde{\mathfrak{E}} \lrcorner,$$

where P and Q are object terms. We adjoin the formula $\widetilde{\mathfrak{P}}$ to the list Σ of formulas of the language $Я_\Sigma$ and denote by $\tilde{\theta}$ the corresponding subordinate generic letter. Obviously P is a word of type $\tilde{\theta}$ if and only if we can realize a word U of type ν such that P is a word of type θ^U. It is easy to see that in this case $+$ is also an algorithm of type $(\tilde{\theta}\tilde{\theta} \to \tilde{\theta})$, and $\cdot$ is an algorithm of type $(д\tilde{\theta} \to \tilde{\theta})$.

It is possible to construct a formula $\widetilde{\mathfrak{Q}}$ such that, for any words ι_1 of type ι and $\tilde{\theta}_1$ of type $\tilde{\theta}$, the condition $\widetilde{\mathfrak{Q}}(\iota_1, \tilde{\theta}_1)$ is satisfied if and only if one can realize a positive natural number m, words $U_1, \cdots, U_m$ of type ν, words P_1 of type $\theta^{U_1}, \cdots, P_m$ of type θ^{U_m}, and real duplexes $\alpha_1, \cdots, \alpha_m$ such that

$$\tilde{\theta}_1 \approx \alpha_1 \cdot P_1 + \cdots + \alpha_m \cdot P_m, \quad |\alpha_1| + \cdots + |\alpha_m| = 1,$$

and

$$\mathfrak{Q}(U_1, R_1, P_1) \,\&\, \ldots \,\&\, \mathfrak{Q}(U_m, R_m, P_m),$$

where $R_i \leftrightharpoons \langle \iota_1 \rangle_s (U_i) \quad (i = 1, \cdots, m)$.

It is possible to show that the list

$$\text{(19)} \qquad \mathrm{A}_m, \widetilde{\mathfrak{P}}, \widetilde{\mathfrak{E}}, +, \cdot, \mathfrak{O}, Ч_0, \mathfrak{I}^*, \widetilde{\mathfrak{Q}}.$$

is a constructive locally convex space. We will call this space the *F-inductive limit* of the expanding inductive spectrum of the locally convex spaces (18).

Obviously, if the list (18) is an expanding inductive spectrum of locally inductive spaces, then, for any point P of its F-inductive limit, one can realize a word U of type ν such that $U \tau P$ is a point of its FR-inductive limit. On the other hand, the following assertion holds.

3.2.1. *It is possible to construct an expanding inductive spectrum of locally convex spaces, having the property that there cannot exist an algorithm which constructs relative to each point P of the F-inductive limit a word U of type ν such that $U \tau P$ is a point of the FR-inductive limit.*

This assertion will be proved by examples in Chapters IV and VI.

If the list (18) defines an expanding inductive spectrum of locally convex spaces, while the directed set $(\mathscr{A}, \mathfrak{X})$ is the set of natural numbers with the usual order relation, then we will say that this list defines an *expanding sequence of locally convex spaces.*

Suppose that the list (18) defines an expanding sequence of locally convex spaces and satisfies in addition the following conditions:

(i) $$\forall k \theta_1^{k+1} (\exists \theta_2^k (\theta_1^{k+1} \underset{k+1}{=} \theta_2^k) \supset \theta_1^{k+1} \in \mathfrak{P}^k).$$

(ii) $$\forall k j \beta (\mathfrak{J}(k, \beta) \equiv \mathfrak{J}(j, \beta));$$

$$\forall k \iota_1^k \theta_1^k (\mathfrak{Q}(k, \iota_1^k, \theta_1^k) \equiv \mathfrak{Q}(k+1, \iota_1^k, \theta_1^k)),$$

where β is a variable of genus t_{II}. Here the variables u and v for words of type ν are replaced by variables k and j for natural numbers. The first formula of condition (ii) means that for any natural numbers k and j the sets $\mathfrak{J}^k$ and $\mathfrak{J}^j$ coincide.

(iii) For any natural numbers k and j with $k<j$, $\mathfrak{P}^k$ is a closed set in the space $\mathfrak{M}^j$. Here by a closed set we understand a set which is simultaneously $\mathfrak{Z}$-closed and $\mathfrak{Z}'$-closed.

Under all these assumptions the inductive spectrum (18) will be called a *strictly expanding sequence of locally convex spaces*. The F-inductive limit and the FR-inductive limit of a strictly expanding sequence of locally convex spaces will be called respectively the *F-strict inductive limit* and the *FR-strict inductive limit.*

We note that in a strictly expanding sequence of locally convex spaces each space $\mathfrak{M}^k$ is a regular closed subspace of the space $\mathfrak{M}^j$ for $k<j$. In this connection see the definition of regular subspace in §2.3, Chapter I.

The concepts of "F-strict inductive limit" and "FR-strict inductive limit" are regarded as the constructive analogues of the concept of "strict inductive limit" in classical mathematics (see **[39]** and **[38]**, Chapter II).

3.3. Suppose given a strictly expanding sequence of locally convex spaces, defined by the list

(20) $$\mathrm{Ч}_0, N, \mathrm{A}_{\mathfrak{m}}, \mathfrak{P}, \mathfrak{E}, +, \cdot, \mathfrak{D}, \mathrm{A}_{\mathfrak{n}}, \mathfrak{J}, \mathfrak{Q}.$$

Here the directed set $(\mathscr{A}, \mathfrak{X})$ is replaced by the directed set N of natural numbers, and for simplicity we delete the algorithm ω of identity mappings from the list defining the expanding sequence of spaces.

Suppose that the FR-strict inductive limit and the F-strict inductive limit of the sequence (20) are defined respectively by the following lists:

$$\mathrm{A}_r, \mathfrak{P}^*, \mathfrak{E}^*, \underset{\theta}{+}, \underset{\theta}{\cdot}, \mathfrak{D}^*, \mathrm{Ч}_0, \mathfrak{J}^*, \mathfrak{Q}^*,$$

$$\mathrm{A}_{\mathfrak{m}}, \tilde{\mathfrak{P}}, \tilde{\mathfrak{E}}, +, \cdot, \mathfrak{D}, \mathrm{Ч}_0, \mathfrak{J}^*, \tilde{\mathfrak{Q}}.$$

Here we employ the notation introduced in §§3.1 and 3.2.

In view of condition (ii) it is obvious that the sets $\mathfrak{J}^k$ and $\mathfrak{J}^j$ coincide for any natural numbers k and j. Denote by δ an algorithm which transforms each element U of the set $\mathfrak{J}^k$ (k arbitrary) into a word $\delta(U)$ of type ι (i.e. an element of the set $\mathfrak{J}^*$) such that

$$\forall k (\langle \delta(U) \rangle_s (k) \doteq U).$$

We shall prove the following formula:

$$(21)\qquad \forall k\iota_1^k\theta_1^k(\tilde{\mathfrak{Q}}(\delta(\iota_1^k), \theta_1^k) \equiv \mathfrak{Q}(k, \iota_1^k, \theta_1^k)).$$

Suppose given a natural number k, a word U of type ι^k and a word P of type θ^k. By definition of the formula $\mathfrak{Q}$ it obviously follows from $\mathfrak{Q}(k, U, P)$ that $\tilde{\mathfrak{Q}}(\delta(U), P)$. We shall prove the converse. Suppose that $\tilde{\mathfrak{Q}}(\delta(U), P)$. Then we can realize a natural number $m \geqq 1$, natural numbers $k_1, \cdots, k_m$, words P_1 of type $\theta^{k_1}, \cdots, P_m$ of type θ^{k_m}, and real duplexes $\alpha_1, \cdots, \alpha_m$ such that

$$(22)\qquad P \underset{}{\approx} \alpha_1 \cdot P_1 + \cdots + \alpha_m \cdot P_m,\ |\alpha_1| + \cdots + |\alpha_m| = 1,$$

$$(23)\qquad \mathfrak{Q}(k_1, U, P_1) \& \ldots \& \mathfrak{Q}(k_m, U, P_m).$$

We write $l \leftrightharpoons \max(k \square k_1 \square \cdots \square k_m)$. Using condition (ii), we conclude from (23) that

$$\mathfrak{Q}(l, U, P_1) \& \ldots \& \mathfrak{Q}(l, U, P_m),$$

and so we find that $\mathfrak{Q}(l, U, \alpha_1 \cdot P_1 + \cdots + \alpha_m \cdot P_m)$, since $|\alpha_1| + \cdots + |\alpha_m| = 1$. We note that for arbitrary natural numbers j and h, if $j \leqq h$, the identity mapping of the space $\mathfrak{M}^j$ into the space $\mathfrak{M}^h$ is one-to-one. Therefore from (22) we find that

$$P \underset{l}{=} \alpha_1 \cdot P_1 \underset{l}{+} \ldots + \alpha_m \cdot P_m,$$

so that $\mathfrak{Q}(l, U, P)$. Since P belongs to the space $\mathfrak{M}^k$, in view of (ii) it follows from $\mathfrak{Q}(l, U, P)$ that $\mathfrak{Q}(k, U, P)$, as we were required to prove.

In a quite analogous way one proves the following formula:

$$(24)\qquad \forall k\iota_1^k\theta_1^k(\mathfrak{Q}^*(\delta(\iota_1^k), k\tau\theta_1^k) \equiv \mathfrak{Q}(k, \iota_1^k, \theta_1^k)).$$

Using (21) and (24), one easily proves the following assertion.

3.3.1. *An FR- (respectively F-) limit of a strictly expanding sequence of T-separated (T′-separated) locally convex spaces is a T-separated (T′-separated) space.*

REMARK. In the proof of (21) and (24) no use was made of condition (iii) on a strongly expanding sequence of locally convex spaces. Therefore assertion 3.3.1 holds also for the case of an FR- (F-) limit of such an expanding sequence of locally convex spaces which satisfies conditions (i) and (ii). In other words, if the list (20) defines an expanding sequence of T-separated (T'-separated) spaces and if this sequence satisfies conditions (i) and (ii), then its FR- (F-) inductive limit is T-separated (T'-separated).

§4. Dimension of constructive locally convex spaces

4.1. Suppose that $\mathfrak{M}$ is a constructive locally convex space given by the list

$$(1)\qquad \text{A}_{\text{m}},\ \mathfrak{P},\ \mathfrak{E},\ \underset{\theta}{+},\ \underset{\theta}{\cdot},\ \mathfrak{D},\ \text{A}_{\text{n}},\ \mathfrak{J},\ \mathfrak{Q}.$$

Suppose that $X_1, \cdots, X_n$ are n points of $\mathfrak{M}$ ($n \geqq 1$). We will say that the points $X_1, \cdots, X_n$ are *linearly dependent in* $\mathfrak{M}$ if we can realize real duplexes

$\alpha_1, \cdots, \alpha_n$ such that

$$(\alpha_1 \neq 0 \vee \ldots \vee \alpha_n \neq 0) \,\&\, (\alpha_1 \cdot X_1 + \ldots + \alpha_n \cdot X_n = \mathfrak{D}).$$

We will say that the points $X_1, \cdots, X_n$ are *linearly independent in* $\mathfrak{M}$ if they are not linearly dependent in $\mathfrak{M}$. It is easy to prove that $X_1, \cdots, X_n$ are linearly independent in $\mathfrak{M}$ if and only if one of the following conditions is satisfied:

$$\forall x_1 \ldots x_n (x_1 \cdot X_1 + \ldots + x_n \cdot X_n = \mathfrak{D} \supset x_1 = 0 \,\&\, \ldots \,\&\, x_n = 0);$$

$$\forall x_1 \ldots x_n (x_1 \neq 0 \vee \ldots \vee x_n \neq 0 \supset \neg (x_1 \cdot X_1 + \ldots + x_n \cdot X_n = \mathfrak{D})),$$

where $x_1, \cdots, x_n$ are variables for real duplexes.

Suppose that n is a natural number $\geqq 1$. We will say that $\mathfrak{M}$ is *n-dimensional in the weak sense* if we can realize n linearly independent points $X_1, \cdots, X_n$ with the property that

$$\forall \theta_1 \exists x_1 \ldots x_n (\theta_1 = x_1 \cdot X_1 + \ldots + x_n \cdot X_n).$$

The space $\mathfrak{M}$ is said to be *n-dimensional in the strong sense* if we can realize n linearly independent points $X_1, \cdots, X_n$ and algorithms $\lambda_1, \cdots, \lambda_n$ of type $(\theta \rightarrow \text{д})$ such that

(2) $$\forall \theta_1 (\theta_1 = \lambda_1(\theta_1) \cdot X_1 + \ldots + \lambda_n(\theta_1) \cdot X_n).$$

In both cases we will say that the points $X_1, \cdots, X_n$ form a *basis for the space* $\mathfrak{M}$.

The space $\mathfrak{M}$ is said to be *finite-dimensional in the weak sense* (*the strong sense*) if we can realize a natural number n such that $\mathfrak{M}$ is n-dimensional in the weak (strong) sense.

The following properties are valid:

1) If $X_1, \cdots, X_n$ are linearly independent points of $\mathfrak{M}$, then

$$\neg (X_1 = \mathfrak{D}) \,\&\, \ldots \,\&\, \neg (X_n = \mathfrak{D}).$$

2) If $\mathfrak{M}$ is n-dimensional in the strong sense, then

$$\forall \theta_1 \theta_2 ((\theta_1 = \theta_2) \equiv (\lambda_1(\theta_1) = \lambda_1(\theta_2) \,\&\, \ldots \,\&\, \lambda_n(\theta_1) = \lambda_n(\theta_2)));$$

$$\forall \theta_1 \theta_2 x y (\lambda_i (x \cdot \theta_1 + y \cdot \theta_2) = x \cdot \lambda_i(\theta_1) + y \cdot \lambda_i(\theta_2)) \quad (i = 1, \ldots, n),$$

$\lambda_1, \cdots, \lambda_n$ being algorithms satisfying condition (2).

3) If $\mathfrak{M}$ is n-dimensional (in the weak or strong sense), then any n linearly independent points in $\mathfrak{M}$ form a basis for it.

Obviously every space n-dimensional in the strong sense is n-dimensional in the weak sense. However, the converse assertion is false. Indeed, the following theorem holds.

4.1.1. *For any natural number $n \geqq 1$, it is possible to construct a locally convex space which is n-dimensional in the weak sense but not n-dimensional in the strong sense.*

PROOF. For the case $n = 1$ we consider the space $\mathscr{F}$ of all F-numbers. This space is one-dimensional in the weak sense, and the point 1 forms a basis for it. Suppose that $\mathscr{F}$ is one-dimensional in the strong sense. Then we can realize an F-number $X \neq 0$ and an algorithm λ of the type (в $\rightarrow$ д) such that

$$\forall r\, (r \underset{\text{в}}{=} \lambda(r) \cdot X).$$

Here r is a variable for F-numbers. Suppose γ is an algorithm such that

$$\forall r\, (\gamma(r) \doteqdot \lambda(r) X).$$

Then γ is applicable to any F-number and transforms each F-number into a real duplex equal to it. But such an algorithm is impossible (see [15]). Therefore $\mathscr{F}$ cannot be one-dimensional in the strong sense.

For any $n \geqq 1$ the space $\mathscr{F}^{(n)}$, the n-times product of the space $\mathscr{F}$ (see §1.3), is n-dimensional in the weak sense but not in the strong sense.

In what follows, by an n-dimensional (finite-dimensional) space we will always understand a space which is n-dimensional (finite-dimensional) in the strong sense.

Concerning the space $\mathfrak{M}$, we will say that it is an *infinite-dimensional space* if for any natural number $n \geqq 1$ there are n linearly independent points realizable in $\mathfrak{M}$.

REMARK. The concepts of n-dimensional, finite-dimensional and infinite-dimensional spaces are defined not only for constructive locally convex spaces, but also for constructive linear spaces.

4.2. Suppose that э is a letter not entering into the alphabet $Ч_3$. We adjoin the alphabet $Ч_3 \cup \{э\}$ to the list of basic alphabets of the language Я and denote its index by r: $A_r = Ч_3 \cup \{э\}$.

Suppose that n is a positive natural number. Then one can construct a one-parameter formula $\mathfrak{P}$ which defines the set of all possible words in the alphabet A_r having the form $\alpha_1 э \cdots э\, \alpha_n$, where $\alpha_1, \cdots, \alpha_n$ are real duplexes. We denote by θ a subordinate generic letter assigned to the formula $\mathfrak{P}$.

It is possible to construct an algorithm И with the following property: for any word P in A_r and any natural number j, if P is an n-term э-system of real duplexes and $1 \leqq j \leqq n$, then И$(j \square P)$ is the jth э-term of the word P.

Now we construct a two-parameter formula $\mathfrak{E}$, an algorithm $\underset{\theta}{+}$ of type $(\theta\theta \rightarrow \theta)$ and an algorithm $\underset{\theta}{\cdot}$ of type (д$\theta \rightarrow \theta$) such that, for any words θ_1 and θ_2 of type θ and any real duplex x,

$$\mathfrak{E}^*(\theta_1, \theta_2) \equiv \forall j\, (1 \leqslant j \leqslant n \supset И(j \square \theta_1) \underset{\text{в}}{=} И(j \square \theta_2));$$

$$\underset{\theta}{+}(\theta_1 \square \theta_2) \simeq (И(1 \square \theta_1) \underset{\text{в}}{+} И(1 \square \theta_2))\, э \ldots э\, (И(n \square \theta_1) \underset{\text{в}}{+} И(n \square \theta_2));$$

$$\underset{\theta}{\cdot}(x \square \theta_1) \simeq (x \underset{\text{в}}{\cdot} И(1 \square \theta_1))\, э \ldots э\, (x \underset{\text{в}}{\cdot} И(n \square \theta_1)).$$

We denote the word $0 э \cdots э 0$ by $\mathfrak{O}$.

It is not hard to see that the list

(3) $$\mathrm{A}_r.\ \mathfrak{P},\ \mathfrak{C},\ \underset{\theta}{+},\ \underset{\theta}{\cdot},\ \mathfrak{D}$$

defines a constructive linear space. We write

$$X_1 \leftrightharpoons 1 \ni 0 \ni \ldots \ni 0.\ X_2 \leftrightharpoons 0 \ni 1 \ni \ldots \ni 0, \ldots,\ X_n \leftrightharpoons 0 \ni 0 \ni \ldots \ni 1.$$

Obviously

(4) $$\forall \theta_1 (\theta_1 \underset{\theta}{=} \text{И}(1 \square \theta_1) \cdot X_1 + \ldots + \text{И}(n \square \theta_1) \cdot X_n).$$

Thus the space (3) is an n-dimensional linear space. Therefore every locally convex space constructed on the basis of the linear space (3) is also an n-dimensional space, while the points $X_1, \cdots, X_n$ form its basis.

Suppose that $\mathfrak{J}$ is the set of all natural numbers. Suppose that m is a natural number such that $0 \leqq m \leqq n$. We denote by $\mathfrak{Q}_{n,m}$ a two-parameter formula such that, for any natural number i and any word θ_1 of type θ,

$$\mathfrak{Q}_{n,m}(i, \theta_1) \equiv \forall j ((1 \leqslant j \leqslant m \supset |\text{И}(j \square \theta_1)| \leqslant 2^{-i}) \,\&\, (m < j \leqslant n \supset \mathfrak{Q}_s(i, \text{И}(j \square \theta_1)))),$$

where $\mathfrak{Q}_s$ is the formula defined in Chapter I, §1.4, Example 2.

We recall that, for any natural number i and any real duplex x,

$$\mathfrak{Q}_s(i, x) \equiv \exists j \forall kl (k, l \geqslant j \supset |x| \cdot |S(k) - S(l)| < 2^{-i}),$$

where S is the Specker algorithm.

Consider the following list:

(5) $$\mathrm{A}_r,\ \mathfrak{P},\ \mathfrak{C},\ \underset{\theta}{+},\ \underset{\theta}{\cdot},\ \mathfrak{D},\ \text{Ч}_0,\ \mathfrak{J},\ \mathfrak{Q}_{n,m}.$$

This list defines an n-dimensional locally convex space, which we shall denote by $\mathscr{T}_{n,m}$. It is easy to see that $\mathscr{T}_{n,m}$ coincides with the space

$$\underbrace{R \times \ldots \times R}_{m \text{ times}} \times \underbrace{\mathscr{S} \times \ldots \times \mathscr{S}}_{n-m \text{ times}}$$

up to equivalence of topologies. Here R and $\mathscr{S}$ are the spaces defined in Chapter I, §1.4, Examples 1 and 2. Therefore we may suppose that $\mathscr{T}_{n,m}$ is the product of the space R taken m times and the space $\mathscr{S}$ taken $n-m$ times. For $m = n$ it coincides with $R^{(n)}$ and for $m = 0$ it coincides with $\mathscr{S}^{(n)}$. We note that

$$\mathfrak{Q}_{n,n}(i, \theta_1) \equiv \forall j (1 \leqslant j \leqslant n \supset |\text{И}(j \square \theta_1)| \leqslant 2^{-i}).$$

Therefore the space $\mathscr{T}_{n,n}$ $(= R^{(n)})$ is a normable space; it is normable with respect to the norm $\mathscr{N}$ defined as follows:

$$\forall \theta_1 (\mathscr{N}(\theta_1) \simeq \max(|\text{И}(1 \square \theta_1)| \square \ldots \square |\text{И}(n \square \theta_1)|)).$$

4.2.1. a) *The space* $\mathscr{T}_{n,n}$ $(= R^{(n)})$ *is an* n*-dimensional normable space, so that it is* T*-separated and* T'*-separated.*

b) *For $m < n$ the space $\mathscr{T}_{n,m}$ is n-dimensional and T'-separated, but not T-separated.*

Part a) follows from what was said above, and part b) follows from Theorems 1.2.1 and 1.2.3.

It is evident from 4.2.1 that, if $m < n$, the space $\mathscr{T}_{n,m}$ is not isomorphic to the space $R^{(n)}$. Thus we have the following corollary:

4.2.2. *For any natural number $n \geqq 1$, one can realize an n-dimensional T'-separated locally convex space which is not isomorphic to the space $R^{(n)}$.*

As we know, in the classical theory of locally convex spaces one proves that every n-dimensional, separated, locally convex space is isomorphic to the product of the space of real numbers taken n times as factors. Thus, Theorem 4.2.2 shows that this assertion does not carry over to constructive mathematics if as the constructive analogue of the concept of separation one takes the constructive concept of "T'-separation". The following question arises: does this assertion carry over to constructive mathematics if as the constructive analogue of the concept of separation one adopts not the concept of T'-separation, but rather the concept of T-separation? Copying the proof of the assertion indicated above in the classical theory, we may prove the following assertions.

1) Every one-dimensional, T-separated, locally convex space is isomorphic to the space R.

2) If $\mathfrak{M}$ is an n-dimensional, T-separated, locally convex space, and if in $\mathfrak{M}$ each $\mathfrak{Z}'$-closed subspace is $\mathfrak{Z}$-closed, then $\mathfrak{M}$ is isomorphic to the space $R^{(n)}$.

However the author has not been able to solve the question: if $\mathfrak{M}$ is n-dimensional ($n > 1$) and T-separated, is $\mathfrak{M}$ then isomorphic to the space $R^{(n)}$?

4.3. In this subsection we shall construct some infinite-dimensional spaces which will be employed in the following chapter.

Suppose that $\mathfrak{J}$ is the set of all natural numbers, $\mathfrak{P}$ the set of all real duplexes, $\mathfrak{E}$ a two-parameter formula defining the equality relation for real duplexes, and $+$ and $\cdot$ the addition and multiplication algorithms for real duplexes.

Suppose that m is a natural number. We shall construct a three-parameter formula $\mathfrak{Q}$ such that, for any natural numbers k and i and any real duplex x,

$$\mathfrak{Q}(k, i, x) \equiv (k < m \supset |x| \leqslant 2^{-i}) \,\&\, (k \geqslant m \supset \mathfrak{Q}_s(i, x)).$$

We consider the list

$$(6) \qquad \mathsf{Ч}_0, \mathfrak{J}, \mathsf{Ч}_3, \mathfrak{P}, \mathfrak{E}, +, \cdot, 0, \mathsf{Ч}_0, \mathfrak{J}, \mathfrak{Q}.$$

We may suppose that this list yields a sequence of locally convex spaces, each space $\mathfrak{M}^k$ of which coincides either with R, if $k < m$, or with $\mathscr{S}$, if $k \geqq m$.

We denote by $\mathscr{T}_{\infty,m}$ the product of the sequence (6) of spaces. By the definition of the product in §1.2 the space $\mathscr{T}_{\infty,m}$ is constructed as follows.

First we may construct a formula $\mathfrak{P}^*$, defining the set of transcriptions relative to the alphabet $Ч_3^{sa}$ of all possible algorithms of the type (н → д). We denote by θ the subordinate generic letter assigned to the formula $\mathfrak{P}^*$. It is possible to construct a two-parameter formula $\mathfrak{E}^*$, an algorithm $\underset{\theta}{+}$ of the type $(\theta\theta \to \theta)$ and an algorithm $\underset{\theta}{\cdot}$ of the type $(д\theta \to \theta)$ such that, for any words θ_1 and θ_2 of type θ and any real duplex x,

$$\mathfrak{E}^*(\theta_1, \theta_2) \equiv \forall k (\langle\theta_1\rangle_3(k) = \langle\theta_2\rangle_3(k));$$

$$\forall k (\langle \underset{\theta}{+} (\theta_1 \square \theta_2)\rangle_3(k) \simeq \langle\theta_1\rangle_3(k) + \langle\theta_2\rangle_3(k));$$

$$\forall k (\langle \underset{\theta}{\cdot} (x \square \theta_1)\rangle_3(k) \simeq x \cdot \langle\theta_1\rangle_3(k)).$$

We denote by $\mathfrak{O}^*$ the transcription of an algorithm which transforms every natural number into the number 0.

Suppose that τ and э are letters not entering into $Ч_0$. We denote by s the index of the alphabet $Ч_0 \cup \{\tau, э\}$, i.e. $A_s = Ч_0 \cup \{\tau, э\}$. It is possible to construct a normal one-parameter formula $\mathfrak{I}^*$ defining the set of all possible words in A_s of the following kind:

$$n_1 э\, i_1 \tau \cdots \tau n_k э\, i_k, \tag{7}$$

where $k \geq 1$, and where $n_1, \cdots, n_k$ and $i_1, \cdots, i_k$ are natural numbers. We denote by ι a subordinate generic letter put into correspondence with the formula $\mathfrak{I}^*$. We denote by L_τ, $И_1$ and $И_2$ algorithms satisfying the following conditions: for any word U of type ι and natural number i, if U has the form (7) and $1 \leq j \leq k$, then $L_\tau(U) \mathrel{\dot{=}} k$, $И_1(j \square U) \mathrel{\dot{=}} n_j$ and $И_2(j \square U) \mathrel{\dot{=}} i_j$ (see §1.2).

Finally, we construct a two-parameter formula $\mathfrak{Q}^*$ such that, for any words ι_1 of type ι and θ_1 of type θ,

$$\mathfrak{Q}^*(\iota_1, \theta_1) \equiv \forall j (1 \leqslant j \leqslant L_\tau(\iota_1) \supset \mathfrak{Q}(И_1(j \square \iota_1), И_2(j \square \iota_1), \langle\theta_1\rangle_3(И_1(j \square \iota_1)))),$$

where $\mathfrak{Q}$ is the formula defined above.

The space $\mathscr{T}_{\infty,m}$ is defined by the following list:

$$Ч_0, \mathfrak{P}^*, \mathfrak{E}^*, \underset{\theta}{+}, \underset{\theta}{\cdot}, \mathfrak{O}^*, A_s, \mathfrak{I}^*, \mathfrak{Q}^*. \tag{8}$$

Suppose that X is a sequence of points of $\mathscr{T}_{\infty,m}$, defined in such a way that, for any natural numbers i and k,

$$\langle X(i)\rangle_3(k) \simeq \begin{cases} 1, & \text{if } k = i, \\ 0, & \text{if } k \neq i. \end{cases}$$

We write $X_i \leftrightharpoons X(i)$. It is easy to see that, for any n, the points $X_1, \cdots, X_n$ are linearly independent in $\mathscr{T}_{\infty,m}$. Hence, and from Theorems 1.2.1 and 1.2.3, we obtain

4.3.1. *For any natural number m, $\mathscr{T}_{\infty,m}$ is an infinite-dimensional, T'-separated, but not T-separated space.*

Chapter III

SPACES ADJOINT TO CONSTRUCTIVE LOCALLY CONVEX SPACES

§1. Bounded sets in a constructive locally convex space

1.1. Suppose that $\mathfrak{M}$ is a locally convex space given by the list

$$\mathrm{A}_{\mathrm{m}},\ \mathfrak{P},\ \mathfrak{C},\ \underset{\theta}{+},\ \underset{\theta}{\cdot},\ \mathfrak{O},\ \mathrm{A}_{\mathrm{n}},\ \mathfrak{J},\ \mathfrak{Q}. \tag{1}$$

We denote the indices of the alphabets $\mathrm{A}_{\mathrm{m}} \cup \mathrm{Ч}_3$ and $\mathrm{A}_{\mathrm{n}} \cup \mathrm{Ч}_3$ by η and ζ respectively: $\mathrm{A}_\eta = \mathrm{A}_{\mathrm{m}} \cup \mathrm{Ч}_3$, $\mathrm{A}_\zeta = \mathrm{A}_{\mathrm{n}} \cup \mathrm{Ч}_3$.

Suppose that $\mathfrak{A}$ is some set of points of $\mathfrak{M}$. We will say that $\mathfrak{A}$ is a bounded set in $\mathfrak{M}$ if

$$\forall \iota_1 \exists x\,(x > 0 \;\&\; \forall \theta_1\,(\theta_1 \in \mathfrak{A} \supset \mathfrak{Q}\,(\iota_1, x \cdot \theta_1))). \tag{2}$$

This means that $\mathfrak{A}$ is absorbed by any neighborhood of zero in $\mathfrak{M}$. This coincides with the definition of a bounded set introduced by von Neumann in classical mathematics [**50**].

The set $\mathfrak{A}$ is said to be a п*-bounded set in* $\mathfrak{M}$ if for any sequence λ_1 of real duplexes and sequence λ_2 of points of $\mathfrak{M}$ we have

$$\forall n\,(\lambda_2(n) \in \mathfrak{A}) \;\&\; (0 \underline{\ \lim\ }_R \lambda_1) \supset (\mathfrak{O} \underline{\ \lim\ }_{\mathfrak{M}} \lambda_3),$$

where λ_3 is a sequence of points of $\mathfrak{M}$, defined as follows:

$$\forall n\,(\lambda_3(n) \simeq \lambda_1(n) \cdot \lambda_2(n)).$$

This definition corresponds to the definition of a bounded set introduced by Kolmogorov in classical mathematics [**44**].

It is easy to verify that *every bounded set in* $\mathfrak{M}$ *is also a* п*-bounded set in* $\mathfrak{M}$. However, the inverse is not true. In [**20**] an example of a set which was п-bounded but not bounded in the space of real duplexes was constructed.

We note that condition (2) is equivalent to the following condition:

$$\forall \iota_1 \exists a\,(a > 0 \;\&\; \forall \theta_1\,(\theta_1 \in \mathfrak{A} \supset \mathfrak{Q}\,(\iota_1, a \cdot \theta_1))), \tag{2'}$$

where a is a variable for rational numbers. Since $\mathfrak{J}$ is a normal formula, condition (2′) is satisfied if and only if we can realize an algorithm λ of the type $(\iota \to \mathrm{p})$ satisfying the conditions:

$$\forall \iota_1\,(\lambda(\iota_1) > 0); \qquad \forall \iota_1 \theta_1\,(\theta_1 \in \mathfrak{A} \supset \mathfrak{Q}\,(\iota_1, \lambda(\iota_1) \cdot \theta_1)).$$

Thus we see that in a certain sense every bounded set in $\mathfrak{M}$ corresponds to some algorithm of the type $(\iota \to \mathrm{p})$.

We introduce the following formula:

$$\mathfrak{D} \leftrightarrows (\langle c_{0,1}\rangle_\zeta \in (\iota \to \mathrm{p})) \,\&\, \forall \iota_1 (\langle c_{0,1}\rangle_\zeta (\iota_1) > 0).$$

$\mathfrak{D}$ is a one-parameter formula, whose parameter is the variable $c_{0,1}$. From the normality of the formula $\mathfrak{J}$ it follows that $\mathfrak{D}$ is also a normal formula. We adjoin this formula to the list Σ of the language $\mathcal{A}_\Sigma$ and denote by σ the corresponding subordinate generic letter.

By definition, the set $\mathfrak{A}$ is bounded if and only if we can realize a word σ_1 of type σ such that

$$\forall \iota_1 \theta_1 (\theta_1 \in \mathfrak{A} \supset \mathfrak{Q} (\iota_1, \langle \sigma_1 \rangle_\zeta (\iota_1) \cdot \theta_1)),$$

or

$$\forall \theta_1 (\theta_1 \in \mathfrak{A} \supset \forall \iota_1 \mathfrak{Q} (\iota_1, \langle \sigma_1 \rangle_\zeta (\iota_1) \cdot \theta_1)).$$

Suppose that A_q is an alphabet, and $\mathfrak{K}$ a normal one-parameter formula defining some set of words in A_q. We suppose that the formula $\mathfrak{K}$ lies in the list of formulas Σ of the language $\mathcal{A}_\Sigma$, and we denote by κ the subordinate generic letter set into correspondence with the formula $\mathfrak{K}$ in this language. Suppose further that $\mathfrak{R}$ is a two-parameter formula, whose parameters are a variable ν of genus κ and a variable α of genus t_{m}. We write

$$\mathfrak{R} (U, P) \leftrightarrows F^{\nu, \alpha}_{U, P} \llcorner \mathfrak{R} \lrcorner,$$

where U and P are arbitrary object terms. We suppose that

$$\forall \varkappa_1 \alpha (\mathfrak{R} (\varkappa_1, \alpha) \supset \alpha \in \mathfrak{P}).$$

Then we will say that the list $(A_q, \mathfrak{K}, \mathfrak{R})$ defines a system of sets of points of $\mathfrak{M}$.

Suppose that $(A_q, \mathfrak{K}, \mathfrak{R})$ is some system of sets of points of $\mathfrak{M}$. We will say that this list defines a *system of bounded sets in* $\mathfrak{M}$ if

(3) $$\forall \varkappa_1 \exists \sigma_1 \forall \theta_1 (\mathfrak{R} (\varkappa_1, \theta_1) \supset \forall \iota_1 \mathfrak{Q} (\iota_1, \langle \sigma_1 \rangle_\zeta (\iota_1) \cdot \theta_1)).$$

This means that each set of the system $(A_q, \mathfrak{K}, \mathfrak{R})$ is a bounded set in $\mathfrak{M}$.

Suppose that $\mathfrak{S}$ is a system of bounded sets in $\mathfrak{M}$, defined by a list $(A_q, \mathfrak{K}, \mathfrak{R})$. $\mathfrak{S}$ is said to be *normal* if $\mathfrak{K}$ is a normal formula. It is said to be *enumerable* if the set $\mathfrak{K}$ is enumerable.

A system $\mathfrak{S}$ of bounded sets is said to be *covering* if

(4) $$\forall \theta_1 \exists \varkappa_1 \mathfrak{R} (\varkappa_1, \theta_1).$$

We will say that the system $\mathfrak{S}$ is a *fundamental system of bounded sets in* $\mathfrak{M}$ if, in addition to condition (3), it satisfies also the following condition:

(5) $$\forall \sigma_1 \exists \varkappa_1 \forall \theta_1 (\forall \iota_1 \mathfrak{Q} (\iota_1, \langle \sigma_1 \rangle_\zeta (\iota_1) \cdot \theta_1) \supset \mathfrak{R} (\varkappa_1, \theta_1)).$$

Obviously, every fundamental system of bounded sets in $\mathfrak{M}$ is a covering system.

The following assertion will play a fundamental role in the construction of the adjoint space.

1.1.1. *For any constructive locally convex space $\mathfrak{M}$ it is possible to construct a normal fundamental system of bounded sets in $\mathfrak{M}$.*

Indeed, we construct a formula $\mathfrak{R}$ in the following way:

$$\mathfrak{R} \leftrightharpoons (\alpha \in \mathfrak{P}) \,\&\, \forall \iota_1 \mathfrak{Q}(\iota_1, \langle \sigma_1 \rangle_{\zeta}(\iota_1) \cdot \alpha),$$

where α is a variable of genus t_{m}.

It is easy to see that the list $(\text{Ч}_0, \mathfrak{D}, \mathfrak{R})$ defines a system of sets of points of the set $\mathfrak{M}$ and possesses properties (3) and (5). Moreover, $\mathfrak{D}$ is a normal formula. Assertion 4.1.1 is proved.

Suppose $\mathfrak{S}$ and $\tilde{\mathfrak{S}}$ are systems of bounded sets in $\mathfrak{M}$, defined respectively by the lists $(\mathrm{A}_q, \mathfrak{K}, \mathfrak{R})$ and $(A_{\tilde{q}}, \tilde{\mathfrak{K}}, \tilde{\mathfrak{R}})$. We denote by κ and $\tilde{\kappa}$ subordinate generic letters whose characteristic formulas are $\mathfrak{R}$ and $\tilde{\mathfrak{R}}$ respectively. We will say that the system $\mathfrak{S}$ majorizes the system $\tilde{\mathfrak{S}}$, or that $\tilde{\mathfrak{S}}$ is majorized by $\mathfrak{S}$, if

$$\forall \tilde{\varkappa}_1 \exists \varkappa_1 \forall \theta_1 (\tilde{\mathfrak{R}}(\tilde{\varkappa}_1, \theta_1) \supset \mathfrak{R}(\varkappa_1, \theta_1)).$$

Two systems $\mathfrak{S}$ and $\tilde{\mathfrak{S}}$ are said to be *equivalent* if each of them majorizes the other.

Obviously, if $\mathfrak{S}$ is a fundamental system of bounded sets in $\mathfrak{M}$, then $\mathfrak{S}$ majorizes any system of bounded sets in $\mathfrak{M}$.

From the definition it is also obvious that any two fundamental systems of bounded sets in $\mathfrak{M}$ are equivalent to one another, and each system of bounded sets in $\mathfrak{M}$ equivalent to some fundamental system of bounded sets in $\mathfrak{M}$ is itself fundamental.

1.2. In the proof of 1.1.1 we indicated a method for constructing a normal fundamental system of bounded sets in any locally convex space. However, in certain cases we may construct a normal fundamental system of bounded sets in a simpler way.

1) We begin with a normed space $\mathfrak{M}$ given by the list

$$\mathrm{A}_{\mathrm{m}}, \mathfrak{P}, \mathfrak{E}, +, \cdot, \mathfrak{D}, \mathcal{N}.$$

We denote by $\mathfrak{K}$ the set of all natural numbers, and by $\mathfrak{R}$ the following formula:

$$\mathfrak{R} \leftrightharpoons (\alpha \in \mathfrak{P}) \,\&\, (\mathcal{N}(\alpha) \leqslant k + 1),$$

where k is a variable for the natural numbers and α is a variable of genus t_{m}.

It is easy to verify that the list $(\text{Ч}_0, \mathfrak{K}, \mathfrak{R})$ defines a normal fundamental system of bounded sets in $\mathfrak{M}$.

2) Now suppose that $\mathfrak{M}$ is an enumerably normed space given by the list

$$\mathrm{A}_{\mathrm{m}}, \mathfrak{P}, \mathfrak{E}, +, \cdot, \mathfrak{D}, \mathcal{N},$$

where $\mathcal{N}$ is an enumerable norm in $\mathfrak{M}$ (see §§4.2 and 4.4 in Chapter I).

Suppose that $\mathfrak{U}$ is an arithmetic universal algorithm, i.e. an algorithm of the type ($\text{HH} \xrightarrow{\cdot} \text{H}$) having the property that for any algorithm φ of type ($\text{H} \xrightarrow{\cdot} \text{H}$) we can realize a natural number m such that

$$\forall n\, (\varphi\, (n) \simeq \mathfrak{U}\, (m \,\Box\, n))$$

(see Theorem 1.1 of [4] or Theorem XIX of [**43**]).

Suppose that A is a set of natural numbers defined in the following way:

$$(m \in A) \leftrightarrows \forall n!\, \mathfrak{U}\, (m \,\Box\, n).$$

This means that A is a set of "indices" of all possible algorithms of type ($\text{H} \to \text{H}$), or a set of indices of all general recursive functions. Obviously the set A is defined by a normal formula. We denote by e a subordinate generic letter assigned to the set A in the language Я_Σ.

It is possible to construct a two-parameter formula $\mathfrak{R}$ such that the list ($\text{Ч}_0, A, \mathfrak{R}$) forms a system of sets of points of the space $\mathfrak{M}$ and such that, for any words e_1 of type e and θ_1 of type θ,

$$\mathfrak{R}\, (e_1, \theta_1) \equiv \forall n\, (\mathcal{N}\, (n \,\Box\, \theta_1) \leqslant \mathfrak{U}\, (e_1 \,\Box\, n) + 1).$$

It is not difficult to prove that each set of the system ($\text{Ч}_0, A, \mathfrak{R}$) is bounded in $\mathfrak{M}$, and that each bounded set in $\mathfrak{M}$ is contained in some set of the system ($\text{Ч}_0, A, \mathfrak{R}$). In other words, *the list* ($\text{Ч}_0, A, \mathfrak{R}$) *defines a normal fundamental system of bounded sets in the enumerably normed space* $\mathfrak{M}$. In what follows this fundamental system of bounded sets will be used several times.

We introduce the following notation. Suppose that e_1 is a word of type e. We denote by $\{e_1\}$ an algorithm of type ($\text{H} \to \text{H}$), defined as follows:

$$\forall n\, (\{e_1\}\, (n) \simeq \mathfrak{U}\, (e_1 \,\Box\, n)).$$

REMARK. For the general case of a multinormed space it is possible to construct a normal fundamental system of bounded sets in an analogous way. The difference consists only in that, in the general case of a multinormed space, in place of the set A one has to choose the sets of transcriptions of algorithms of the type ($\nu \to \text{H}$), where ν is a generic letter assigned to the set of the multinorm $\mathscr{A}$ of the given space.

3) Now suppose given an $\mathscr{A}$-family of constructive locally convex spaces, defined by the list

(6) $$\text{A}_q,\ \mathscr{A},\ \text{A}_{\mathfrak{m}},\ \mathfrak{P},\ \mathfrak{E},\ +,\ \cdot,\ \mathfrak{D},\ \text{A}_{\mathfrak{n}},\ \mathfrak{J},\ \mathfrak{Q}.$$

Here we are using the terminology and notation introduced in §§1.1 and 1.2 of Chapter II.

Suppose that the product of an $\mathscr{A}$-family of spaces (6) is defined by the list

(7) $$\text{Ч}_0,\ \mathfrak{P}^\bullet,\ \mathfrak{E}^\bullet,\ \underset{\theta}{+},\ \underset{\theta}{\cdot},\ \mathfrak{D}^\bullet,\ \text{A}_s,\ \mathfrak{J}^\bullet,\ \mathfrak{Q}^\bullet.$$

We denote the space with index U of the $\mathscr{A}$-family (6) by $\mathfrak{M}^U$, and the product (7) by $\mathfrak{M}$.

Suppose that A_u is an alphabet, $\mathfrak{K}$ a normal two-parameter formula whose parameters are a variable γ of genus ν and a variable α of genus t_u, and $\mathfrak{R}$ a three-parameter formula whose parameters are a variable γ of genus ν, a variable β of genus t_u and a variable α of genus t_{m}. Here ν is a subordinate generic letter assigned to the formula $\mathscr{A}$. We suppose that the formula $\mathfrak{R}$ lies in the list Δ of formulas of the language $Я_\Sigma^\Delta$, and denote by κ the pregeneric letter assigned to the formula $\mathfrak{R}$ in this language. We write

$$\mathfrak{K}^U \leftrightharpoons F_U^\gamma \llcorner \mathfrak{K} \lrcorner; \quad \mathfrak{R}^U \leftrightharpoons F_U^\gamma \llcorner \mathfrak{R} \lrcorner; \quad \mathfrak{K}(U, P) \leftrightharpoons F_{U, P}^{\gamma, \beta} \llcorner \mathfrak{K} \lrcorner;$$

$$\mathfrak{R}(U, P, Q) \leftrightharpoons F_{U, P, Q}^{\gamma, \beta, \alpha} \llcorner \mathfrak{R} \lrcorner,$$

where U, P and Q are arbitrary object terms.

We suppose that for any word U of type ν the list $(A_u, \mathfrak{K}^U, \mathfrak{R}^U)$ defines a normal fundamental system of bounded sets in the space $\mathfrak{M}^U$. On the basis of this hypothesis we may construct a normal fundamental system of bounded sets in $\mathfrak{M}$ as follows.

Suppose that v is the index of the alphabet $A_q \cup A_u$, i.e. $A_v = A_q \cup A_u$. We introduce the following formulas:

$$\mathfrak{K}^* \leftrightharpoons \forall \mathrm{v}_1 ((\langle c_{0,1} \rangle_v (\mathrm{v}_1) \overset{+}{\in} t_u) \,\&\, \mathfrak{K}(\mathrm{v}_1, \langle c_{0,1} \rangle_v (\mathrm{v}_1)));$$

$$\mathfrak{R}^* \leftrightharpoons (c_{0,2} \in \mathfrak{P}^*) \,\&\, \forall \mathrm{v}_1 \mathfrak{R}(\mathrm{v}_1, \langle c_{0,1} \rangle_v (\mathrm{v}_1), \langle c_{0,2} \rangle_r (\mathrm{v}_1)).$$

The parameter of the formula $\mathfrak{K}^*$ is the variable $c_{0,1}$, and the parameters of $\mathfrak{R}^*$ are the variables $c_{0,1}$ and $c_{0,2}$. $\mathfrak{K}^*$ is a normal formula. It is possible to prove that the list $(Ч_0, \mathfrak{K}^*, \mathfrak{R}^*)$ defines a normal fundamental system of bounded sets in the space $\mathfrak{M}$. This is proved on the basis of the following assertion: for any set $\mathfrak{A}$ of points of the product $\mathfrak{M}$, $\mathfrak{A}$ is a bounded set in $\mathfrak{M}$ if and only if for any word U of type ν the set of points X of the space $\mathfrak{M}^U$ such that $\exists\, \theta_1 (\theta_1 \in \mathfrak{A} \,\&\, \langle \theta_1 \rangle_r (U) \doteq X)$ is bounded in $\mathfrak{M}^U$. The proof of this last assertion is analogous to the proof of the corresponding assertion in classical mathematics.

4) For the projective limit of the projective spectrum of locally convex spaces it is possible to construct a normal fundamental system of bounded sets in the projective limit, starting from the normal fundamental systems of bounded sets in the spaces of the given projective spectrum, in the same way as was used in 3) for the construction of a normal fundamental system of bounded sets in the product of an $\mathscr{A}$-family of locally convex spaces.

1.3. In this subsection we use the terminology and notation introduced in §3 of Chapter II.

Consider now bounded sets in the inductive limit of an expanding sequence of constructive locally convex spaces. Suppose given an expanding sequence of locally convex spaces, defined by the list

$$Ч_0,\ N,\ A_{\mathrm{m}},\ \mathfrak{P},\ \mathfrak{C},\ +,\ \cdot,\ \mathfrak{D},\ A_{\mathfrak{n}},\ \mathfrak{J},\ \mathfrak{Q} \tag{8}$$

(see §3.2, Chapter II). For each natural number k the list

(9) $$\mathrm{A}_m,\ \mathfrak{P}^k,\ \mathfrak{E}^k,\ +,\ \cdot,\ \mathfrak{O},\ \mathrm{A}_n,\ \mathfrak{I}^k,\ \mathfrak{Q}^k$$

defines the space with index k from the sequence (8), which we shall denote by $\mathfrak{M}^k$.

Suppose that the FR-inductive limit and the F-inductive limit of the sequence (8) are defined respectively by the following lists:

(10) $$\mathrm{A}_r,\ \mathfrak{P}^*,\ \mathfrak{E}^*,\ \underset{\theta}{+},\ \underset{\theta}{\cdot},\ \mathfrak{O}^*,\ \mathrm{Ч}_0,\ \mathfrak{I}^*,\ \mathfrak{Q}^*;$$

(11) $$\mathrm{A}_m,\ \tilde{\mathfrak{P}},\ \tilde{\mathfrak{E}},\ +,\ \cdot,\ \mathfrak{O},\ \mathrm{Ч}_0,\ \mathfrak{I}^*,\ \tilde{\mathfrak{Q}}.$$

We denote the FR-inductive limit by $\mathfrak{M}$, and the F-inductive limit (11) by $\tilde{\mathfrak{M}}$.

We recall that points of $\mathfrak{M}$ (words of type θ) are words of the type $k \tau P$, where k is a natural number and P is a point of $\mathfrak{M}^k$. The points of $\tilde{\mathfrak{M}}$ (words of type $\tilde{\theta}$) are those words P in A_m which satisfy the condition $\exists k\ (P \in \mathfrak{M}^k)$.

Suppose that $\mathfrak{A}$ is some set of points of $\mathfrak{M}$. We denote by $\tilde{\mathfrak{A}}$ the set of points of $\tilde{\mathfrak{M}}$ defined as follows:

$$(\tilde{\theta}_1 \in \tilde{\mathfrak{A}}) \leftrightharpoons \exists k\,(k\tau\tilde{\theta}_1 \in \mathfrak{A}).$$

The set $\tilde{\mathfrak{A}}$ is taken to be the *image* of the set $\mathfrak{A}$ in $\tilde{\mathfrak{M}}$. For convenience of exposition we introduce the following convention: if the set $\tilde{\mathfrak{A}}$ is contained, and bounded, in some space $\mathfrak{M}^k$, we also will say that the set $\mathfrak{A}$ is contained, and bounded, in $\mathfrak{M}^k$.

Now we introduce the following definitions: the FR-inductive limit $\mathfrak{M}$ (F-inductive limit $\tilde{\mathfrak{M}}$) is said to be *α-regular* if for any bounded set $\mathfrak{A}$ in $\mathfrak{M}$ ($\tilde{\mathfrak{M}}$) we can realize a natural number k such that $\mathfrak{A}$ is contained in the set $\mathfrak{M}^k$. The space $\mathfrak{M}$ ($\tilde{\mathfrak{M}}$) is said to be *β-regular*, if for any set $\mathfrak{A}$ and any natural number k such that $\mathfrak{A}$ is bounded in $\mathfrak{M}$ ($\tilde{\mathfrak{M}}$) and contained in $\mathfrak{M}^k$ we can realize a natural number j such that $\mathfrak{A}$ is contained in and bounded in $\mathfrak{M}^j$. The space $\mathfrak{M}$ ($\tilde{\mathfrak{M}}$) is said to be *regular* if for any bounded set $\mathfrak{A}$ in $\mathfrak{M}$ ($\tilde{\mathfrak{M}}$) we can realize a natural number k such that $\mathfrak{A}$ is contained in and bounded in $\mathfrak{M}^k$. One finds the corresponding definitions for classical mathematics in, for example, [7, **10**].

Obviously an FR-inductive limit $\mathfrak{M}$ (F-inductive limit $\tilde{\mathfrak{M}}$) is regular if and only if it is α-regular and β-regular.

1.3.1. *Suppose that $\mathfrak{A}$ is a set of points of an F-inductive limit $\tilde{\mathfrak{M}}$ (FR-inductive limit $\mathfrak{M}$). If a natural number k is realizable such that $\mathfrak{A}$ is contained in and bounded in $\mathfrak{M}^k$, then $\mathfrak{A}$ is bounded in $\mathfrak{M}$.*

Proof. The assertion need only be proved for the case when $\mathfrak{A}$ is a set of points of $\tilde{\mathfrak{M}}$. Suppose that $\mathfrak{A}$ is contained in and bounded in a space $\mathfrak{M}^k$, where k is a natural number. Then

(12) $$\forall \iota_1^k \exists a\,(a > 0\ \&\ \forall \theta_1^k\,(\theta_1^k \in \mathfrak{A} \supset \mathfrak{Q}\,(k,\ \iota_1^k,\ a \cdot \theta_1^k))).$$

We note that the identity mapping of $\mathfrak{M}^k$ into $\mathfrak{M}$ is continuous, i.e. that

(13) $$\forall \iota_1 \exists \iota^k \forall \theta_1^k (\mathfrak{Q}(k, \iota_1^k, \theta_1^k) \supset \tilde{\mathfrak{Q}}(\iota_1, \theta_1^k)).$$

Since the set $\mathfrak{A}$ is contained in $\mathfrak{M}^k$, it follows from (12) and (13) that

$$\forall \iota_1 \exists a (a > 0 \,\&\, \forall \tilde{\theta}_1 (\tilde{\theta}_1 \in \mathfrak{A} \supset \tilde{\mathfrak{Q}}(\iota_1, a \cdot \tilde{\theta}_1))).$$

Thus $\tilde{\mathfrak{A}}$ is bounded in $\tilde{\mathfrak{M}}$.

Now we suppose that the list (8) defines a strictly expanding sequence of locally convex spaces, and that $\mathfrak{M}$ and $\tilde{\mathfrak{M}}$ are its strict FR-inductive limit and its strict F-inductive limit respectively (see the definitions in §3.2, Chapter II).

1.3.2. *The F-strict inductive limit* (*FR-strict inductive limit*) *of a strictly expanding sequence of locally convex spaces is* α-*regular.*

Proof. Suppose that $\mathfrak{A}$ is a set of points of the F-strict inductive limit $\tilde{\mathfrak{M}}$, bounded in $\tilde{\mathfrak{M}}$ and contained in the space $\mathfrak{M}^k$, where k is some natural number. Then the following holds:

(14) $$\forall \iota_1 \exists a (a > 0 \,\&\, \forall \theta_1^k (\theta_1^k \in \mathfrak{A} \supset \tilde{\mathfrak{Q}}(\iota, a \cdot \theta_1^k))).$$

In view of formula (21) in §3.3, Chapter II, it is obvious that

(15) $$\forall \iota_1^k \exists \iota_1 \forall \theta_1^k (\tilde{\mathfrak{Q}}(\iota_1, \theta_1^k) \supset \mathfrak{Q}(k, \iota_1^k, \theta_1^k)).$$

It follows from (14) and (15) that

$$\forall \iota_1^k \exists a (a > 0 \,\&\, \forall \theta_1^k (\theta_1^k \in \mathfrak{A} \supset \mathfrak{Q}(k, \iota_1^k, a \cdot \theta_1^k))),$$

i.e. that the set $\mathfrak{A}$ is bounded in $\mathfrak{M}^k$. Thus the strict inductive limit $\tilde{\mathfrak{M}}$ is β-regular. For the case of the FR-inductive limit the proof is carried out in a quite analogous fashion.

1.3.3. *Suppose that the list* (8) *defines a strictly expanding sequence of locally convex spaces. Suppose that* $\tilde{\mathfrak{M}}$ *is its F-strict inductive limit and that* $\mathfrak{A}$ *is a set of points of the space* $\tilde{\mathfrak{M}}$. *If* $\mathfrak{A}$ *is a bounded set in* $\tilde{\mathfrak{M}}$, *then*

$$\neg \forall k \exists \tilde{\theta}_1 ((\tilde{\theta}_1 \in \mathfrak{A}) \,\&\, \neg (\tilde{\theta}_1 \in \mathfrak{M}^k)).$$

Proof. Suppose that $\mathfrak{A}$ is a set of points of $\tilde{\mathfrak{M}}$, and that

(16) $$\forall k \exists \tilde{\theta}_1 (\tilde{\theta}_1 \in \mathfrak{A} \,\&\, \neg (\tilde{\theta}_1 \in \mathfrak{M}^k)).$$

We shall prove that the set $\mathfrak{A}$ is not bounded in $\tilde{\mathfrak{M}}$.

First of all, it follows from (16) that we can realize an algorithm φ_1 of type $(\text{H} \to \tilde{\theta})$ such that

$$\forall k (\varphi_1(k) \in \mathfrak{A} \,\&\, \neg (\varphi_1(k) \in \mathfrak{M}^k)).$$

Suppose that φ_2 is a sequence of real indices defined as follows:

$$\forall k \left(\varphi_2(k) = \frac{1}{k+1}\right).$$

We construct a sequence φ of points of $\widetilde{\mathfrak{M}}$ such that

$$\forall k\,(\varphi(k) \simeq \varphi_2(k) \cdot \varphi_1(k)).$$

Obviously $\forall k \urcorner (\varphi(k) \in \mathfrak{M}^k)$. On the other hand, φ is a sequence of points of $\widetilde{\mathfrak{M}}$, so that

$$\forall k \exists m\,(\varphi(k) \in \mathfrak{M}^m).$$

This means that we can realize an algorithm ψ of type $(\mathrm{H} \to \mathrm{H})$ for which

$$\forall k\,(\varphi(k) \in \mathfrak{M}^{\psi(k)}).$$

We construct an algorithm γ of the type $(\mathrm{H} \to \tilde{\theta})$ according to the following schema:

$$\begin{cases} \gamma(0) \rightleftharpoons 0, \\ \gamma(k+1) \simeq \psi(\gamma(k)), \end{cases}$$

and then we construct an algorithm λ of the type $(\mathrm{H} \to \mathrm{H})$, as follows:

$$\forall k\,(\lambda(k) \simeq \varphi(\gamma(k))).$$

It is easy to see that

$$\forall k\,(\urcorner(\lambda(k) \in \mathfrak{M}^{\gamma(k)})\ \&\ (\lambda(k) \in \mathfrak{M}^{\gamma(k+1)})). \tag{17}$$

It follows from (17) that the sequence γ is strictly increasing.

Since the list (8) defines a strictly expanding sequence of spaces, $\mathfrak{M}^{\gamma(k)}$ is a closed subspace of $\mathfrak{M}^{\gamma(k+1)}$ for each k (see condition (iii) in §3.2, Chapter II). Hence it follows, using (17), that

$$\forall k \exists \iota_1^{\gamma(k+1)}\, \forall \tilde{\theta}_1\,(\tilde{\theta}_1 \in \mathfrak{M}^{\gamma(k)} \supset \urcorner \mathfrak{Q}(\gamma(k+1), \iota_1^{\gamma(k+1)}, \lambda(k) - \tilde{\theta}_1)). \tag{18}$$

We note that in view of condition (ii) (see §3.2, Chapter II) the sets $\mathfrak{J}^k$ (k runs through the natural numbers) coincide with one another, and accordingly coincide with a set $\mathfrak{J}^0$. We denote by ν a subordinate generic letter assigned to this set $\mathfrak{J}^0$. Thus words of type ν are elements of the set $\mathfrak{J}^0$, and so they are elements of the set $\mathfrak{J}^k$ for any k.

It follows from (18) that we can realize an algorithm α of the type $(\mathrm{H} \to \nu)$ such that

$$\forall k \tilde{\theta}_1\,(\tilde{\theta}_1 \in \mathfrak{M}^{\gamma(k)} \supset \urcorner \mathfrak{Q}(\gamma(k+1), \alpha(k), \lambda(k) - \tilde{\theta}_1)). \tag{19}$$

In view of axioms III2* and III5* of a family of locally convex spaces (see §1.1, Chapter II) one can realize algorithms δ of type $(\mathrm{H}\nu\nu \to \nu)$ and β of type $(\mathrm{H}\nu \to \nu)$ such that

$$\forall k \mathrm{v}_1 \mathrm{v}_2 \theta_1^k\,(\mathfrak{Q}(k, \delta(k \square \mathrm{v}_1 \square \mathrm{v}_2), \theta_1^k) \supset \mathfrak{Q}(k, \mathrm{v}_1, \theta_1^k)\ \&\ \mathfrak{Q}(k, \mathrm{v}_2, \theta_1^k));$$

$$\forall k \mathrm{v}_1 \theta_1^k \theta_2^k\,(\mathfrak{Q}(k, \beta(k \square \mathrm{v}_1), \theta_1^k)\ \&\ \mathfrak{Q}(k, \beta(k \square \mathrm{v}_1), \theta_2^k) \supset \mathfrak{Q}(k, \mathrm{v}_1, \theta_1^k + \theta_2^k)).$$

Denote by H the set of natural numbers enumerated by the algorithm γ. Since γ is strictly increasing, H is algorithmically solvable. We denote by ξ

the algorithm inverse to the algorithm γ relative to the natural numbers (see [21]). Obviously

$$\forall n\,(n \in H \supset \gamma(\xi(n)) \doteq n) \text{ and } \forall k\,(\xi(\gamma(k)) \doteq k).$$

Now we shall construct algorithms η of the type $(\mathrm{H} \to \nu)$, χ of the type $(\mathrm{H} \to \mathrm{H})$, ζ of the type $(\mathrm{H} \to \nu)$, and ρ of the type $(\mathrm{H} \to \nu)$, according to the following schemas respectively:

$$\eta(n) \simeq \begin{cases} Y, & \text{if } n = 0 \text{ or } n > 0 \,\&\, \neg(n-1 \in H); \\ \alpha(\xi(n-1)), & \text{if } n > 0 \,\&\, (n-1 \in H) \end{cases}$$

(Y some fixed word of type ν);

$$\chi(n) \simeq \begin{cases} \gamma(0) & \text{if } n = 0; \\ \chi(n-1), & \text{if } n > 0 \,\&\, \neg(n-1 \in H); \\ \gamma(\xi(n-1)+1), & \text{if } n > 0 \,\&\, (n-1 \in H). \end{cases}$$

$$\begin{cases} \zeta(0) \doteq \eta(0); \\ \zeta(n+1) \simeq \delta(\chi(n+1) \square \eta(n+1) \square \beta(\chi(n+1) \square \zeta(n))). \end{cases}$$

$$\rho(n) \simeq \begin{cases} \beta(\chi(n) \square \zeta(n)), & \text{if } \neg(n \in H); \\ \beta(\gamma(\xi(n)+1) \square \zeta(n)), & \text{if } n \in H. \end{cases}$$

We denote by ι_1 the transcription of the algorithm ρ. Then ι_1 is a word of type ι, i.e. an element of the set $\mathfrak{Z}^*$. We shall show that

$$\forall k \neg \tilde{\mathfrak{Q}}(\iota_1, \lambda(k)). \tag{20}$$

Suppose that k is a natural number, and that $\tilde{\mathfrak{Q}}(\iota_1, \lambda(k))$. It follows that a natural number m and points $X_0, \cdots, X_m$ of the space $\mathfrak{M}$ are realizable such that

$$X_i \in \mathfrak{M}^i,\ \mathfrak{Q}(i, \rho(i), X_i) \quad (i = 0, \ldots, m) \tag{21}$$

$$\lambda(k) \eqsim X_0 + \ldots + X_m. \tag{22}$$

We find a natural number h such that $m \leq \gamma(k+h)$. It is obvious from (17) that $h > 0$. Without loss of generality we may suppose that $m = \gamma(k+h)$.

By means of induction on j $(0 \leq j \leq h-1)$ we may prove the following formula:

$$\mathfrak{Q}(\gamma(k+h-j),\ \zeta(\gamma(k+h-j-1)+1),\ X_{\gamma(k+h-j-1)+1} + \ldots + X_m). \tag{23}$$

In the proof of (23) use was made of formulas (21), (22), the definitions of the algorithms η, χ, ζ and ρ, and also conditions (i) and (ii) of §3.2, Chapter II.

Substituting $j = h-1$ into (23), we find that

$$\mathfrak{Q}(\gamma(k+1),\ \zeta(\gamma(k)+1),\ X_{\gamma(k)+1} + \ldots + X_m). \tag{24}$$

Since $\chi(\gamma(k)+1) = \gamma(k+1)$, by the definition of the algorithm ζ we find from (24) that

$$\mathfrak{Q}(\gamma(k+1), \eta(\gamma(k)+1), X_{\gamma(k)+1}+\ldots+X_m),$$

i.e.

(25) $$\mathfrak{Q}(\gamma(k+1), \alpha(k), X_{\gamma(k)+1}+\ldots+X_m).$$

We write $X \leftrightharpoons X_0 + \cdots + X_{\gamma(k)}$. Then $X \in \mathfrak{M}^{\gamma(k)}$ and

$$X_{\gamma(k)+1}+\ldots+X_m \approx \lambda(k) - X.$$

Both parts of this equation are points of the space $\mathfrak{M}^{\gamma(k+1)}$, so that they are equal to one another in that space also. Therefore we find from (25) that

$$\mathfrak{Q}(\gamma(k+1), \alpha(k), \lambda(k) - X).$$

This contradicts property (19). Therefore the hypotheses $\tilde{\mathfrak{Q}}(\iota_1, \lambda(k))$ is rejected, and formula (20) is proved. It is obvious from (20) that

$$\neg(\mathfrak{D}\ \underline{\lim}_{\widetilde{\mathfrak{M}}}\ \varphi).$$

Thus the set $\mathfrak{A}$ is not п-bounded in $\widetilde{\mathfrak{M}}$ (see the definition of a п-bounded set in §1.1), so that $\mathfrak{A}$ is not bounded in $\widetilde{\mathfrak{M}}$. Theorem 1.3.3 is proved.

In a quite analogous way one proves the following theorem.

1.3.4. *Suppose that the list* (8) *defines a strictly expanding sequence of locally convex spaces. Let* $\mathfrak{M}$ *be its FR-strict inductive limit and* $\mathfrak{A}$ *a set of points of the space* $\mathfrak{M}$. *If* $\mathfrak{A}$ *is a bounded set in* $\mathfrak{M}$, *then*

$$\neg\forall k \exists \theta_1 ((\theta_1 \in \mathfrak{A}) \,\&\, \neg([\theta_1]_2^{\tau} \in \mathfrak{M}^k)).$$

As we know, in classical mathematics one proves that the strict inductive limit of any strictly expanding sequence of locally convex spaces is regular (see [**39**], and also [**7**], Chapter XI, and [**38**], Chapter III). One may consider Theorems 1.3.2–1.3.4 to be the results of a partial carryover of this theorem to constructive mathematics.

In Chapter VI the following theorem will be proved.

1.3.5. *It is possible to construct a strictly increasing sequence of locally convex spaces such that in its F-strict inductive limit (FR-strict inductive limit) one can realize a bounded set* $\mathfrak{A}$ *for which*

$$\neg\exists k \forall \tilde{\theta}_1 (\tilde{\theta}_1 \in \mathfrak{A} \supset \tilde{\theta}_1 \in \mathfrak{M}^k)$$

(*respectively* $\neg\exists k \forall \theta_1 \ (\theta_1 \in \mathfrak{A} \supset [\theta_1]_2^{\tau} \in \mathfrak{M}^k)$).

In other words, that bounded set $\mathfrak{A}$ is not contained in any of the spaces $\mathfrak{M}^k$. That F-strict inductive limit (FR-strict inductive limit) is not α-regular, and so not regular.

§2. The space adjoint to a constructive locally convex space

2.1. Suppose that $\mathfrak{M}$ and $\widetilde{\mathfrak{M}}$ are locally convex spaces, given respectively by the following lists:

(1) $$\mathrm{A}_{\mathfrak{m}},\ \mathfrak{P},\ \mathfrak{E},\ +,\ \bullet,\ \mathfrak{D},\ \mathrm{A}_{\mathfrak{n}},\ \mathfrak{J},\ \mathfrak{Q},$$

(2) $$\mathrm{A}_{\tilde{\mathfrak{m}}},\ \tilde{\mathfrak{P}},\ \tilde{\mathfrak{E}},\ \tilde{+},\ \tilde{\bullet},\ \tilde{\mathfrak{D}},\ \mathrm{A}_{\tilde{\mathfrak{n}}},\ \tilde{\mathfrak{J}},\ \tilde{\mathfrak{Q}}.$$

Suppose that λ is a linear operator of the type $(\theta \rightarrow \theta)$ (see §1.5 of Chapter I). By definition, a linear operator is said to be *continuous* if

$$\forall \tilde{\iota}_1 \exists \iota_1 \forall \theta_1 (\mathfrak{Q}(\iota_1\ \theta_1) \supset \tilde{\mathfrak{Q}}(\tilde{\iota}_1,\ \lambda(\theta_1))).$$

We will say that the linear operator λ is *bounded* if for any bounded set $\mathfrak{A}$ in $\mathfrak{M}$ the set of those points $\tilde{\theta}_1$ such that $\exists \theta_1 (\theta_1 \in \mathfrak{A} \,\&\, \tilde{\theta}_1 \cong \lambda(\theta_1))$ is bounded in $\tilde{\mathfrak{M}}$. In other words, the linear operator λ is said to be bounded if it transforms any bounded set in $\mathfrak{M}$ into a bounded set in $\tilde{\mathfrak{M}}$. Using the notation of the fundamental systems of bounded sets in $\mathfrak{M}$ and $\tilde{\mathfrak{M}}$, we may formulate the definition in the following way: suppose that $(\mathrm{A}_q, \mathfrak{K}, \mathfrak{N})$ and $(\mathrm{A}_q, \tilde{\mathfrak{K}}, \tilde{\mathfrak{N}})$ are fundamental systems of bounded sets in $\mathfrak{M}$ and $\tilde{\mathfrak{M}}$ respectively. The linear operator λ is said to be *bounded* if

$$\forall \varkappa_1 \exists \tilde{\varkappa}_1 \forall \theta_1 (\mathfrak{N}(\varkappa_1,\ \theta_1) \supset \tilde{\mathfrak{N}}(\tilde{\varkappa}_1,\ \lambda(\theta_1))),$$

where κ_1 and $\tilde{\kappa}_1$ are variables for the sets $\mathfrak{K}$ and $\tilde{\mathfrak{K}}$ respectively.

It is not hard to verify that every continuous linear operator is bounded. However, not every bounded linear operator is continuous.

Concerning a linear operator λ of the type $(\theta \rightarrow \tilde{\theta})$, we shall say that it is п-*continuous* if the following condition is satisfied: for any sequence φ of points of $\mathfrak{M}$, if φ converges to zero in $\mathfrak{M}$, then the sequence ψ defined by the formula

$$\forall n (\psi(n) \simeq \lambda(\varphi(n))),$$

converges to zero in $\tilde{\mathfrak{M}}$.

The concept of a "п-continuous operator" is the constructive analog of the concept of a "countably continuous operator" in classical mathematics. In classical mathematics one proves that if $\mathfrak{M}$ is a countably normed space and $\tilde{\mathfrak{M}}$ is any locally convex space, then any linear operator mapping $\mathfrak{M}$ into $\tilde{\mathfrak{M}}$ is continuous if and only if it is bounded or countably continuous. In Chapter VI it will be proved that the constructive analog of this assertion does not hold. In fact, we can construct a constructive countably normed space $\mathfrak{M}$, a constructive locally convex space $\tilde{\mathfrak{M}}$, and a linear operator λ mapping $\mathfrak{M}$ into $\tilde{\mathfrak{M}}$, such that λ is bounded and п-continuous, but not continuous.

2.2. Now we consider linear operators mapping a locally convex space $\mathfrak{M}$ into the space R of real duplexes. Such operators will be called *linear functionals* in the space $\mathfrak{M}$.

We denote by η and ζ the indices of the alphabets $\mathrm{A}_{\mathfrak{m}} \cup \mathrm{Ч}_3$ and $\mathrm{A}_{\mathfrak{n}} \cup \mathrm{Ч}_3$, respectively: $\mathrm{A}_\eta = \mathrm{A}_{\mathfrak{m}} \cup \mathrm{Ч}_3$ and $\mathrm{A}_\zeta = \mathrm{A}_{\mathfrak{n}} \cup \mathrm{Ч}_3$.

By definition, the linear functional $\lambda_{\eta,1}$ in the space $\mathfrak{M}$ is said to be *continuous* if

$$\forall n \exists \iota_1 \forall \theta_1 (\mathfrak{Q}(\iota_1, \theta_1) \supset |\lambda_{\eta,1}(\theta_1)| \leqslant 2^{-n}).$$

This means that in the alphabet A_ζ^{sa} we can realize an algorithm $\lambda_{\zeta,1}$ of the type $(\mathrm{H} \to \iota)$ satisfying the condition

$$\forall n \theta_1 (\mathfrak{Q}(\lambda_{\zeta,1}(n), \theta_1) \supset |\lambda_{\eta,1}(\theta_1)| \leqslant 2^{-n}). \tag{3}$$

Concerning the algorithm $\lambda_{\zeta,1}$, we will say that it is a *regulator of continuity of the linear functional* $\lambda_{\eta,1}$, and write

$$(\lambda_{\zeta,1} \, \underline{\mathbf{reg\ cont}}_{\mathfrak{M}} \, \lambda_{\eta,1}),$$

if $\lambda_{\zeta,1}$ is an algorithm of the type $(\mathrm{H} \to \iota)$ and satisfies condition (3).

We suppose that in the list of basic alphabets of the language $\mathcal{Я}_\Sigma$ one finds the alphabet $\mathrm{Ч}_0 \cup \{\otimes\}$. We denote its index by π, i.e. $A_\pi = \mathrm{Ч}_0 \cup \{\otimes\}$. Concerning the word $t_{\pi,1}$ in the alphabet A_π, we shall say that it is a *canonical word* in A_π if $t_{\pi,1}$ contains one and only one entrance of the sign $\otimes$. We shall denote by $\overline{\mathrm{lef}}_\pi$ and $\overline{\mathrm{rig}}_\pi$ algorithms in A_π^{sa} which are applicable to any canonical word in A_π and transform each such word into the left and right wings of the unique appearance of the sign $\otimes$ in the word P ([26], §9).

Suppose that Θ is an object term. We introduce the notation

$$\underline{\Theta}_\pi \leftrightharpoons \langle \overline{\mathbf{lef}}_\pi(\Theta) \rangle_\eta; \quad \overline{\Theta}^\pi \leftrightharpoons \langle \overline{\mathbf{rig}}_\pi(\Theta) \rangle_\zeta.$$

Concerning the word $t_{\pi,1}$ in the alphabet A_π, we shall say that it is a *complete cipher of a continuous linear functional in* $\mathfrak{M}$, and write

$$(t_{\pi,1} \in \underline{\mathrm{comp\ \ ciph}}_{\mathfrak{M}}), \tag{4}$$

if, first, $t_{\pi,1}$ is a canonical word in A_π, secondly, $\underline{t_{\pi,1}}_\pi$ is a linear functional in $\mathfrak{M}$, and, thirdly, $\overline{t_{\pi,1}}^\pi$ is a regulator of continuity of the functional $\underline{t_{\pi,1}}_\pi$.

It is easy to see that if $\mathfrak{P}$ is a normal formula, then the condition (4) may be expressed also by a normal formula.

We denote the formula (4) by $\mathfrak{P}^{\text{ш}}$, and adjoin it to the list Σ of formulas of the language $\mathcal{Я}_\Sigma$. In this connection we understand that in the formula (4) $t_{\pi,1}$ is a variable of genus t_π and plays the role of a parameter of this formula.

The formula $\mathfrak{P}^{\text{ш}}$ will be assigned some subordinate generic letter, which we shall denote by ш.

We introduce the following formula:

$$\mathfrak{C}^{\text{ш}} \leftrightharpoons \forall \theta_1 (\underline{t_{\pi,1}}_\pi(\theta_1) \underset{\mathrm{B}}{=} \underline{t_{\pi,1}}_\pi(\theta_2))$$

and write

$$(P \underset{\text{ш}}{=} Q) \leftrightharpoons F_P^{t_{\pi,1}} \, {}_Q^{t_{\pi,2}} \llcorner \mathfrak{C}^{\text{ш}} \lrcorner,$$

where P and Q are arbitrary object terms.

We fix on some word Y of type ι. We write $\mathfrak{O}^{\text{ш}} \leftrightharpoons U \otimes V$, where U is the transcription of an algorithm which transforms any word of type θ into the number 0, and V is the transcription of an algorithm which transforms any natural number into the word Y. Obviously $\mathfrak{D}^{\text{щ}}$ is a word of type щ.

In view of Axiom III2 and property 1.3.7 of Chapter I, we can realize algorithms δ of the type $(\iota\iota \to \iota)$ and γ of the type $(\iota\,\text{д} \to \iota)$ with the following properties:

$$\forall \iota_1 \iota_2 \theta_1 \, (\mathfrak{Q}\,(\delta\,(\iota_1 \square \iota_2),\ \theta_1) \supset \mathfrak{Q}\,(\iota_1,\ \theta_1) \,\&\, \mathfrak{Q}\,(\iota_2,\ \theta_1));$$

$$\forall \iota_1 x \theta_1 \, (\mathfrak{Q}\,(\gamma\,(\iota_1 \square x),\ \theta_1) \supset \mathfrak{Q}\,(\iota_1,\ x \cdot \theta_1)).$$

We construct in $A_\pi^{sa} \cup A_\eta$ algorithms α and ξ, and in $A_\pi^{sa} \cup A_\zeta$ algorithms β and μ, such that

$$\forall \text{щ}_1 \text{щ}_2 \theta_1 \, (\alpha\,(\text{щ}_1 \square \text{щ}_2 \square \theta_1) \simeq \underbrace{\text{щ}_1}_{\pi}\,(\theta_1) \underset{\text{в}}{+} \underbrace{\text{щ}_2}_{\pi}\,(\theta_1));$$

$$\forall \text{щ}_1 \text{щ}_2\, n \, (\beta\,(\text{щ}_1 \square \text{щ}_2 \square n) \simeq \delta\,(\overline{\text{щ}_1}^{\pi}\,(n+1) \square \overline{\text{щ}_2}^{\pi}\,(n+1)));$$

$$\forall x \text{щ}_1 \theta_1 \, (\xi\,(x \square \text{щ}_1 \square \theta_1) \simeq x \underset{\text{в}}{\cdot} \underbrace{\text{щ}_1}_{\pi}\,(\theta_1));$$

$$\forall x \text{щ}_1 n \, (\mu\,(x \square \text{щ}_1 \square n) \simeq \gamma\,(\overline{\text{щ}_1}^{\pi}\,(n) \square x)).$$

Then we construct in A_π^{sa} an algorithm $\underset{\text{щ}}{+}$, and in $A_\pi^{sa} \cup \text{Ч}_3$ an algorithm $\underset{\text{щ}}{\cdot}$, as follows:

$$\forall \text{щ}_1 \text{щ}_2 \, (\underset{\text{щ}}{+}\,(\text{щ}_1 \square \text{щ}_2) \simeq \mathcal{E}\,(\alpha \uparrow \text{щ}_1 \square \text{щ}_2 \square),\ A_\eta^{sa} З \otimes \mathcal{E}\,(\beta \uparrow \text{щ}_1 \square \text{щ}_2 \square),\ A_\zeta^{sa} З).$$

$$\forall x \text{щ}_1 \, (\underset{\text{щ}}{\cdot}\,(x \square \text{щ}_1) \simeq \mathcal{E}\,(\xi \uparrow x \square \text{щ}_1 \square),\ A_\eta^{sa} З \otimes \mathcal{E}\,(\mu \uparrow x \square \text{щ}_1 \square),\ A_\zeta^{sa} З).$$

It is not hard to prove that $\underset{\text{щ}}{+}$ is an algorithm of the type (щ щ → щ), and that $\underset{\text{щ}}{\cdot}$ is an algorithm of the type (д щ → щ). Moreover, it is possible to prove that the list

$$A_\pi,\ \mathfrak{P}^{\text{щ}},\ \mathfrak{E}^{\text{щ}},\ \underset{\text{щ}}{+},\ \underset{\text{щ}}{\cdot},\ \mathfrak{D}^{\text{щ}} \tag{5}$$

defines a constructive linear space.

2.3. Along with the concept of complete cipher of a continuous linear functional in $\mathfrak{M}$, we introduce further the concept of an F-cipher of a continuous linear functional in $\mathfrak{M}$, as follows. Suppose that $t_{\pi,1}$ is a word in the alphabet A_π. We will say that $t_{\pi,1}$ is the *F-cipher of a continuous linear functional in* $\mathfrak{M}$ if $t_{\pi,1}$ has the form $P\otimes$, where P is the transcription relative to the alphabet A_η^{sa} of some linear functional in $\mathfrak{M}$. In other words, the word $t_{\pi,1}$ is said to be an F-cipher of a continuous linear functional in $\mathfrak{M}$ if, first, $t_{\pi,1}$ has the form $P\otimes$, where P is a word in Ч_0; second, $\underbrace{t_{\pi,1}}_{\pi}$ is a linear functional in $\mathfrak{M}$; and, third, one can realize an algorithm $\lambda_{\zeta,1}$ of the type $(\iota \to \text{н})$ which is a regulator of convergence of the functional $\underbrace{t_{\pi,1}}_{\pi}$.

Obviously, if the word $P \otimes Q$ is a complete cipher of a continuous linear functional in $\mathfrak{M}$, then the word $P\otimes$ is the F-cipher of a continuous linear functional in $\mathfrak{M}$. Conversely, if the word $P\otimes$ is the F-cipher of a continuous linear functional in $\mathfrak{M}$, then a word Q in $Ч_0$ is realizable such that $P\otimes Q$ is the complete cipher of a continuous linear functional in $\mathfrak{M}$. At the same time, the following theorem holds.

2.3.1. *It is possible to construct a normed space $\mathfrak{M}$ having the following property: there does not exist an algorithm which transforms each F-cipher $P\otimes$ of a continuous linear functional in $\mathfrak{M}$ into a word Q such that $P\otimes Q$ is the complete cipher of a continuous linear functional in $\mathfrak{M}$.*

Proof. We denote by $\mathfrak{m}$ the index of the alphabet $Ч_3\cup\{\tau\}$, i.e. $А_{\mathfrak{m}} = Ч_3\cup\{\tau\}$, and by $\mathfrak{P}$ the set of all τ-systems of real duplexes. The set $\mathfrak{P}$ may be defined by a normal formula. We denote by θ a subordinate generic letter assigned to the set $\mathfrak{P}$. It is possible to construct algorithms $\underset{\theta}{+}$ of the type $(\theta\theta\to\theta)$ and $\underset{\theta}{\cdot}$ of the type $(д\theta\to\theta)$ such that for any words θ_1 and θ_2 of type θ and any real duplex x, if $\theta_1 \overline{\overset{\circ}{=}} \alpha_1\tau\cdots\tau\alpha_k$ and $\theta_2 \overline{\overset{\circ}{=}} \beta_1\tau\cdots\tau\beta_l$, then

$$\underset{\theta}{+}(\theta_1\Box\theta_2)\simeq\begin{cases}(\alpha_1+\beta_1)\tau\ldots\tau(\alpha_k+\beta_k)\tau\beta_{k+1}\tau\ldots\tau\beta_l, & \text{if } l>k;\\ (\alpha_1+\beta_1)\tau\ldots\tau(\alpha_l+\beta_l)\tau\alpha_{l+1}\tau\ldots\tau\alpha_k, & \text{if } k>l.\\ (\alpha_1+\beta_1)\tau\ldots\tau(\alpha_k+\beta_k), & \text{if } k=l.\end{cases}$$

$$\underset{\theta}{\cdot}(x\Box\theta_1)\simeq x\cdot\alpha_1\tau\ldots\tau x\cdot\alpha_k.$$

We denote by L_τ an algorithm calculating the number of τ-terms of words of type θ. Suppose that θ_1 is a word of type θ and that $1\leqq i\leqq L_\tau(\theta_1)$. We denote the ith τ-term of the word θ_1 by $[\theta_1]_i^\tau$.

We shall construct an algorithm $\mathscr{N}$ of the type $(\theta\to д)$ as follows:

$$\forall\theta_1\left(\mathscr{N}(\theta_1)\simeq\left(\sum_{i=1}^{L_\tau(\theta_1)}|[\theta_1]_i^\tau|^2\right)^{1/2}.\right.$$

We define a two-parameter formula $\mathfrak{E}$ such that, for any words θ_1 and θ_2 of type θ,

$$\mathfrak{E}(\theta_1,\theta_2)\equiv(\mathscr{N}(\theta_1-\theta_2)=0).$$

Here we set $\theta_1-\theta_2\leftrightharpoons\theta_1+(-1)\cdot\theta_2$.

Now we consider the following list:

(6) $$А_{\mathfrak{m}},\ \mathfrak{P},\ \mathfrak{E},\ \underset{\theta}{+},\ \underset{\theta}{\cdot},\ 0,\ \mathscr{N}.$$

This list defines a normed space. This space will be denoted by R_2. We note that the space (6) differs from the space R_2 constructed in [**26**], §13, only in that the points of the space (6) are τ-systems of real duplexes, and in [**26**] they are τ-systems of rational numbers.

Suppose that $\mathscr{A}$ is an algorithm of the type $(н\dot{\to}н)$, the problem of whose applicability to the natural numbers is undecidable (see, for example, [**26**], [**28**] or [**4**]). This means that

(7) $$\neg\forall n\,(!\mathcal{A}(n) \vee \neg !\mathcal{A}(n)).$$

It is possible to construct an algorithm G which is applicable to any word of the type $m \square n$, where m and n are natural numbers, such that $G(m \square n) \doteq \Lambda$ if and only if the process of application of the algorithm $\mathcal{A}$ to the number m ends exactly at n steps.

We construct an algorithm $\mathfrak{B}$ of the type (нн → д) such that, for any natural numbers m and n,

$$\mathfrak{B}(m \square n) \simeq \begin{cases} 2^n, & \text{if } G(m \square n) \doteq \Lambda; \\ 0, & \text{if } G(m \square n) \not\doteq \Lambda. \end{cases}$$

The following formulas are obvious:

(8) $$\forall mn\,(\mathfrak{B}(m \square n) = 0 \vee \mathfrak{B}(m \square n) \doteq 2^n);$$

(9) $$\forall m\,((!\mathcal{A}(m) \equiv \exists n\,(\mathfrak{B}(m \square n) \doteq 2^n))\,\&\,(\neg !\mathcal{A}(m) \equiv \forall n\,(\mathfrak{B}(m \square n) \doteq 0)));$$

(10) $$\forall mn\,(\mathfrak{B}(m \square n) \doteq 2^n \supset \forall l\,(l \neq n \supset \mathfrak{B}(m \square l) \doteq 0)).$$

Now we shall construct in the alphabet A_m^{sa} an algorithm $\mathfrak{C}$ of the type (нθ → д) such that, for any natural number m and any word θ_1 of type θ,

(11) $$\mathfrak{C}(m \square \theta_1) \simeq \sum_{i=0}^{H}{}_! \gamma_{i+1} \cdot \mathfrak{B}(m \square i),$$

where

$$H \leftrightharpoons L_\tau(\theta_1) - 1,\ \gamma_j \leftrightharpoons [\theta_1]_j^\tau \qquad (j = 1, \ldots, H+1).$$

It is not hard to verify that, for each number m, $(\mathfrak{C} \uparrow m \square)$ is a linear functional in R_2. We prove further that

(12) $$\forall m \neg\neg\exists \lambda_{\zeta,1}\,(\lambda_{\zeta,1}\ \underline{\text{reg cont}}_{R_2}\,(\mathfrak{C} \uparrow m \square)).$$

Suppose that m is a natural number. We suppose that $\neg !\mathcal{A}(m)$. Then it follows from (9) and (11) that $\forall \theta_1(\mathfrak{C}(m \square \theta_1) = 0)$. Therefore any algorithm of the type (н → н) is a regulator of continuity of the functional $(\mathfrak{C} \uparrow m \square)$.

Now we suppose that $!\mathcal{A}(m)$. Then we can realize a natural number n_0 such that $\mathfrak{B}(m \square n_0) = 2^{n_0}$. We construct an algorithm $\lambda_{\zeta,1}$ as follows:

$$\forall k\,(\lambda_{\zeta,1}(k) \simeq n_0 + k).$$

We shall show that $(\lambda_{\zeta,1}\ \underline{\text{reg cont}}_{R_2}(\mathfrak{C} \uparrow m \square))$, i.e. that

(13) $$\forall k\theta_1\,(\mathcal{N}(\theta_1) \leqslant 2^{-(n_0+k)} \supset |\mathfrak{C}(m \square \theta_1)| \leqslant 2^{-k}).$$

For any word θ_1 of type θ we have

(14) $$|\mathfrak{C}(m \square \theta_1)| \leqslant \Big(\sum_{i=0}^{H}{}_! \gamma_{i+1}^2\Big)^{1:2} \cdot \Big(\sum_{i=1}^{H}{}_! \mathfrak{B}(m \square i)^2\Big)^{1:2},$$

where H and γ_j are the same as in (11). From (10) and from $\mathfrak{B}(m \square n_0) = 2^{n_0}$

it follows that

$$(15)\qquad \Big(\sum_{i=0}^{H} \mathfrak{B}(m\,\square\, i)^2_?\Big)^{1:2} \leqslant 2^{n_0}.$$

And then formula (13) follows from (14) and (15). Thus we have proved that

$$(16)\qquad !\mathcal{A}(m) \vee \neg !\mathcal{A}(m) \supset \exists \lambda_{\zeta,1}(\lambda_{\zeta,1}\ \underline{\text{reg cont}}_{R_2}(\mathfrak{C}\uparrow m\,\square)).$$

Obviously

$$(17)\qquad \forall m \neg\neg(!\mathcal{A}(m) \vee \neg !\mathcal{A}(m)).$$

(12) follows from (16) and (17) directly. Thus (12) is proved.

We introduce the following concept. The linear functional $\lambda_{\eta,1}$ in R_2 will be called *quasicontinuous* if

$$\neg\neg \exists \lambda_{\zeta,1}(\lambda_{\zeta,1}\ \underline{\text{reg cont}}_{R_2}\,\lambda_{\eta,1}).$$

We preface the proof of Theorem 2.3.1 by the proof of the following assertion.

1) *It is not true that every quasicontinuous linear functional in R_2 is a continuous linear functional in R_2.*

In order to prove this we suppose that every quasicontinuous linear functional in R_2 is a continuous linear functional in R_2. It is clear from (12) that $(\mathfrak{C}\uparrow m\,\square)$ is a quasicontinuous linear functional in R_2 for any m. Hence in particular it follows that

$$\forall m \exists n \forall \theta_1 (\mathcal{N}(\theta_1) \leqslant 2^{-n} \supset |\mathfrak{C}(m\,\square\,\theta_1)| \leqslant 2^{-1}).$$

This means that we can realize an algorithm δ of the type $(\mathrm{H}\to\mathrm{H})$ such that

$$(18)\qquad \forall m \theta_1 (\mathcal{N}(\theta_1) \leqslant 2^{-\delta(m)} \supset |\mathfrak{C}(m\,\square\,\theta_1)| \leqslant 2^{-1}).$$

Suppose that m is a natural number such that $!\mathscr{A}(m)$. Then we can realize a natural number n_0 such that $G(m\,\square\, n_0) \doteq \Lambda$. We construct a word X of type θ as follows:

$$X \leftrightharpoons \underbrace{0\tau \ldots \tau 0}_{n_0\ \text{zeros}}\tau 2^{-n_0}.$$

Obviously $|\mathfrak{C}(m\,\square\, X)| = 1 > 2^{-1}$. Therefore it follows from (18) that $\mathcal{N}(X) > 2^{-\delta(m)}$. On the other hand, from the definition of the word X it is obvious that $\mathcal{N}(X) = 2^{-n_0}$. Therefore we have obtained $n_0 < \delta(m)$. Thus we have proved that for any natural number m, if $!\mathscr{A}(m)$, then the process of application of the algorithm $\mathscr{A}$ to the number m ends at less than $\delta(m)$ steps. Therefore it follows that the problem of applicability of $\mathscr{A}$ to the natural numbers is decidable. But this is impossible. The assertion 1) is proved.

Now we turn to the proof of Theorem 2.3.1. In fact, we shall prove the following assertion.

2) *It is not possible to have an algorithm which transforms each F-cipher $P\otimes$ of a continuous linear functional in R_2 into a word Q such that $P\otimes Q$ is the complete cipher of a continuous linear functional in R_2.*

In order to prove this, we suppose that an algorithm $\mathfrak{L}$ which has the property indicated in proposition 2) is realizable. Then, if the word $P\otimes$ is an F-cipher of a continuous linear functional in R_2, it follows that $\mathfrak{L}$ is applicable to the word $P\otimes$, $\langle\mathfrak{L}(P\otimes)\rangle_\zeta$ is an algorithm of the type $(\text{н}\to\text{н})$, and

$$(19) \qquad \forall\theta_1 n\,(\mathcal{N}(\theta_1)\leqslant 2^{-\langle\mathfrak{L}(P\otimes)\rangle_\zeta(n)}\supset|\langle P\rangle_\eta(\theta_1)|\leqslant 2^{-n}).$$

Suppose that λ is any quasicontinuous linear functional in R_2. Write $P\leftrightharpoons\xi\lambda,\mathrm{A}^{\text{sa}}_\eta\mathfrak{Z}$. Then the word $P\otimes$ cannot fail to be the F-cipher of a continuous linear functional in R_2. Hence (i) $\neg\neg!\,\mathfrak{L}(P\otimes)$. Therefore $!\,\mathfrak{L}(P\otimes)$. (ii) $\langle\mathfrak{L}(P\otimes)\rangle_\zeta$ cannot fail to be an algorithm of the type $(\text{н}\to\text{н})$, so that $\langle\mathfrak{L}(P\otimes)\rangle_\zeta$ is an algorithm of the type $(\text{н}\to\text{н})$. This holds since the condition "$\langle\mathfrak{L}(P\otimes)\rangle_\zeta$ is an algorithm of type $(\text{н}\to\text{н})$" is expressed by a normal formula. And, (iii), condition (19) cannot fail to be satisfied, so that (19) is satisfied, since (19) is a normal formula. Thus the algorithm $\mathfrak{L}$ is applicable to any word of the type $P\otimes$, where P is the transcription of a quasicontinuous linear functional in R_2, and transforms every such word into the transcription of the regulator of continuity of the functional $\langle P\rangle_\eta$. Hence it follows that every quasicontinuous linear functional in R_2 is continuous. This contradicts assertion 1). Therefore an algorithm $\mathfrak{L}$ with the indicated property is not possible, and assertion 2) is proved. Thus, Theorem 2.3.1 is proved as well.

2.4. Now we turn to the consideration of the space (5). This is the linear space of all complete ciphers of continuous linear functionals in the space $\mathfrak{M}$. Now we shall define locally convex topologies on the space (5).

Suppose that $\mathfrak{S}$ is a normal system of bounded sets in $\mathfrak{M}$, defined by the list $(\mathrm{A}_q, \mathfrak{K}, \mathfrak{R})$ (see §1.2). We suppose that $\mathfrak{S}$ satisfies the following conditions:

$$(20) \qquad \forall\varkappa_1\varkappa_2\exists\varkappa_3\forall\theta_1\,(\mathfrak{R}(\varkappa_1,\theta_1)\vee\mathfrak{R}(\varkappa_2,\theta_1)\supset\mathfrak{R}(\varkappa_3,\theta_1));$$

$$(21) \qquad \forall\varkappa_1\exists\varkappa_2\forall\theta_1\,(\mathfrak{R}(\varkappa_1.\,\theta_1)\supset\mathfrak{R}(\varkappa_2, 2\cdot\theta_1)).$$

Now we construct a two-parameter formula $\mathfrak{Q}$ such that, for any words κ_1 of type κ and щ_1 of type щ,

$$(22) \qquad \mathfrak{Q}^{\text{ш}}(\varkappa_1,\text{ш}_1)\equiv\forall\theta_1\,(\mathfrak{R}(\varkappa_1,\theta_1)\supset|\underline{\text{ш}_1}_\pi(\theta_1)|\leqslant 1).$$

Here we recall that words of type κ are elements of the set $\mathfrak{K}$, and words of type щ are elements of the set $\mathfrak{P}^{\text{ш}}$.

We shall prove that *the list* $(\mathrm{A}_q, \mathfrak{K}, \mathfrak{Q})$ *defines a locally convex topology on the linear space* (5). In order to prove this, it is sufficient for us to verify for the list $(\mathrm{A}_q, \mathfrak{K}, \mathfrak{Q}^{\text{ш}})$ the topological axioms of a locally convex space. We shall verify some of these, in fact the following:

III 2. $\forall\varkappa_1\varkappa_2\exists\varkappa_3\forall\text{ш}_1\,(\mathfrak{Q}^{\text{ш}}(\varkappa_3,\text{ш}_1)\supset\mathfrak{Q}^{\text{ш}}(\varkappa_1,\text{ш}_1)\,\&\,\mathfrak{Q}^{\text{ш}}(\varkappa_2,\text{ш}_1));$

III 4. $\forall\varkappa_1\text{ш}_1\exists x\,(x>0\,\&\,\mathfrak{Q}^{\text{ш}}(\varkappa_1, x\cdot\text{ш}_1));$

III 5. $\forall\varkappa_1\exists\varkappa_2\forall\text{ш}_1\text{ш}_2\,(\mathfrak{Q}^{\text{ш}}(\varkappa_2,\text{ш}_1)\,\&\,\mathfrak{Q}^{\text{ш}}(\varkappa_2,\text{ш}_2)\supset\mathfrak{Q}^{\text{ш}}(\varkappa_1,\text{ш}_1+\text{ш}_2)).$

III2 follows directly from (20) and (22). III4 is proved as follows. Suppose given words κ_1 of type κ and щ_1 of type щ. In view of property (3) it is possible to find a word σ_1 of type σ such that

$$\forall\theta_1 (\mathfrak{R}(\varkappa_1, \theta_1) \supset \forall\iota\mathfrak{Q}(\iota_1, \langle\sigma_1\rangle_\zeta(\iota_1)\cdot\theta_1)).$$

Then we put

$$x \leftrightharpoons \langle\sigma_1\rangle_\zeta(\overline{\text{щ}_1^\pi}(0)).$$

We have

$$\forall\theta_1 (\mathfrak{Q}(\overline{\text{ш}_1^\pi}(0), \theta_1) \supset |\underline{\text{ш}_{1_\pi}}(\theta_1)| \leqslant 1).$$

Therefore

$$\forall\theta_1 (\mathfrak{R}(\varkappa_1, \theta_1) \supset x\cdot|\underline{\text{ш}_{1_\pi}}(\theta_1)| \leqslant 1),$$

i.e.

$$\mathfrak{Q}^{\text{ш}}(\kappa^1, x\cdot\text{щ}^1).$$

Condition III4 is verified.

Now we prove III5. Suppose given a word κ_1 of type κ. In view of (21) it is possible to find a word κ_2 of type κ such that

$$(\forall\theta_1 (\mathfrak{R}(\varkappa_1, \theta_1) \supset \mathfrak{R}(\varkappa_2, 2\cdot\theta_1)).$$

Suppose that щ_1 and щ_2 are words of type щ such that $\mathfrak{Q}^{\text{ш}}(\kappa_2, \text{щ}_1)$ and $\mathfrak{Q}^{\text{ш}}(\kappa_2, \text{щ}_2)$. Then

$$\forall\theta_1 (\mathfrak{R}(\varkappa_2, \theta_1) \supset (|\underline{\text{ш}_{1_\pi}}(\theta_1)| \leqslant 1) \& (|\underline{\text{ш}_{2_\pi}}(\theta_1)| \leqslant 1)).$$

Hence we find that

$$\forall\theta_1 \left(\mathfrak{R}(\varkappa_1, \theta_1) \supset \left(|\underline{\text{ш}_{1_\pi}}(\theta_1)| \leqslant \frac{1}{2}\right) \& \left(|\underline{\text{ш}_{2_\pi}}(\theta_1)| \leqslant \frac{1}{2}\right)\right),$$

so that

$$\forall\theta_1 (\mathfrak{R}(\varkappa_1, \theta_1) \supset |\underline{\text{ш}_1 + \text{ш}_{2_\pi}}(\theta_1) \leqslant 1),$$

i.e.

$$\mathfrak{Q}^{\text{ш}}(\kappa_1, \text{щ}_1 + \text{щ}_2),$$

as we were required to prove.

Thus we have found that the list $(\mathrm{A}_q, \mathfrak{K}, \mathfrak{Q}^{\text{ш}})$ defines a locally convex topology on the space (5). This topology will be called the $\mathfrak{S}$-*topology* on the space (5).

Suppose that $\mathfrak{S}$ and $\tilde{\mathfrak{S}}$ are normal systems of bounded sets in $\mathfrak{M}$, satisfying conditions (20) and (21). The following assertion is obvious:

2.4.1. *If the system $\mathfrak{S}$ of bounded sets majorizes the system $\tilde{\mathfrak{S}}$ of bounded sets* (*see* §1.1), *then the $\mathfrak{S}$-topology majorizes the $\tilde{\mathfrak{S}}$-topology. Therefore, if the systems $\mathfrak{S}$ and $\tilde{\mathfrak{S}}$ are equivalent, then the $\mathfrak{S}$-topology and the $\tilde{\mathfrak{S}}$-topology are equivalent.*

Now we prove the following property:

2.4.2. *If the system* $\mathfrak{S}$ *of bounded sets is a covering and satisfies conditions* (20) *and* (21), *then the* $\mathfrak{S}$*-topology is* T'*-separated.*

PROOF. Suppose that the system $\mathfrak{S}$ is defined by the list $(\mathrm{A}_q, \mathfrak{K}, \mathfrak{R})$. Since $\mathfrak{S}$ is a covering system,

$$\forall \theta_1 \exists \varkappa_1 \mathfrak{R}(\varkappa_1, \theta_1). \tag{23}$$

We have to prove that

$$\forall \text{ш}_1 (\forall \varkappa_1 \mathfrak{Q}^{\text{ш}}(\varkappa_1, \text{ш}_1) \supset \text{ш}_1 = \mathfrak{O}^{\text{ш}}). \tag{24}$$

Suppose that щ_1 is a word of type щ such that $\forall \kappa_1 \mathfrak{Q}^{\text{ш}}(\kappa_1, \text{щ}_1)$. Then

$$\forall \varkappa_1 \theta_1 (\mathfrak{R}(\varkappa_1, \theta_1) \supset |\underline{\text{ш}_{1_\pi}}(\theta_1)| \leqslant 1). \tag{25}$$

It is easy to deduce from (25) and (21) that

$$\forall \varkappa_1 n \theta_1 (\mathfrak{R}(\varkappa_1, \theta_1) \supset |\underline{\text{ш}_{1_\pi}}(\theta_1)| \leqslant 2^{n}). \tag{26}$$

Now it follows from (26) and (23) that

$$\forall n \theta_1 (|\underline{\text{ш}_{1_\pi}}(\theta_1)| \leqslant 2^{-n}).$$

Thus

$$\forall \theta_1 (|\underline{\text{ш}_{1_\pi}}(\theta_1)| = 0), \text{ т. е. } \text{ш}_1 = \mathfrak{O}^{\text{ш}}.$$

Formula (24) is proved, and so the $\mathfrak{S}$-topology is T'-separated.

It is not hard to prove that *if* $\mathfrak{S}$ *is a fundamental family of boundary sets in* $\mathfrak{M}$, *then* $\mathfrak{S}$ *satisfies conditions* (20) *and* (21). Therefore, if $\mathfrak{S}$ is some normal fundamental system of bounded sets in $\mathfrak{M}$, then we may define an $\mathfrak{S}$-topology on the space (5). This $\mathfrak{S}$-topology will be called the *strong topology* on the space (5).

We recall that if $\mathfrak{S}$ is a fundamental system of bounded sets in $\mathfrak{M}$, then $\mathfrak{S}$ is a covering system and majorizes any system of bounded sets in $\mathfrak{M}$. Therefore we have

2.4.3. *The strong topology is T-separated and majorizes any* $\mathfrak{S}$*-topology, where* $\mathfrak{S}$ *is a normal system of bounded sets in* $\mathfrak{M}$ *satisfying conditions* (20) *and* (21).

Suppose that $\mathfrak{S}$ is some normal fundamental system of bounded sets in $\mathfrak{M}$, defined by the list $(\mathrm{A}_q, \mathfrak{K}, \mathfrak{R})$. Then the $\mathfrak{S}$-topology is defined by the list $(\mathrm{A}_q, \mathfrak{K}, \mathfrak{Q}^{\text{ш}})$. This $\mathfrak{S}$-topology is a strong topology. The list

$$\mathrm{A}_\pi, \mathfrak{P}^{\text{щ}}, \mathfrak{C}^{\text{щ}}, \underset{\text{щ}}{+}, \underset{\text{щ}}{\cdot}, \mathfrak{O}^{\text{щ}}, \mathrm{A}_q, \mathfrak{K}, \mathfrak{Q}^{\text{щ}} \tag{27}$$

defines a constructive locally convex space. This space will be said to be *adjoint to the space* $\mathfrak{M}$.

Thus the adjoint space to a locally convex space $\mathfrak{M}$ is the space of all complete ciphers of linear functionals in $\mathfrak{M}$, equipped with the strong topology. The adjoint space is T'-separated.

2.5. Suppose given a locally convex space $\mathfrak{M}$ defined by the list (1). Suppose that the space adjoint to $\mathfrak{M}$ is defined by the list (27). We denote the space (27) by $\mathfrak{M}^{\text{ш}}$.

The topology $(\mathrm{A}_{\text{п}}, \mathfrak{J}, \mathfrak{Q})$ of the space $\mathfrak{M}$ will also be called the *original topology in the space* $\mathfrak{M}$. Using the adjoint space $\mathfrak{M}^{\text{ш}}$, we now define in $\mathfrak{M}$, or, more precisely, on the linear space

$$(28) \qquad \mathrm{A}_{\text{m}}, \mathfrak{P}, \mathfrak{C}, \underset{\theta}{+}, \underset{\theta}{\cdot}, \mathfrak{D},$$

the so-called strong topology, as follows. $\mathfrak{M}^{\text{ш}}$ is a locally convex space, so that it is possible to construct in $\mathfrak{M}^{\text{ш}}$ a normal fundamental system of bounded sets. Suppose that $(\mathrm{A}_r, \mathfrak{K}_1, \mathfrak{R}_1)$ is any normal fundamental system of bounded sets in $\mathfrak{M}^{\text{ш}}$. We construct a two-parameter formula $\mathfrak{Q}_1$ such that, for any element ν_1 of the set $\mathfrak{K}_1$ and any point θ_1 of the set $\mathfrak{M}$,

$$\mathfrak{Q}_1(\nu_1, \theta_1) \equiv \forall \text{ш}_1 (\mathfrak{R}_1(\nu_1, \text{ш}_1) \supset |\underline{\text{ш}}_{1_\pi} \cdot (\theta_1)| \leqslant 1).$$

One may prove that the list

$$(29) \qquad \mathrm{A}_r, \mathfrak{K}_1, \mathfrak{Q}_1$$

defines a locally convex topology on the linear space (28). The proof of this assertion is analogous to the proof presented above of the analogous assertion for the list $(\mathrm{A}_q, \mathfrak{K}, \mathfrak{Q}^{\text{ш}})$.

The topology (29) will be called the *strong topology in the space* $\mathfrak{M}$.

As we know, in classical mathematics (under the classical understanding of concepts and assertions), in any locally convex space the strong topology majorizes the original topology (see for example [7], Chapter IX). We shall show that this assertion does not carry over in constructive mathematics. Indeed, we shall prove the following theorem.

2.5.1. *It is possible to realize a constructive locally convex space in which the strong topology does not majorize the original topology.*

Proof. We consider the space $\mathscr{S}$ constructed in Chapter I, §1.4, Example 2. First of all we show that every continuous linear functional in $\mathscr{S}$ is identically equal to zero. Indeed, suppose that λ is a continuous linear functional in $\mathscr{S}$, and that $\lambda(1) \neq 0$. Then it is possible to find a natural number n such that $|\lambda(1)| > 2^{-n}$. Since the functional λ is continuous, we can realize a natural number i such that

$$\forall x (\mathfrak{Q}_s(i, x) \supset |\lambda(x)| \leqslant 2^{-n}).$$

Hence it follows that $\neg \mathfrak{Q}_s(i, 1)$. Thus we have proved that $\exists i \neg \mathfrak{Q}_s(i, 1)$, i.e. that

$$\exists i \neg \exists j \forall kl (k, l \geqslant j \supset |S(k) - S(l)| < 2^{-i}).$$

But this contradicts the following property of the algorithm S:

$$\forall i \neg \neg \exists j \forall kl (k, l \geqslant j \supset |S(k) - S(l)| < 2^{-i}).$$

Therefore we have found that $\lambda(1)=0$, and hence that $\forall x(\lambda(x)=0)$.

Since every continuous linear functional in $\mathcal{S}$ is identically equal to zero, the space adjoint to $\mathcal{S}$ contains only the zero functional. Therefore a fundamental system of bounded sets in the space adjoint to $\mathcal{S}$ consists of a unique set, whose unique element is the zero functional. Hence a fundamental system of neighborhoods of the origin in the strong topology of the space $\mathcal{S}$ consists of a unique neighborhood, containing all the points of the space $\mathcal{S}$.

We suppose that the strong topology in $\mathcal{S}$ majorizes the original topology. Then any neighborhood of the origin in the original topology must contain all the points of the space $\mathcal{S}$, i.e. $\forall ix\mathfrak{Q}_s(i,x)$. Hence it follows that $\forall i\,\mathfrak{Q}_s(i,1)$, i.e.

$$\forall i\exists j\forall kl\,(k,\,l\geqslant j\supset|S\,(k)-S\,(l)|<2^{-i}).$$

This is not possible. Therefore the strong topology in $\mathcal{S}$ cannot majorize the original topology. The theorem is proved.

In Chapter VI we shall construct a constructive enumerably normed space for which the strong topology does not majorize the original topology. Along with Theorem 2.5.1, this is also a point of essential divergence between the constructive and classical theories of locally convex spaces.

§3. The space adjoint to a constructive normed space

3.1. Suppose given a constructive normed space $\mathfrak{M}$ defined by the list

$$(1)\qquad A_{\text{ш}},\ \mathfrak{P},\ \mathfrak{C},\ \underset{\theta}{+},\ \underset{\theta}{\cdot},\ \mathfrak{D},\ \mathcal{N}$$

(see §4.2, Chapter I). As was proved in §1.2, a normal fundamental system of bounded sets in $\mathfrak{M}$ may be defined by a list $(\text{Ч}_0,\mathfrak{K},\mathfrak{N})$, where $\mathfrak{K}$ is the set of all natural numbers and the formula $\mathfrak{N}$ is defined in such a way that, for any natural number k and any point θ_1 of $\mathfrak{M}$,

$$\mathfrak{N}\,(k,\,\theta_1)\equiv(\mathcal{N}\,(\theta_1)\leqslant k+1).$$

Considering $\mathfrak{M}$ as a locally convex space, we may define a space adjoint to $\mathfrak{M}$, which is a locally convex space.

In the space $\mathfrak{M}^{\text{ш}}$ adjoint to the normed space $\mathfrak{M}$, the locally convex topology is defined by the list $(\text{Ч}_0,\mathfrak{K},\mathfrak{Q}^{\text{ш}})$, where $\mathfrak{K}$ is the set of all natural numbers and the formula $\mathfrak{Q}^{\text{ш}}$ is defined in such a way that, for any natural number k and any complete cipher ш_1 of a continuous linear functional in $\mathfrak{M}$,

$$\mathfrak{Q}^{\text{ш}}\,(k,\,\text{ш}_1)\equiv\forall\theta_1\,(\mathcal{N}\,(\theta_1)\leqslant k+1\supset|\underline{\text{ш}_{1_\pi}}\,(\theta_1)\,|\leqslant 1).$$

As we know, in the classical theory of locally convex spaces one proves that the adjoint space to any normed space is a normed space. The question arises: does this assertion carry over into constructive mathematics? We shall consider this question in this section.

Suppose that λ is a linear functional in $\mathfrak{M}$. We shall say that the real duplex x is the *norm* of the functional λ, and write

$$(x \,\underline{\text{norm}}_{\mathfrak{M}}\, \lambda),$$

if the following two conditions are satisfied:

(2) $$\forall \theta_1 (|\lambda(\theta_1)| \leqslant x \cdot \mathcal{N}(\theta_1));$$

(3) $$\forall n \exists \theta_1 (|\lambda(\theta_1)| > (x - 2^{-n}) \cdot \mathcal{N}(\theta_1)).$$

This definition is equivalent to the definition introduced in [**26**], §12.3.

3.1.1. *Suppose that $\mathfrak{M}$ is a normed space. If one can realize an algorithm $\mathfrak{N}$ of the type* $(\text{щ} \to \text{д})$ *such that*

(4) $$\forall \text{щ}_1 (\mathfrak{N}(\text{щ}_1) \,\underline{\text{norm}}_{\mathfrak{M}}\, \underline{\text{щ}_1}_\pi),$$

then the adjoint space $\mathfrak{M}^{\text{щ}}$ *is normable.*

PROOF. Suppose that we can realize an algorithm $\mathfrak{N}$ of the type $(\text{щ} \to \text{д})$ which satisfies condition (4). It is not hard to verify that $\mathfrak{N}$ satisfies the axioms of a norm. In order to prove that $\mathfrak{M}^{\text{щ}}$ is normable, we need to prove the following formulas:

(5) $$\forall k \exists n \forall \text{щ}_1 (\mathfrak{N}(\text{щ}_1) \leqslant 2^{-n} \supset \mathfrak{Q}^{\text{щ}}(k, \text{щ}_1));$$

(6) $$\forall n \exists k \forall \text{щ}_1 (\mathfrak{Q}^{\text{щ}}(k, \text{щ}_1) \supset \mathfrak{N}(\text{щ}_1) \leqslant 2^{-n}).$$

Let us prove (5). Suppose given a number k. We choose a number n such that $2^n \geqq k+1$. Obviously, if $\mathfrak{N}(\text{щ}_1) \leqq 2^{-n}$, then in view of (2) we get

$$\forall \theta_1 (|\underline{\text{щ}_1}_\pi(\theta_1)| \leqslant 2^{-n} \cdot \mathcal{N}(\theta_1)),$$

so that

$$\forall \theta_1 (\mathcal{N}(\theta_1) \leqslant k+1 \supset |\underline{\text{щ}_1}_\pi(\theta_1)| \leqslant 1),$$

i.e. $\mathfrak{Q}^{\text{щ}}(k, \text{щ}_1)$. Formula (5) is proved.

Now let us prove (6). Suppose given a number n. We choose a number k such that $k+1 = 2^n$. Suppose that щ_1 is a word of type щ such that $\mathfrak{Q}^{\text{щ}}(k, \text{щ}_1)$, i.e. such that

(7) $$\forall \theta_1 (\mathcal{N}(\theta_1) \leqslant k+1 \supset |\underline{\text{щ}_1}_\pi(\theta_1)| \leqslant 1).$$

Suppose that m is any natural number. In view of (3) it is possible to find a word θ_1 of type θ such that

(8) $$(\mathfrak{N}(\text{щ}_1) - 2^{-m}) \cdot \mathcal{N}(\theta_1) < |\underline{\text{щ}_1}_\pi(\theta_1)|.$$

Obviously $\neg(\theta_1 = \mathfrak{O})$. Hence $\mathcal{N}(\theta_1) \neq 0$ and $\mathcal{N}((2^n : \mathcal{N}(\theta_1)) \cdot \theta_1) = 2^n \leqq k+1$, so that in view of (7) we get

$$|\underline{\text{щ}_1}_\pi((2^n : \mathcal{N}(\theta_1)) \cdot \theta_1)| \leqslant 1,$$

i.e.

(9) $$|\underline{\text{щ}_1}_\pi(\theta_1)| \leqslant 2^{-n} \cdot \mathcal{N}(\theta_1).$$

It follows from (8) and (9) that $\mathfrak{N}(\text{щ}_1) - 2^{-m} \leq 2^{-n}$. This is true for any m, so that we obtain $\mathfrak{N}(\text{щ}_1) \leq 2^{-n}$. Formula (6) is proved.

In [26], §11, N. A. Šanin introduced the concept of a totally bounded metric space. Using that concept, we now define the concept of a totally bounded set in a normed space. Suppose that $\mathfrak{A}$ is a set of points of the space $\mathfrak{M}$. Denote by ρ an algorithm of the type $(\theta\theta \to \text{д})$, defined as follows:

$$\forall \theta_1\theta_2 (\rho(\theta_1 \square \theta_2) \simeq \mathcal{N}(\theta_1 - \theta_2)).$$

Obviously the list $(\text{A}_{\text{ш}}, \mathfrak{A}, \rho)$ defines a constructive metric space (see [26], §9). This is the metric space of the normed space $\mathfrak{M}$. We will say that $\mathfrak{A}$ is a *totally bounded set* in the normed space $\mathfrak{M}$ if the metric space $(\text{A}_{\text{ш}}, \mathfrak{A}, \rho)$ is totally bounded.

3.1.2. *Suppose that $\mathfrak{M}$ is a normed space and $\mathfrak{A}$ is the set of points θ_1 of the space $\mathfrak{M}$ such that $\mathcal{N}(\theta_1) = 1$. If the set $\mathfrak{A}$ is totally bounded in $\mathfrak{M}$, then the adjoint space $\mathfrak{M}^{\text{ш}}$ is normable.*

PROOF. By the hypothesis of the theorem, the metric space $(\text{A}_{\text{ш}}, \mathfrak{A}, \rho)$ is totally bounded. Obviously every word щ_1 of type щ is a complete cipher of a uniformly continuous functional in this metric space. In view of Theorem 11.2.1 of [26], it is possible to construct an algorithm $\mathfrak{N}$ of the type $(\text{щ} \to \text{д})$ such that, for any word щ_1 of type щ, $\mathfrak{N}(\text{щ}_1)$ is the least upper bound of the functional $\underline{\text{щ}_1}_{\pi}$ in the space $(\text{A}_{\text{ш}}, \mathfrak{A}, \rho)$. We shall prove that

$$\forall \text{щ}_1 (\mathfrak{N}(\text{щ}_1) \underset{\mathfrak{M}}{\underline{\text{norm}}} \underline{\text{щ}_1}_{\pi}). \tag{10}$$

First of all, we shall show that for any word щ_1 of type щ we have

$$\forall \theta_1 (|\underline{\text{щ}_1}_{\pi}(\theta_1)| \leqslant \mathfrak{N}(\text{щ}_1) \cdot \mathcal{N}(\theta_1)). \tag{11}$$

Suppose that θ_1 is a word of type θ. If $\theta_1 = \mathfrak{O}$, then, obviously,

$$|\underline{\text{щ}_1}_{\pi}(\theta_1)| \leqslant \mathfrak{N}(\text{щ}_1) \cdot \mathcal{N}(\theta_1).$$

Now we suppose that $\urcorner(\theta_1 = \mathfrak{O})$. Then $\mathcal{N}(\theta_1) \neq 0$. We put $X \leftrightharpoons (1 : \mathcal{N}(\theta_1)) \cdot \theta_1$. Obviously $\mathcal{N}(X) = 1$, i.e. X belongs to the set $\mathfrak{A}$. Therefore, by the definition of the algorithm $\mathfrak{N}$, we have $\underline{\text{щ}_1}_{\pi}(X) \leq \mathfrak{N}(X)$, whence it follows that

$$-\underline{\text{щ}_1}_{\pi}(\theta_1) \leqslant \mathfrak{N}(\text{щ}_1) \cdot \mathcal{N}(\theta_1). \tag{12}$$

Analogously one proves that

$$-\underline{\text{щ}_1}_{\pi}(\theta_1) \leqslant \mathfrak{N}(\text{щ}_1) \cdot \mathcal{N}(\theta_1). \tag{13}$$

It follows from (12) and (13) that

$$|\underline{\text{щ}_1}_{\pi}(\theta_1)| \leqslant \mathfrak{N}(\text{щ}_1) \cdot \mathcal{N}(\theta_1).$$

Thus we have proved that

(14) $$(\theta_1 = \mathfrak{O}) \vee \neg(\theta_1 = \mathfrak{O}) \supset | \underset{\pi}{\underbrace{\text{ш}_1}} (\theta_1) | \leqslant \mathfrak{N}(\text{ш}_1) \cdot \mathcal{N}(\theta_1).$$

Since $\forall \theta_1 \neg\neg(\theta_1 = \mathfrak{O} \vee \neg(\theta_1 = \mathfrak{O}))$, from (14) we find that

$$\forall \theta_1 \neg\neg(| \underset{\pi}{\underbrace{\text{ш}_1}} (\theta_1) | \leqslant \mathfrak{N}(\text{ш}_1) \cdot \mathcal{N}(\theta_1)),$$

so that we obtain formula (11) as well.

It remains to be proved that, for any word ш$_1$ of type ш,

(15) $$\forall n \exists \theta_1 (| \underset{\pi}{\underbrace{\text{ш}_1}} (\theta_1) | > (\mathfrak{N}(\text{ш}_1) - 2^{-n}) \cdot \mathcal{N}(\theta_1)).$$

Suppose given a number n. It follows from the definition of the algorithm $\mathfrak{N}$ that one can realize a point θ_1, belonging to the set $\mathfrak{A}$, and such that

$$\underset{\pi}{\underbrace{\text{ш}_1}} (\theta_1) > \mathfrak{N}(\text{ш}_1) - 2^{-n}.$$

Since θ_1 belongs to $\mathfrak{A}$, we have $\mathcal{N}(\theta_1) = 1$, so that θ_1 satisfies the condition

$$| \underset{\pi}{\underbrace{\text{ш}_1}} (\theta_1) | > (\mathfrak{N}(\text{ш}_1) - 2^{-n}) \cdot \mathcal{N}(\theta_1).$$

Formula (15), and hence also (10), are proved. In view of Theorem 3.1.1 it follows from (10) that $\mathfrak{M}$ is normable.

3.2. Theorems 3.1.1 and 3.1.2 give us conditions sufficient for the adjoint space to a normed space to be normable. However, not every normed space satisfies these conditions. In this subsection we show that there exists a normed space whose adjoint is not normable.

Consider the space R_2, constructed in the proof of Theorem 2.3.1. R_2 is a normed space. Now we shall prove the following proposition.

3.2.1. *The space* $R_2^{\text{ш}}$ *adjoint to* R_2 *is not metrizable* (*in fact, not even semi-metrizable*).

Proof. By definition the space R_2 is defined by the list (6) of §2.3. $R_2^{\text{ш}}$ is the adjoint to R_2. The points of the space $R_2^{\text{ш}}$ will be called *words of type* ш. A locally convex topology in $R_2^{\text{ш}}$ is defined by the list $(\text{Ч}_0, \mathfrak{K}, \mathfrak{Q}^{\text{ш}})$, where $\mathfrak{K}$ is the set of all natural numbers, and the formula $\mathfrak{Q}^{\text{ш}}$ is defined in such a way that, for any natural number k and any word ш$_1$ of type ш,

$$\mathfrak{Q}^{\text{ш}}(k, \text{ш}_1) \equiv \forall \theta_1 (\mathcal{N}(\theta_1) \leqslant k + 1 \supset | \underset{\pi}{\underbrace{\text{ш}_1}} (\theta_1) | \leqslant 1).$$

Suppose that $\mathscr{A}$ is an algorithm of the type $(\text{н} \dot{\rightarrow} \text{н})$, the problem of whose applicability to the natural numbers is undecidable, and G an algorithm applicable to any pair of natural numbers of the form $m \,\Box\, n$ and such that $G(m \,\Box\, n) \,\overline{\underline{\circ}}\, \Lambda$ if and only if the process of the application of $\mathscr{A}$ to the number m stops after exactly n steps (see the proof of 2.3.1).

We shall construct an algorithm $\mathfrak{B}$ of the type $(\text{нн} \rightarrow \text{д})$ such that, for any natural numbers m and n,

$$\mathfrak{B}(m \square n) \simeq \begin{cases} 2, & \text{if} \quad G(m \square n) \mathrel{\overline{\overline{\circ}}} \Lambda; \\ 0, & \text{if} \quad G(m \square n) \not\equiv \Lambda. \end{cases}$$

Now we construct in the alphabet A^{sa}_{m} an algorithm $\mathfrak{C}$ of the type $(\text{н}\theta \to \text{д})$ such that, for any natural number m and any word θ_1 of type θ,

$$\mathfrak{C}(m \square \theta_1) \simeq \sum_{i=0}^{H} \gamma_{i+1} \cdot \mathfrak{B}(m \square i),$$

where

$$H \rightleftharpoons L_\tau(\theta_1) - 1, \gamma_j \rightleftharpoons [\theta_1]_j^\tau \; (j = 1, \ldots, H+1).$$

Using the method used in the proof of 2.3.1, we may prove that, for any natural number m and any word θ_1 of type θ,

$$\neg ! \mathcal{A}(m) \supset \mathfrak{C}(m \square \theta_1) = 0;$$

$$! \mathcal{A}(m) \supset |\mathfrak{C}(m \square \theta_1)| \leqslant 2 \cdot \mathcal{N}(\theta_1).$$

Hence we find that

(16) $$\forall m \theta_1 (|\mathfrak{C}(m \square \theta_1)| \leqslant 2 \cdot \mathcal{N}(\theta_1)).$$

Now we denote by Z a word in the alphabet Ч_0 such that

(17) $$\forall_n (\langle Z \rangle_0 (n) \simeq n + 1).$$

Now we construct an algorithm $\mathfrak{D}$ as follows:

$$\forall m (\mathfrak{D}(m) \simeq \mathcal{E}(\mathfrak{C} \uparrow m \square), A^{sa}_{m} \; 3 \otimes Z).$$

It follows from (16) and (17) that, for any m, $\langle Z_0 \rangle$ is a regulator of convergence of the linear functional $(\mathfrak{C} \uparrow m \square)$, and so, for any m, $\mathfrak{D}(m)$ is a complete cipher of a continuous linear functional in R_2. In other words, $\mathfrak{D}$ defines a sequence of points of the space $R_2^{\text{ш}}$.

Now we suppose that the space $R_2^{\text{ш}}$ is semimetrizable. From this hypothesis it follows that we can realize an algorithm ρ of the type $(\text{щ щ} \to \text{в})$ for which

(18) $$\forall \text{щ}_1 \text{щ}_2 (\rho(\text{щ}_1 \square \text{щ}_2) \geqslant 0);$$

(19) $$\forall n \exists k \forall \text{щ}_1 \text{щ}_2 (\mathfrak{Q}^{\text{ш}}(k, \text{щ}_1 - \text{щ}_2) \supset \rho(\text{щ}_1 \square \text{щ}_2) \leqslant 2^{-n});$$

(20) $$\forall k \exists n \forall \text{щ}_1 \text{щ}_2 (\rho(\text{щ}_1 \square \text{щ}_2) \leqslant 2^{-n} \supset \mathfrak{Q}^{\text{ш}}(k, \text{щ}_1 - \text{щ}_2)).$$

Now we construct an algorithm $\mathfrak{H}$ of the type $(\text{н} \to \text{н})$ as follows:

$$\forall m (\mathfrak{H}(m) \simeq \rho(\mathfrak{D}(m) \square \mathfrak{D}^{\text{ш}})).$$

Since $\mathfrak{H}$ is a sequence of F-numbers,

$$\forall m \exists x (x \underset{\text{в}}{=} \mathfrak{H}(m)),$$

where x is a variable for FR-numbers. Therefore it is possible to construct an algorithm $\mathfrak{F}$ of the type $(\text{н} \to \text{д})$ such that $\forall m(\mathfrak{F}(m) = \mathfrak{H}(m))$, whence

$$\forall m\,(\mathfrak{F}(m) \underset{\text{в}}{=} \rho\,(\mathfrak{D}(m) \,\square\, \mathfrak{D}^{\text{ш}})). \tag{21}$$

It follows from (18) that $\mathfrak{F}$ is a sequence of positive FR-numbers. Moreover, it follows from (19) – (21) that

$$\forall n \exists k \forall m\,(\mathfrak{Q}^{\text{ш}}(k, \mathfrak{D}(m)) \supset \mathfrak{F}(m) \leqslant 2^{-n}); \tag{22}$$

$$\forall k \exists n \forall m\,(\mathfrak{F}(m) \leqslant 2^{-n} \supset \mathfrak{Q}^{\text{ш}}(k, \mathfrak{D}(m))). \tag{23}$$

By (23) we find a natural number n_0 such that

$$\forall m\,(\mathfrak{F}(m) \leqslant 2^{-n_0} \supset \mathfrak{Q}^{\text{ш}}(0, \mathfrak{D}(m))). \tag{24}$$

Suppose that m is a natural number. If $\neg\,!\mathscr{A}(m)$, then $\mathfrak{D}(m) = \mathfrak{O}^{\text{ш}}$, so that $\forall k \mathfrak{Q}^{\text{ш}}(k, \mathfrak{D}(m))$, whence in view of (22) we get $\forall n(\mathfrak{F}(m) \leqq 2^{-n})$, i.e. $\mathfrak{F}(m) = 0$.

Now suppose that $!\mathscr{A}(m)$. Then we can realize a natural number n such that $\mathfrak{B}(m \,\square\, n) \overset{\circ}{=} 2$. We construct a point X of the space R_2 as follows: $X \leftrightharpoons 0\tau \cdots \tau 0 \tau 1$ (n zeros). We have

$$\mathscr{N}(X) = 1, \ |\mathfrak{C}(m \,\square\, X)| = 2.$$

Thus we have proved that

$$\exists \theta_1 (\mathscr{N}(\theta_1) = \& \,|\underbrace{\mathfrak{D}(m)}_{\pi}(\theta_1)| = 2),$$

so that it follows that

$$\neg\,\forall \theta_1 (\mathscr{N}(\theta_1) \leqslant 1) \supset |\underbrace{\mathfrak{D}(m)}_{\pi}(\theta_1)| \leqslant 1),$$

i.e. $\neg\,\mathfrak{Q}^{\text{ш}}(0, \mathfrak{D}(m))$. Hence in view of (24) we find that $\mathfrak{F}(m) > 2^{-n_0}$. Thus for any natural number m we have

$$(!\mathscr{A}(m) \supset \mathfrak{F}(m) > 2^{-n_0}) \& (\neg\,!\mathscr{A}(m) \supset \mathfrak{F}(m) = 0).$$

Hence it is not difficult to show that for any m

$$(!\mathscr{A}(m) \equiv \mathfrak{F}(m) > 0) \& (\neg\,!\mathscr{A}(m) \equiv \mathfrak{F}(m) < 2^{-n_0}). \tag{25}$$

In view of Theorem 1.3 in [4] we have

$$\forall m\,(\mathfrak{F}(m) > 0 \vee \mathfrak{F}(m) < 2^{-n_0}).$$

Therefore we find from (25) that

$$\forall m\,(!\mathscr{A}(m) \vee \neg\,!\mathscr{A}(m)).$$

But this contradicts the property of the algorithm $\mathscr{A}$. Hence the space $R_2^{\text{ш}}$ is not semimetrizable. Theorem 3.2.1 is proved.

The following is a consequence of Theorem 3.2.1.

3.2.2. *The space $R_2^{\text{ш}}$ adjoint to the normed space R_2 is not normable* (*in fact not even seminormable*).

We note further that the space $R_2^{ш}$ contains a bounded neighborhood of the origin. Indeed, we choose for example the neighborhood of the origin with the index 0. It is not hard to prove that

$$\forall k \exists x (x > 0 \,\&\, \forall ш_1 (\mathfrak{Q}^{ш}(0, ш_1) \supset \mathfrak{Q}^{ш}(k, x \cdot ш_1))).$$

This means that the neighborhood of the origin with index 0 is a bounded set in R_2. Thus R_2 is T'-separated and has a bounded neighborhood of the origin. In view of Theorem 4.2.3, Chapter I, if $R_2^{ш}$ is multinormable, it is also normable. Therefore from Theorem 3.2.2 we obtain

3.2.3. *The space $R_2^{ш}$ adjoint to the normed space R_2 is not multinormable (in fact, not multiseminormable).*

Now we consider the space l_2 defined in §13 of [**26**]. We may suppose that l_2 is the standard FR-completion of the space R_2 considered above. l_2 is a complete Hilbert space. Denote by $l_2^{ш}$ the space adjoint to l_2. Using Theorem 3.2.2, we may prove without difficulty that the space $l_2^{ш}$ is also not normable. If l_2 and $l_2^{ш}$ were isomorphic, the normability of l_2 would imply that $l_2^{ш}$ is normable. We thus obtain

3.2.4. *The space $l_2^{ш}$ adjoint to the complete Hilbert space l_2 is not isomorphic to l_2.*

Suppose that P is the complete cipher of a continuous linear functional in l_2. We shall say that P is *the complete cipher of a normed continuous linear functional in l_2*, if there exists a real duplex x which is the norm of the functional $\underline{P}_\pi$ in l_2.

By using the construction in the proof of Theorem 3.2.1 we may prove, as well, the following theorems:

3.2.5. *There does not exist an algorithm which transforms the complete cipher of an arbitrary normable continuous linear functional in l_2 into the norm of that functional in l_2.*

3.2.6. *There does not exist an algorithm $\mathfrak{H}$ such that, for any word P, if P is the complete cipher of a normed continuous linear functional in l_2, then $\mathfrak{H}(P)$ is a point of the space l_2 and*

$$\forall \theta_1 (\underline{P}_\pi(\theta_1) = C(\mathfrak{H}(P) \square \theta_1)),$$

*where C is an algorithm which calculates the scalar product in l_2 (see [**26**], §12).*

Theorems 3.2.5 and 3.2.6 were proved by G. E. Minc in [**16**].

§4. Dimension of the adjoint space

4.1. In this subsection we shall consider the spaces $\mathscr{T}_{n,m}$ constructed in §4.2 of Chapter II. By definition, $\mathscr{T}_{n,m}$ ($n > 0, 0 \leq m \leq n$) is a locally convex space given by the list

$$A_r, \mathfrak{P}, \mathfrak{C}, \underset{\theta}{+}, \underset{\theta}{\cdot}, \mathfrak{D}, Ч_0, \mathfrak{Z}, \mathfrak{Q}_{n,m} \tag{1}$$

(see §4.2, Chapter II). Here all unexplained notation is to be understood as in §4.2, Chapter II.

We fix natural numbers m and n such that $n > 0$ and $0 \leqq m \leqq n$. The space $\mathscr{T}_{n,m}$ is an n-dimensional space whose basis is formed by the points $X_1, \cdots, X_n$.

In the proof of Theorem 2.5.1 we have already proved that every linear functional in the space $\mathscr{S} = \mathscr{T}_{1,0}$ is identically equal to zero. Using an analogous method we now prove that, for any continuous linear functional λ in the space $\mathscr{T}_{n,m}$ and for any h $(m < h \leqq n)$,

$$\lambda(X_h) = 0 \tag{2}$$

(here we recall that $\text{И}(h \square X_h) \doteq 1$ and $\text{И}(j \square X_h) \doteq 0$ for any $j \neq h$, $1 \leqq j \leqq n$).

Suppose that λ is a continuous linear functional in $\mathscr{T}_{n,m}$, and that $m < h \leqq n$. In order to prove (2), we suppose that $\lambda(X_h) \neq 0$. Then it is possible to find a natural number k such that $|\lambda(X_h)| > 2^{-k}$. Since λ is continuous, for that number k we can find a natural number i such that

$$\forall \theta_1 (\mathfrak{Q}_{n,m}(i, \theta_1) \supset |\lambda(\theta_1)| \leqslant 2^{-k}).$$

Hence it follows that $\urcorner \mathfrak{Q}_{n,m}(i, X_h)$, i.e. that

$$\urcorner \forall j ((1 \leqslant j \leqslant m \supset |\text{И}(j \square X_h)| \leqslant 2^{-i}) \& (m < j \leqslant n \supset \mathfrak{Q}_s(i, \text{И}(j \square X_h)))). \tag{3}$$

Obviously

$$\forall j (1 \leqslant j \leqslant m \supset |\text{И}(j \square X_h)| \leqslant 2^{-i}),$$

so that we find from (3) that

$$\urcorner \forall j (m < j \leqslant n \supset \mathfrak{Q}_s(i, \text{И}(j \square X_h))). \tag{4}$$

Formula (4) is equivalent to the following:

$$\urcorner (\forall j (m < j \leqslant n \& j \neq h \supset \mathfrak{Q}_s(i, 0)) \& \mathfrak{Q}_s(i, 1)).$$

It follows directly from this formula that $\urcorner \mathfrak{Q}_s(i, 1)$. Thus we have proved that $\exists i \urcorner \mathfrak{Q}_s(i, 1)$. But this is not possible (see the proof of Theorem 2.5.1). Therefore $\lambda(X_h) = 0$, as we wished to prove.

We note that if $m = 0$, then it follows from (2) that for any continuous linear functional λ in $\mathscr{T}_{n,m}$ we have

$$\lambda(X_h) = 0 \qquad (h = 1, \cdots, n).$$

Hence one easily deduces that every continuous linear functional in $\mathscr{T}_{n,0}$ is identically equal to zero. Therefore the set $\mathscr{T}_{n,0}^{\text{III}}$ adjoint to $\mathscr{T}_{n,0}$ consists only of the null element.

Now we consider the case $m > 0$. We construct algorithms $\mathfrak{B}_i$, $1 \leqq i \leqq m$, such that

$$\forall \theta_1 (\mathfrak{B}_i(\theta_1) \simeq \text{И}(i \square \theta_1)).$$

Then we denote by Z the transcription of an algorithm of the type $(\text{н} \to \text{н})$ such that $\forall k(\langle Z\rangle_0(k) \doteq k)$. We write

$$Y_i \rightleftharpoons \mathcal{E}\, \mathfrak{B}_i,\, \mathrm{A}_r^{\text{за}}\ 3 \otimes Z \quad (i = 1, \ldots, m).$$

It is easy to see that for any i, $1 \leqq i \leqq m$, $\mathfrak{B}_i$ is a continuous linear functional in $\mathscr{T}_{n,m}$, and $\langle Z_0\rangle$ is a regulator of continuity for the functional $\mathfrak{B}_i$. Thus for any i, $1 \leqq i \leqq m$, Y_i is a complete cipher of a continuous linear functional in $\mathscr{T}_{n,m}$. In other words, Y_i is a point of the adjoint space $\mathscr{T}_{n,m}^{\text{ш}}$.

Suppose that λ is any continuous linear functional in $\mathscr{T}_{n,m}$. We shall show that

$$\forall \theta_1 (\lambda(\theta_1) = \lambda(X_1)\cdot \mathfrak{B}_1(\theta_1) + \ldots + \lambda(X_m)\cdot \mathfrak{B}_m(\theta_1)). \tag{5}$$

Indeed, suppose that θ_1 is a point of $\mathscr{T}_{n,m}$. We have

$$\theta_1 \underset{\theta}{=} \text{И}(1 \square \theta_1)\cdot X_1 + \ldots + \text{И}(n \square \theta_1)\cdot X_n.$$

Since λ is a linear functional,

$$\lambda(\theta_1) = \text{И}(1 \square \theta_1)\ \lambda(X_1) + \ldots + \text{И}(n \square \theta_1)\cdot \lambda(X_n). \tag{6}$$

From (6) and (2) we find directly that

$$\lambda(\theta_1) = \lambda(X_1)\cdot \text{И}(1 \square \theta_1) + \ldots + \lambda(X_m)\cdot \text{И}(m \square \theta_1),$$

i.e. formula (5) is proved.

It is clear from (5) that if P is the complete cipher of a continuous linear functional in $\mathscr{T}_{n,m}$, then

$$P \underset{\text{щ}}{=} \underset{\longleftarrow \pi}{P}(X_1)\cdot Y_1 + \ldots + \underset{\longleftarrow \pi}{P}(X_m)\cdot Y_m.$$

It is not hard to verify that in the adjoint space $\mathscr{T}_{n,m}^{\text{ш}}$ the points $Y_1, \cdots, Y_m$ are linearly independent. Thus the points $Y_1, \cdots, Y_m$ form a basis of the adjoint space, and so it is an m-dimensional space. Thus we have proved the following theorem:

4.1.1. *For any natural numbers n and m $(n > 0,\ 0 \leqq m \leqq n)$, it is possible to construct a T'-separated, n-dimensional, locally convex space whose adjoint is an m-dimensional space.*

We shall prove further that for $m \geqq 1$ the adjoint space is normable. To this end we first of all define a locally convex topology in $\mathscr{T}_{n,m}^{\text{ш}}$.

In view of Theorem 1.1.1 we may construct for the space $\mathscr{S}$ a normal fundamental system of bounded sets. Suppose that $(\mathrm{A}_p, \mathfrak{K}, \mathfrak{N})$ is some normal fundamental system of bounded sets in $\mathscr{S}$.

Suppose that the letter э does not enter into $\mathrm{A}_p \cup \text{Ч}_0$, and denote by q the index of the alphabet $\mathrm{A}_p \cup \text{Ч}_0 \cup \{\text{э}\}$, i.e. $\mathrm{A}_q = \mathrm{A}_p \cup \text{Ч}_0 \cup \{\text{э}\}$.

We may construct the set $\mathfrak{K}^*$ of words in A_q satisfying the following condition: a word P in A_q belongs to $\mathfrak{K}^*$ if and only if P is an n-termed э-system of words in $\mathrm{A}_p \cup \text{Ч}_0$ whose m first э-terms are natural numbers, and whose $n - m$ remaining э-terms are elements of the set $\mathfrak{K}$. This set $\mathfrak{K}^*$ may be defined by a normal formula. We denote by κ a subordinate generic letter assigned to the set $\mathfrak{K}^*$ (in the language $Я_\Sigma$).

We construct an algorithm $\widetilde{И}$ having the following property: if P is a n-termed э-system of words in $A_p \cup Ч_0$ and i is a natural number such that $1 \leqq i \leqq n$, then $\widetilde{И}(i \square P)$ has meaning and is the ith э-term of the word P.

Then we construct a two-parameter formula $\mathfrak{K}^*$ such that, for any element κ_1 of the set $\mathfrak{K}^*$ and any point θ_1 of the space $\mathscr{T}_{n,m}$,

$$\mathfrak{R}^*(\varkappa_1, \theta_1) \equiv \forall_i ((1 \leqslant i \leqslant m \supset |И(i \square \theta_1)| \leqslant \widetilde{И}(i \square \varkappa_1) + 1) \& (m < i \leqslant n \supset \mathfrak{R}(\widetilde{И}(i \square \varkappa_1), И(i \square \theta_1)))).$$

The list $(A_q, \mathfrak{K}^*, \mathfrak{R}^*)$ defines a normal fundamental system of bounded sets in $\mathscr{T}_{n,m}$. Recall that $\mathscr{T}_{n,m}$ is the product of the space R taken m times and the space $\mathscr{S}$ taken $n-m$ times.

We construct a two-parameter formula $\mathfrak{Q}^{ш}$ such that, for any element κ_1 of the set $\mathfrak{K}^*$ and any point $ш_1$ of the adjoint space $\mathscr{T}_{n,m}^{ш}$,

$$\mathfrak{Q}^{ш}(\varkappa_1, ш_1) \equiv \forall \theta_1 (\mathfrak{R}^*(\varkappa_1, \theta_1) \supset |\underbrace{ш_1}_{\pi}(\theta_1)| \leqslant 1).$$

The list $(A_q, \mathfrak{K}^*, \mathfrak{Q}^{ш})$ defines a locally convex topology for the adjoint space.

Now we construct an algorithm $\mathscr{N}^{ш}$ such that, for any point $ш_1$ of $\mathscr{T}_{n,m}^{ш}$,

$$\mathscr{N}^{ш}(ш_1) \simeq \max(|\underbrace{ш_1}_{\pi}(X_1)| \square \ldots \square |\underbrace{ш_1}_{\pi}(X_m)|).$$

It is not hard to verify that $\mathscr{N}^{ш}$ satisfies all the axioms for a norm. We shall show that the space $\mathscr{T}_{n,m}^{ш}$ is normable by means of the norm $\mathscr{N}^{ш}$. To this end it suffices for us to prove the following two formulas:

$$\forall \varkappa_1 \exists k \forall ш_1 (\mathscr{N}^{ш}(ш_1) \leqslant 2^{-k} \supset \mathfrak{Q}^{ш}(\varkappa_1, ш_1)); \tag{7}$$

$$\forall k \exists \varkappa_1 \forall ш_1 (\mathfrak{Q}^{ш}(\varkappa_1, ш_1) \supset \mathscr{N}^{ш}(ш_1) \leqslant 2^{-k}). \tag{8}$$

Let us prove (7). Suppose given a word κ_1 of type κ. We find a natural number k such that

$$2^k \geqslant n \cdot (\max(\widetilde{И}(1 \square \varkappa_1) \square \ldots \square \widetilde{И}(m \square \varkappa_1)) + 1).$$

Suppose that $ш_1$ is a point of $\mathscr{T}_{n,m}^{ш}$ such that $\mathscr{N}^{ш}(ш_1) \leqq 2^{-k}$. We have to prove that $\mathfrak{Q}^{ш}(\kappa_1, ш_1)$, i.e. that

$$\forall \theta_1 (\mathfrak{R}^*(\varkappa_1, \theta_1) \supset |\underbrace{ш_1}_{\pi}(\theta_1)| \leqslant 1). \tag{9}$$

Suppose that θ_1 is a point of the space $\mathscr{T}_{n,m}$ such that $\mathfrak{R}^*(\kappa_1, \theta_1)$. Then

$$\forall i (1 \leqslant i \leqslant m \supset |И(i \square \theta_1)| \leqslant \widetilde{И}(i \square \varkappa_1) + 1),$$

so that

$$\forall i (1 \leqslant i \leqslant m \supset |И(i \square \theta_1)| \leqslant 2^k : n).$$

Therefore we have

$$|\underbrace{\text{щ}_1}_{\pi}(\theta_1)| = |\underbrace{\text{щ}_1}_{\pi}(X_1)\text{И}(1\,\Box\,\theta_1) + \ldots + \underbrace{\text{щ}_1}_{\pi}(X_m)\cdot\text{И}(m\,\Box\,\theta_1)|$$
$$\leqslant \mathcal{N}^{\text{щ}}(\text{щ}_1)\cdot(|\text{И}(1\,\Box\,\theta_1)| + \ldots + |\text{И}(m\,\theta_1)|)$$
$$\leqslant 2^{-k}\cdot n\cdot(2^k : n) = 1.$$

Thus formula (9), and formula (7) as well, are proved.

Now we shall prove (8). Suppose given a natural number k. Put $h = 2^k$. Let U be some element of the set $\mathfrak{K}$. Then write

$$\varkappa_1 \rightleftharpoons \underbrace{h\text{э}\ldots\text{э}h}_{m \text{ times}}\text{э}\underbrace{U\text{э}\ldots\text{э}U}_{n-m \text{ times}}.$$

Obviously κ_1 is a word of type κ. We shall prove that

$$\forall \text{щ}_1 (\mathfrak{Q}^{\text{щ}}(\varkappa_1, \text{щ}_1) \supset \mathcal{N}^{\text{щ}}(\text{щ}_1) \leqslant 2^{-k}). \tag{10}$$

Suppose that щ_1 is a point of the space $\mathcal{T}_{n,m}^{\text{щ}}$ such that $\mathfrak{Q}^{\text{щ}}(\kappa_1, \text{щ}_1)$, i.e. that

$$\forall \theta_1 (\mathfrak{R}^*(\varkappa_1, \theta_1) \supset |\underbrace{\text{щ}_1}_{\pi}(\theta_1)| \leqslant 1).$$

It is easy to see that $\mathfrak{R}^*(\kappa_1, h\cdot X_i)$ for any i, $1 \leqq i \leqq m$, so that $|\underbrace{\text{щ}_1}_{\pi}(h\cdot X_i)| \leqq 1$, whence $|\underbrace{\text{щ}_1}_{\pi}(X_i)| \leqq 1 : h$. Hence we obtain $\mathcal{N}^{\text{щ}}(\text{щ}_1) \leqq 2^{-k}$. Thus (10) and (8) are proved.

We may prove further that the space $\mathcal{T}_{n,m}^{\text{щ}}$ is isomorphic to the space $R^{(m)}$ (the m-fold product of the space R of real duplexes).

Thus we obtain the following: *for any natural numbers n and m, $n > 0$, $1 \leqq m \leqq n$, the space $\mathcal{T}_{n,m}^{\text{щ}}$ is isomorphic to the space $R^{(m)}$.*

REMARK. As we know, in the classical theory one proves that the adjoint to any separable, n-dimensional, locally convex space is an n-dimensional space. Theorem 4.1.1 shows that this assertion does not carry over to constructive mathematics if one adopts T'-separation as the constructive analogue of the separation axiom. The following question arises: if $\mathfrak{M}$ is a T-separated, n-dimensional space, is its adjoint n-dimensional? For the case $n = 1$ the answer to this question is positive. For $n > 1$ the answer is not known to the author.

4.2. We shall now consider the spaces $\mathcal{T}_{\infty,m}$ constructed in §4.3, Chapter II.

Suppose that m is a natural number. The space $\mathcal{T}_{\infty,m}$ is the list

$$\text{Ч}_0, \mathfrak{P}^*, \mathfrak{C}^*, \underset{\theta}{+}, \underset{\theta}{\cdot}, \mathfrak{D}^*, A_8, \mathfrak{I}^*, \mathfrak{Q}^*$$

(see §4.3, Chapter II). Here all unexplained notation is the same as in §4.3 of Chapter II.

Using a method analogous to that by which formula (2) was proved, it is possible to show that if λ is a continuous linear functional in $\mathcal{T}_{\infty,m}$ and if $n \geqq m$, then

$$\lambda(X_n) = 0, \tag{11}$$

where X_n is a point of $\mathscr{T}_{\infty,m}$ satisfying the condition

$$(\langle X_n\rangle_3 (n) = 1) \ \& \ \forall k\, (k \neq n \supset \langle X_n\rangle_3 (k) = 0).$$

Now we shall prove that if λ is a continuous linear functional in $\mathscr{T}_{\infty,m}$, then

$$\forall \theta_1 (\forall n (n < m \supset \langle \theta_1\rangle_3 (n) = 0) \supset \lambda(\theta_1) = 0). \tag{12}$$

Suppose that λ is a continuous linear functional in $\mathscr{T}_{\infty,m}$ and θ_1 a point of $\mathscr{T}_{\infty,m}$ such that

$$\forall n\, (n < m \supset \langle \theta_1\rangle_3 (n) = 0).$$

Construct a sequence φ of points of $\mathscr{T}_{\infty,m}$ such that for any k

$$\langle \varphi(k)\rangle_3 (n) \simeq \begin{cases} 0, & \text{if} \quad n \leqslant k, \\ \langle \theta_1\rangle_3 (n), & \text{if} \quad n > k. \end{cases}$$

It is easy to see that for any $k \geqq m$ we have

$$\theta_1 \underset{\theta}{=} \langle \theta_1\rangle_3 (m) \cdot X_m + \ldots + \langle \theta_1\rangle_3 (k) \cdot X_k + \varphi(k).$$

Therefore, using (11), we get

$$\forall k\, (k \geqslant m \supset \lambda(\theta_1) = \lambda(\varphi(k))). \tag{13}$$

We shall prove the following formula:

$$\forall \iota_1 \exists l \forall k\, (k \geqslant l \supset \mathfrak{Q}^* (\iota_1, \varphi(k))). \tag{14}$$

Suppose that ι_1 is a word of type ι. Put

$$l \rightleftharpoons \max (\text{И}_1 (1 \square \iota_1) \square \ldots \square \text{И}_1 (L_\tau (\iota_1) \square \iota_1)).$$

Suppose that $k \geqq l$. Then for any $n \leqq l$ we have $\langle \varphi(k)\rangle_3(n) = 0$, so that $\forall i\, \mathfrak{Q}(n, i, \langle \varphi(k)\rangle_3(n))$. Hence we get $\mathfrak{Q}^*(\iota_1, \varphi(k))$. Thus (14) is proved.

Since λ is continuous,

$$\forall i \exists \iota_1 \forall \theta_1 (\mathfrak{Q}^* (\iota_1, \theta_1) \supset |\lambda(\theta_1)| \leqslant 2^{-i}). \tag{15}$$

It follows from (14) and (15) that

$$\forall i \exists l \forall k\, (k \geqslant l \supset |\lambda(\varphi(k))| \leqslant 2^{-i}). \tag{16}$$

Then it follows from (13) and (16) that

$$\forall i\, (|\lambda(\theta_1)| \leqslant 2^{-i}),$$

so that $\lambda(\theta_1) = 0$. This proves (12).

We denote the space adjoint to $\mathscr{T}_{\infty,m}$ by $\mathscr{T}^{\text{ш}}_{\infty,m}$. If $m = 0$ it follows from (12) that every continuous linear functional in $\mathscr{T}_{\infty,m}$ is identically equal to zero, so that the adjoint space consists of only one point.

Now consider the case $m > 0$. We construct in Ч_3^{sa} an algorithm $\mathfrak{B}_n$ and in A_s^{sa} an algorithm $\mathfrak{C}_n$, $0 \leqq n < m$, such that

$$\forall \theta_1 (\mathfrak{B}_n(\theta_1) \simeq \langle \theta_1 \rangle_3 (n)); \qquad \forall k (\mathfrak{C}_n (k) \simeq n \ni k).$$

Then we write

$$Y_n \leftrightharpoons \mathcal{E}\mathfrak{B}_n, И_3^{sa} \mathcal{3} \otimes \mathcal{E}\mathfrak{C}_n, A_s^{sa} \mathcal{3} \ (n = 0, \ldots, m-1).$$

It is easy to verify that $Y_0, \cdots, Y_{m-1}$ are linearly independent points of the space $\mathscr{T}^{ш}_{\infty,m}$.

Suppose that $ш_1$ is any point of $\mathscr{T}^{ш}_{\infty,m}$. We shall show that

(17) $\forall \theta_1 (\underset{\pi}{\underbrace{ш_1}} (\theta_1) = \underset{\pi}{\underbrace{ш_1}} (X_0) \cdot \langle \theta_1 \rangle_3 (0) + \ldots + \underset{\pi}{\underbrace{ш_1}} (X_{m-1}) \cdot \langle \theta_1 \rangle_3 (m-1)).$

Suppose that θ_1 is a point of $\mathscr{T}^{ш}_{\infty,m}$. One can find a point P of that space such that

$$\theta_1 \underset{\theta}{=} \langle \theta_1 \rangle_3 (0) \cdot X_0 + \ldots + \langle \theta_1 \rangle_3 (m-1) \cdot X_{m-1} + P,$$
$$\forall n (n < m \supset \langle P \rangle_3 (n) = 0).$$

Hence (12) implies (17). So, we find that

$$\forall ш_1 (ш_1 \underset{ш}{=} \underset{\pi}{\underbrace{ш_1}} (X_0) \cdot Y_0 + \ldots + \underset{\pi}{\underbrace{ш_1}} (X_{m-1}) \cdot Y_{m-1}).$$

The space $\mathscr{T}^{ш}_{\infty,m}$ is thus m-dimensional. One may prove further that it is a normed space and that it is isomorphic to the space $R^{(m)}$. Thus we have obtained the following assertion.

4.2.1. *For any natural number m, it is possible to construct a T′-separated, infinite-dimensional, locally convex space whose adjoint space is a normed m-dimensional space.*

CHAPTER IV

CONSTRUCTIVE GENERALIZED FUNCTIONS

§1. Almost uniformly continuous constructive functions

1.1. In this section we introduce the concept of almost uniformly continuous constructive function and consider some of its properties. This concept will be regarded as the constructive analogue of the concept of "everywhere defined and continuous functions" in classical mathematics. Almost uniformly continuous constructive functions and their complete ciphers serve in the present work as the initial data for the formation of constructive generalized functions.

We introduce the following alphabet: $Ч_4 \leftrightharpoons Ч_3 \cup \{\diamondsuit, \sigma\}$. This is the general alphabet of the objects considered in this section.

Let f be a constructive function defined everywhere, i.e. an operation of the type $(д \to д)$. We will say that f is an *almost uniformly continuous constructive function* if we can realize an algorithm g of the type $(\text{н н} \to \text{н})$ such that

$$(1)\ \forall mixy\,(|x| \leqslant m \,\&\, |y| \leqslant m \,\&\, |x-y| < 2^{-g(m\square i)} \supset |f(x) - f(y)| < 2^{-i}).$$

An algorithm g satisfying the condition (2) will be called a *regulator of almost uniform continuity of the function f*. From the definition it is clear that every almost uniformly continuous function[1] f is uniformly continuous in each segment $-m\triangle m$,[2] where m is any natural number. We denote the condition (1) by $\mathfrak{C}\,(f,g)$.

Concerning the word $c_{4,1}$ in the alphabet $Ч_4$, we will say that it is a *canonical word in* $Ч_4$, and write

$$(c_{4,1} \in \underline{\text{can}}_{\,4}),$$

if it has the form $P \diamondsuit Q$, where P and Q are words in $Ч_0$. Suppose that $\overline{\text{lef}}$ and $\overline{\text{rig}}$ are algorithms applicable to any canonical word in $Ч_4$ and working each such word $c_{4,1}$ into the left and right wings respectively of the unique imbedding of the sign $\diamondsuit$ in $c_{4,1}$. We write

$$\underset{\llcorner 4}{\Theta} \leftrightharpoons \langle\, \overline{\textbf{lef}}\,(\Theta)\,\rangle_3; \qquad \overset{\ulcorner 4}{\Theta} \leftrightharpoons \langle\, \overline{\textbf{rig}}\,(\Theta)\rangle_0,$$

where Θ is any object term. In what follows, instead of $\underline{\Theta}_4$ and $\overline{\Theta}^4$ we shall write simply $\underline{\Theta}$ and $\overline{\Theta}$, if this does not give rise to confusion.

---

[1] We frequently will omit the adjective "constructive".

[2] The concepts of "segment" and "interval", and the notation for them, are understood as in [6].

We introduce the following formulas:[3]

$$\mathfrak{P}^{\text{ш}} \leftrightharpoons (c_{4,1} \in \underline{\mathrm{can}_4}) \,\&\, (c_{4,1} \in \underline{\mathrm{funct}}) \,\&\, (\overline{c_{4,1}} \in (\text{нн} \to \text{н})) \,\&\, \mathfrak{C}\,(\overline{c_{4,1}}, \underline{c_{4,1}});$$

$$\mathfrak{C}^{\text{ш}} \leftrightharpoons \forall x\,(\underline{c_{4,1}}(x) \underset{\text{в}}{=} \underline{c_{4,2}}(x));$$

$$\mathfrak{C}^{*} \leftrightharpoons \forall x\,(|x| \leqslant m \supset \underline{c_{4,1}}(x) \underset{\text{в}}{=} \underline{c_{4,2}}(x)),$$

where $(c_{4,1} \in \underline{\mathrm{funct}})$ denotes that $c_{4,1}$ is a function.

The parameter of the formula $\widetilde{\mathfrak{P}}^{\text{ш}}$ is the variable $c_{4,1}$, the parameters of the formula $\mathfrak{C}^{\text{ш}}$ are the variables $c_{4,1}$ and $c_{4,2}$, and the parameters of the formula $\mathfrak{C}^{*}$ are the variables m, $c_{4,1}$ and $c_{4,2}$. We adjoin the formula $\mathfrak{P}^{\text{ш}}$ to the list of formulas Σ of the language $Я_{\Sigma}$, and denote by ш a subordinate generic letter assigned to the formula $\mathfrak{P}^{\text{ш}}$. A word of type ш will be called a *complete cipher of an almost uniformly continuous function*. Obviously, if P is a word of type ш, then $\underline{P}$ is an almost uniformly continuous function and $\overline{P}$ is its regulator of almost uniform continuity.

We introduce the notation

$$(P \underset{\text{ш}}{=} Q) \leftrightharpoons F^{c_{4,1},c_{4,2}}_{P,\,Q} \llcorner \mathfrak{C}^{\text{ш}} \lrcorner;\ (P \underset{M}{=} Q) \leftrightharpoons F^{m,c_{4,1},c_{4,2}}_{M,P,Q} \llcorner \mathfrak{C}^{*} \lrcorner;\ \mathfrak{C}^{M} \leftrightharpoons F^{m}_{M} \llcorner \mathfrak{C}^{*} \lrcorner,$$

where P, Q and M are arbitrary object terms.

We fix an algorithm χ in $Ч_3^{sa}$ transforming any real duplex into the number 0, and an algorithm μ in $Ч_3^{sa}$ transforming any word of the type $m \square i$, where m and i are natural numbers, into the number 0. Then we write

$$\mathfrak{D}^{\text{ш}} \leftrightharpoons \mathcal{E}\chi,\, Ч_3^{sa}\, 3 \diamondsuit \mathcal{E}\mu,\, Ч_0^{sa}\, 3.$$

It is not hard to construct algorithms $\underset{\text{ш}}{+}$ of the type $(\text{шш} \to \text{ш})$ and $\underset{\text{ш}}{\cdot}$ of the type $(\text{дш} \to \text{ш})$, satisfying the following conditions:

$$\forall \text{ш}_1\text{ш}_2 x\,(\underset{\text{ш}}{+} \llcorner \text{ш}_1 \square \text{ш}_2 \lrcorner (x) \simeq \underline{\text{ш}_1}(x) + \underline{\text{ш}_2}(x));$$

$$\forall \alpha \text{ш}_1 x\,(\underset{\text{ш}}{\cdot} \llcorner x \square \text{ш}_1 \lrcorner (x) \simeq \alpha \cdot \underline{\text{ш}_1}(x)).$$

Here α is a variable for real duplexes.

On the basis of the results of §11.2 of [**26**] it is possible to construct an algorithm $\mathcal{N}$ such that, for any natural number m and any word ш_1 of type ш, $\mathcal{N}(m \square \text{ш}_1)$ is the least upper bound of the function ш_1 in the segment $-M\Delta m$.

It is easy to verify that the list

(2) $$Ч_4,\ \mathfrak{P}^{\text{ш}},\ \mathfrak{C}^{\text{ш}},\ \underset{\text{ш}}{+},\ \underset{\text{ш}}{\cdot},\ \mathfrak{D}^{\text{ш}},\ \mathcal{N}$$

---

[3] The sign c_4 will be used as an object generic letter, assigned to the alphabet $Ч_4$. In keeping with this we shall write $c_{4,1}, c_{4,2}, \cdots$ for variables of genus c_4 (i.e. variables for words in the alphabet $Ч_4$).

defines a constructive enumerably normed space, and for each natural number m the list

$$\text{(3)} \qquad \text{Ч}_4,\ \mathfrak{P}^{\text{ш}},\ \mathfrak{E}^{\text{ш}},\ \underset{\text{ш}}{+},\ \underset{\text{ш}}{\cdot},\ \mathfrak{D}^{\text{ш}},\ (\mathcal{N} \uparrow m \square)$$

defines a constructive normed space.

We shall denote the space (2) by $\mathfrak{S}$, and the space (3) by C_m.[4)]

It is possible to prove that $\mathfrak{S}$ is a complete enumerably normed space, and that for each natural number m C_m is a complete normed space.

We introduce further certain operations over words of type ш. We construct an algorithm B of the type $(\text{н ш} \to \text{н})$, satisfying the condition

$$\forall m \text{ш}_1 (\mathcal{N}(m \square \text{ш}_1) < 2^{B(m \square \text{ш}_1)}).$$

Now we construct algorithms γ, δ and $\underset{\text{ш}}{\times}$ such that, for any words ш_1 and ш_2 of type ш,

$$\gamma(\text{ш}_1 \square \text{ш}_2) \overset{+}{\in} c_0; \qquad \delta(\text{ш}_1 \square \text{ш}_2) \overset{+}{\in} c_0;$$

$$\forall x (< \gamma(\text{ш}_1 \square \text{ш}_2) >_3 (x) \simeq \underline{\text{ш}_1}(x) \cdot \underline{\text{ш}_2}(x));$$

$$\forall mi (< \delta(\text{ш}_1 \square \text{ш}_2) >_0 (m \square i)$$

$$\simeq \max(\overline{\text{ш}_1}(B(m \square \text{ш}_2) + i + 1) \square \overline{\text{ш}_2}(B(m \square \text{ш}_1) + i + 1)));$$

$$\underset{\text{ш}}{\times}(\text{ш}_1 \square \text{ш}_2) \simeq \gamma(\text{ш}_1 \square \text{ш}_2) \diamondsuit \delta(\text{ш}_1 \square \text{ш}_2).$$

It is not hard to verify that $\underset{\text{ш}}{\times}$ is an operator of the type $(\text{ш ш} \to \text{ш})$. This operator is called the *operation of multiplication of words of type* ш. In place of $\underset{\text{ш}}{\times}(\text{ш}_1 \square \text{ш}_2)$ we shall write $(\text{ш}_1 \underset{\text{ш}}{\times} \text{ш}_2)$.

Now we construct the operator of linear substitution of a variable for a word of type ш.

We construct algorithms G of type $(\text{д} \dot{\to} \text{н})$ and H of type $(\text{д д н} \to \text{н})$ such that

$$\forall x (x \neq 0 \supset ! G(x) \,\&\, |x| > 2^{-G(x)});$$

$$\forall xym (|x| \cdot m \dotplus |y| \leqslant H(x \square y \square m)).$$

We construct algorithms μ, ν and $L^{\text{ш}}$ such that, for any real duplexes α and β, and any word ш_1 of type ш,

---

4) In the note [32] we introduced the normed space of complete ciphers of all possible functions defined and uniformly continuous on the segment $-m \triangle m$. This space was denoted by C_m. It is easy to see that, for any function f defined and uniformly continuous on $-m \triangle m$, we can realize an almost uniformly continuous function F such that

$$\forall x (|x| \leqslant \supset F(x) = f(x)).$$

One may easily verify that the space C_m introduced in the present monograph is isomorphic to the normed space of complete ciphers of uniformly continuous functions on $-m \triangle m$. For convenience of exposition we will not consider separately uniformly continuous functions on the segment $-m \triangle m$. Such functions are replaced here by the corresponding almost uniformly continuous functions.

$$\mu(\alpha\square\beta\square\text{ш}_1)\overset{+}{\in} c_0;\quad \nu(\alpha\square\beta\square\text{ш}_1)\overset{+}{\in} c_0;$$
$$\forall x(<\mu(\alpha\square\beta\square\text{ш}_1)>_3(x)\simeq \underline{\text{ш}_1}(\alpha\cdot x+\beta));$$
$$\forall mi(<\nu(\alpha\square\beta\square\text{ш}_1)>_0(m\square i)\simeq G(\alpha)\dot{+}\overline{\text{ш}_1}(H(\alpha\square\beta\square m)\square i)));$$
$$L^{\text{ш}}(\alpha\square\beta\square\text{ш}_1)\simeq\mu(\alpha\square\beta\square\text{ш}_1)\lozenge\nu(\alpha\square\beta\square\text{ш}_1).$$

One may verify that if $\alpha\neq 0$, then $!L^{\text{ш}}(\alpha\square\beta\square\text{ш}_1)$, and $L^{\text{ш}}(\alpha\square\beta\square\text{ш}_1)$ is a word of type ш. Moreover, for any duplexes α and β, if $\alpha\neq 0$, then $(L^{\text{ш}}\uparrow\alpha\square\beta\square)$ is an operator of the type $(\text{ш}\to\text{ш})$. Obviously

$$\forall\alpha\beta\text{ш}_1 x(\alpha\neq 0\supset \underline{L^{\text{ш}}(\alpha\square\beta\square\text{ш}_1)}(x)\underset{\text{B}}{=}\underline{\text{ш}_1}(\alpha x+\beta)).$$

1.2. In [**26**], §15.5, the concept of the integral of the complete cipher of a uniformly continuous function in any given segment $\alpha\Delta\beta$ was introduced (in the context: in the space $\mathfrak{L}(\alpha,\beta)$ of *FR*-constructs summable in $\alpha\Delta\beta$). It is possible to construct an algorithm ξ such that, for any real duplexes α and β with $\alpha\leqq\beta$, and any word ш_1 of type ш, ξ transforms the word $\alpha\square\beta\square\text{ш}_1$ into the complete cipher of a function which is uniformly continuous in the segment $\alpha\Delta\beta$ and satisfies the condition

$$\forall x(x\in\alpha\Delta\beta\supset\underline{\xi(\alpha\square\beta\square\text{ш}_1)}(x)=\underline{\text{ш}_1}(x)).$$

The integral of the complete cipher $\xi(\alpha\square\beta\square\text{ш}_1)$ in the segment $\alpha\Delta\beta$ will also be called *the integral of the word* ш_1 *in the segment* $\alpha\Delta\beta$.

On the basis of the algorithms constructed in [**26**], §15, one may construct an algorithm I_1 such that, for any real duplexes α and β with $\alpha\leqq\beta$, and any word ш_1 of type ш, $I_1(\alpha\square\beta\square\text{ш}_1)$ is the integral of the complete cipher ш_1 in the segment $\alpha\Delta\beta$. This integral will be denoted by $\int_\alpha^\beta(\text{ш}_1)$. Starting from the algorithm I_1, we may construct an algorithm I_2 such that, for any duplexes α and β, and any word ш_1 of type ш,

$$I_2(\alpha\square\beta\square\text{ш}_1)\simeq\int_\gamma^\beta(\text{ш}_1)-\int_\gamma^\alpha(\text{ш}_1),$$

where $\gamma=\min(\alpha\square\beta)$. $I_2(\alpha\square\beta\square\text{ш}_1)$ is called the *integral of the complete cipher of* ш_1 *with the limits* α *and* β, and is denoted by $\int_\alpha^\beta(\text{ш}_1)$.

Suppose that Ω is some interval or segment, and let f and φ be functions defined in Ω. By the definition introduced, for example, in [**28**], the function φ is the *derivative* of the function f in Ω if

$$\forall xn(x\in\Omega\supset\exists m\forall y(y\in\Omega\,\&\,|y-x|<2^{-m}\supset|f(y)-f(x)-(y-x)\cdot\varphi(x)|\leqslant|y-x|\cdot 2^{-n})).\tag{4}$$

If this condition is satisfied we will also say that f is the *antiderivative* of the function φ in Ω.

Suppose that ш$_1$ and ш$_2$ are words of type ш. We will say that the word ш$_1$ is the *derivative* of the word ш$_2$ in Ω (or that ш$_2$ is the *antiderivative* of the word ш$_1$ in Ω) if the function $\underbrace{\text{ш}_1}$ is the derivative of the function $\underbrace{\text{ш}_2}$ in Ω. The condition "the word ш$_1$ is the derivative of the word ш$_2$ in Ω" will be denoted by $Д_\Omega(\text{ш}_1, \text{ш}_2)$.

If $\Omega \mathrel{\overline{\underline{\circ}}} -\infty \nabla \infty$, then in place of the statement "the word ш$_1$ is the derivative of the word ш$_2$ in $-\infty \nabla \infty$", we will say simply that ш$_1$ is the derivative of the word ш$_2$, and in place of $Д_{-\infty\nabla\infty}(\text{ш}_1, \text{ш}_2)$ we will write $Д(\text{ш}_1, \text{ш}_2)$.

Suppose further that k is any natural number. We shall say that ш$_1$ *is the derivative of order k of the word* ш$_2$ (ш$_2$ *is the antiderivative of order k of the word* ш$_1$), and write $Д^k(\text{ш}_1, \text{ш}_2)$, if we can realize a word $P_0, \cdots, P_k$ of type ш which satisfies the following conditions:

1) $P_0 \underset{\text{ш}}{=} \text{ш}_1,\ P_k \underset{\text{ш}}{=} \text{ш}_2$;

2) for any i, $1 \leq i \leq k$, P_{i-1} is the derivative of P_i.

It is possible to construct an algorithm I of the type (дш → ш) such that

$$\forall \alpha \text{ш}_1 x\, (\underbrace{I(\alpha \square \text{ш}_1)}(x) = \int_\alpha^x (\text{ш}_1)).$$

It is not hard to show that, for any real duplex α and any word ш$_1$ of type ш, ш$_1$ is the derivative of $I(\alpha \square \text{ш}_1)$. Hence it follows that, for any words ш$_1$ and ш$_2$ of type ш,

$$Д(\text{ш}_1, \text{ш}_2) \equiv \forall x\, (\underbrace{\text{ш}_2}(x) - \underbrace{\text{ш}_2}(\alpha) = \underbrace{I(\alpha \square \text{ш}_1)}(x)), \tag{5}$$

where α is some fixed duplex. Thus the condition $Д(\text{ш}_1, \text{ш}_2)$ may be expressed by a normal formula.

Now we shall construct an algorithm J of the type (нш → ш) satisfying the following recursive schema:

$$\begin{cases} J(0 \square \text{ш}_1) = \text{ш}_1; \\ J(k+1 \square \text{ш}_1) \simeq I(0 \square J(k \square \text{ш}_1)). \end{cases}$$

For any natural number k and any word ш$_1$ of type ш, ш$_1$ is the derivative of order k of $J(k \square \text{ш}_1)$, i.e. $J(k \square \text{ш}_1)$ is the antiderivative of order k of the word ш$_1$. The algorithm J will be used in what follows.

Following the definition introduced by N. A. Šanin in [**26**], §16.3, we introduce the following concept: a *polynomial frame*, or a *word of type* ю, is any σ-system of real duplexes.[5] We denote by dg the algorithm computing the number of entrances of the sign σ into words of type ю. Suppose that ю$_1$ is a word of type ю. The number dg(ю) will be called the (*formal*) *degree*

[5] A *σ-system* of real duplexes is a word of the type $\alpha_0 \sigma \cdots \sigma \alpha_k$, where $\alpha_0, \cdots, \alpha_k$ are real duplexes.

of the word ю$_1$. It is possible to construct an algorithm H of type (ю → ш) such that for any word ю$_1$ of type ю, if ю$_1 \overline{\overset{\circ}{=}} \alpha_0 \sigma \cdots \sigma \alpha_k$, where $\alpha_1, \cdots, \alpha_k$ are real duplexes, then

$$\forall x (\underline{H(\text{ю}_1)}(x) = \alpha_0 + \alpha_1 \cdot x + \ldots + \alpha_k \cdot x^k).$$

It is possible to construct an algorithm $D^{\text{ю}}$ of the type (ю → ю) such that: a) if σ does not enter into the word ю$_1$, then $D^{\text{ю}}(\text{ю}_1) = 0$; b) if σ enters into the word ю$_1$ and ю$_1 \overline{\overset{\circ}{=}} \alpha_0 \sigma \alpha_1 \sigma \cdots \sigma \alpha_i \sigma \cdots \sigma \alpha_k$ ($\alpha_0, \cdots, \alpha_k$ are duplexes), then $D^{\text{ю}}(\text{ю}_1) \overline{\overset{\circ}{=}} \alpha_1 \sigma \cdots \sigma i \alpha_i \sigma \cdots \sigma k \cdot \alpha_k$. It is easy to see that, for any word ю$_1$ of type ю, $H(D^{\text{ю}}(\text{ю}_1))$ is the derivative of $H(\text{ю}_1)$. We shall also say that $D^{\text{ю}}(\text{ю}_1)$ is the *derivative of* ю$_1$.

On the basis of the definition of the relation Д^k and property (5), one may prove by induction that, for any positive natural number k and any words ш$_1$ and ш$_2$ of type ш, ш$_1$ is the derivative of order k of ш$_2$ if and only if a word ю$_1$ of type ю is realizable such that dg(ю$_1$) $< k$ and

$$\text{ш}_2 \underset{\text{ш}}{=} J(k \square \text{ш}_1) \underset{\text{ш}}{+} H(\text{ю}_1).$$

In what follows we shall use some elementary functions. In particular we shall use the exponential function and the trigonometric function. It is possible to define these functions in constructive mathematics in a way analogous to what was done by Goodstein in [41, 42]. For example, we define the exponential function as follows: suppose that ξ is an algorithm of the type (д → н) and η an algorithm of the type (н → н), having the properties

$$\forall x (\xi(x) \geqslant |x|); \quad \forall n (\eta(n) \simeq \mu j (((2n)^{2n} : (2n)!) < 2^j)).$$

We construct algorithms φ and χ as follows:

$$\forall xn \left(\varphi(x \square n) \simeq \sum_{i=0}^{n} \frac{x^i}{i!} \right); \qquad \forall xk (\chi(x \square k) \simeq \eta(\xi(x)) + k).$$

One may verify that

$$\forall xkmn (m, n \geqslant \chi(x \square k) \supset |\varphi(x \square m) - \varphi(x \square n)| < 2^{-k}).$$

Thus, for any duplex x, $(\varphi \uparrow x \square)$ is a fundamental sequence of real duplexes and $(\chi \uparrow x \square)$ is its regulator of convergence in itself. We denote by L the algorithm of passage to the limit in the space of real duplexes, and construct the algorithm exp as follows:

$$\forall x (\exp(x) \simeq L(\mathcal{E}(\varphi \uparrow x \square), \text{Ч}_3^{\text{sa}}\, 3 \odot \mathcal{E}(\chi \uparrow x \square), \text{Ч}_0^{\text{sa}}\, 3)).$$

It turns out that exp is an almost everywhere uniformly continuous function. Accordingly, one may construct for it a corresponding word of type ш. In what follows we shall use without reference the properties of the function exp proved in [41, 42].

§2. Preparatory remarks. Topology in the space of words of the type ш

2.1. Starting from the simple space $\mathscr{C}$ of complete ciphers of almost uniformly continuous functions, we shall construct the space of constructive generalized functions. In this construction we shall base ourselves on the definition of the concept of generalized function in classical mathematics introduced by Mikusiński, Sikorski and Korevaar (see [**48**], [**45**], and also [**19**]). In the same way the concept of constructive generalized function was introduced in [**32**]. However, if the concept of constructive generalized function introduced in [**32**] is the result of an almost verbatim carry-over of the definition of Mikusiński and Sikorski to constructive mathematics, in this paper we introduce the concept of generalized function using a method having a different character. We first introduce the concept of Mikusiński sequence, and then consider the topological structure of such sequences. Then we construct the space of constructive generalized functions using completions of the space of complete ciphers of almost uniformly continuous functions relative to appropriate topologies. These topologies are enumerably normable, and therefore metrizable.

It is clear that within the limits of the space $\mathscr{C}$ one cannot define the operation of differentiation, which would construct relative to each word of type ш another word of the same type ш, its derivative. We extend the space $\mathscr{C}$ to a new space of "constructive generalized functions", within whose limits one can make a reasonable definition of the operation of differentiation, this operation being, moreover, continuous in this new space.

First of all we note the following circumstance. Suppose that ϵ is some sequence of positive real duplexes converging to zero. Suppose that P is any word of type ш. It is possible to construct a sequence λ of words of type ш such that[6)]

$$\forall nx\,(\underset{\smile}{\lambda_n}(x) \simeq (\underset{\smile}{P}(x+\varepsilon_n) - \underset{\smile}{P}(x)) : \varepsilon_n).$$

It is easy to verify that if we can realize a word Q of type ш such that Q is the derivative of the word P, then the sequence λ converges as well in the space $\mathscr{C}$. Moreover, the sequence λ has the following property:

1) *We can realize a sequence φ of words of type* ш *such that λ_n is the derivative of φ_n for any natural number n, and such that the sequence φ converges in itself in $\mathscr{C}$.*

We put off the proof of this property to §4, where a stronger property will be proved.

Starting from the sequence λ, we may construct a sequence ψ of words of type ш such that

---

[6)] If φ is a sequence of objects of some type, then instead of $\varphi(n)$ we shall write φ_n. Here $\epsilon_n \leftrightharpoons \epsilon(n)$ and $\lambda_n \leftrightharpoons \lambda(n)$.

$$\forall nx\,(\underline{\psi_n}(x) \simeq (\underline{\lambda_n}(x+\varepsilon_n) - \underline{\lambda_n}(x)) : \varepsilon_n).$$

Then, starting from the sequence ψ, using the same method we may construct a new sequence χ, and this process of construction may be continued without end. On continuing this process we obtain a sequence F of words of type ш which has the following general property (whose proof we put off to §4):

(2) *One can realize a natural number k and a sequence of words of type* ш *such that F_n is the derivative of order k of G_n for any natural number n, and the sequence G converges in itself in the space $\mathscr{C}$.*

We see that if one takes the sequence λ as the "derivative" of the word P, and the sequence ψ as the "derivative" of the sequence λ, and so forth, then for any sequence F having the property (2) one can realize a sequence having the same property (2) and being the derivative of the sequence F.

A sequence F of words of the type ш having the property (2) will be called a *Mikusiński sequence of finite order.*

We may introduce a more general concept in the following way: a *Mikusiński sequence* is any sequence F of words of type ш having the following property:

(3) *For any natural number $m \geqq 1$, one can realize a natural number k and a sequence G of words of type* ш *such that F_n is the derivative of order k of G_n for any n, and the sequence G converges in itself in the space C_m.*[7)]

Obviously every Mikusiński sequence of finite order is a Mikusiński sequence.

Suppose that ю$_1$ is a word of type ю (see §1.2) and that k is a natural number. We introduce the notation

$$(\mathrm{dg}(ю_1) \prec\cdot\, k) \rightleftharpoons (k \geqslant 1 \supset \mathrm{dg}(ю_1) < k)\,\&\,(k = 0 \supset ю_1 \mp 0).$$

As we noted in §1.2, the condition "F_n is the derivative of order k of G_n" is equivalent to the condition "one can realize a word P of type ю such that $\mathrm{dg}(P) \prec\cdot\, k$ and $G_n = J(k \,\square\, F_n) + H(P)$." Thus the definition of Mikusiński sequence presented above is equivalent to the following definition: a sequence F of words of type ш will be called a *Mikusiński sequence* if one can realize a sequence ξ of natural numbers and an algorithm φ of the type (н н $\rightarrow$ ю) such that

a) $\forall mn\,(\mathrm{dg}(\varphi(m \,\square\, n)) \prec\cdot\, \xi(m))$;

b) for any $m \geqq 1$ the sequence G defined by the formula

$$\forall n\,(G(n) \simeq J(\xi(m) \,\square\, F(n)) + H(\varphi(n))),$$

converges in itself in the space C_m.

2.2. Suppose that φ is a sequence of words of type ю. We shall say that that sequence converges in itself (converges to a duplex y) to the real duplex x if the sequence of numbers $\bar{\varphi}$ defined by the formula

---

7) We note that the concept "convergence in C_m" corresponds to the classical concept of "uniform convergence in the segment $[-m, m]$", and the concept of "convergence in $\mathscr{C}$" corresponds to the classical concept of "almost uniform convergence" (see [48]).

$$\forall n\,(\overline{\varphi}\,(n) \simeq \underline{H\,(\varphi\,(n))\,(x)}),$$

converges in itself (converges to y). We shall say that the sequence φ converges in itself in the space $\mathscr{C}$ (in C_m) if the sequence ψ of words of type ш defined by the formula

$$\forall n\,(\psi\,(n) \simeq H\,(\varphi\,(n))),$$

converges in itself in $\mathscr{C}$ (in C_m).

We note that, for any sequence φ of words of type ш, φ converges in itself in $\mathscr{C}$ if and only if it converges in itself in each space C_m. This follows from the Remark in §4.4, Chapter I.

We introduce the following convention: in what follows (up to the end of Chapter VI), in each case when the sign m is used as the notation for a variable, we will suppose that it denotes a variable for *positive* natural numbers.

Fix a sequence X of pairwise distinct real duplexes in the segment $-1\triangle 1$. For definiteness we suppose that the sequence X is defined as follows:

$$\forall\, n\,(X\,(n) \simeq -1 + 2^{-n}).$$

We write $x_n \leftrightharpoons X(n)$. Then $x_0 = 0$, $x_1 = -\frac{1}{2}, \cdots$.

We construct an algorithm A such that, for any $k \geqq 1$, $A(k)$ is the value of the following determinant:

$$A\,(k) = \begin{vmatrix} 1\; x_0 \ldots x_0^{k-1} \\ \ldots\ldots\ldots \\ 1\; x_{k-1} \ldots x_{k-1}^{k-1} \end{vmatrix}.$$

We have $A(k) \neq 0$ for any $k \geqq 1$. We construct an algorithm B such that for any natural numbers k, i and j, if $k \geqq 1$ and $0 \leqq i, j \leqq k-1$, then $B(k \square i \square j)$ is the cofactor of the element x_i^j of the determinant $A(k)$.

Suppose that P is a word of type ю and $P \stackrel{\circ}{=} \alpha_0 \sigma \cdots \sigma \alpha_{k-1}$ $(k \geqq 1)$. Then, if for any i, $0 \leqq i \leqq k-1$, we have

$$\alpha_0 + \alpha_1 \cdot x_i + \ldots + \alpha_{k-1} \cdot x_i^{k-1} = y_i$$

($y_0, \cdots, y_{k-1}$ are certain duplexes), it follows that

$$\alpha_j = \frac{1}{A(k)}\,[B\,(k \square 0 \square j) \cdot y_0 + \ldots + B\,(k \square k-1 \square j) \cdot y_{k-1}]$$

$(j = 0, \cdots, k-1)$. We construct an algorithm C such that for any $k \geqq 1$

$$C\,(k) = \frac{k}{|A\,(k)|} \cdot \max_{0 \leqslant i,\, j \leqslant k-1} |B\,(k \square i \square j)|.$$

Then we construct an algorithm α of the type (HH → H) such that for any natural numbers m and k, if $m, k \geqq 1$, then

$$C\,(k) \cdot \sum_{j=0}^{k-1} m^j \leqslant 2^{\alpha(m \square k)}.$$

We note the following assertion: for any word P of type ю and any natural number k, if $\mathrm{dg}(P) \prec \cdot\, k$ and

$$|\underline{H(P)(x_i)}| < 2^D$$

for any i, $0 \leqq i \leqq k-1$, D being some integer, then

$$\forall mx(|x| \leqslant m \supset |\underline{H(P)(x)}| < 2^{D+\alpha(m\square k)}).$$

We shall prove the following lemma.

2.2.1. *Suppose that φ is a sequence of words of type* ю *and k is a natural number such that $\forall n(\mathrm{dg}(\varphi(n)) \prec \cdot\, k)$. If the sequence φ converges in itself at k points $x_0, \cdots, x_{k-1}$, then φ converges in itself in the space C_m for any $m \geqq 1$, and so φ converges in itself in the space $\mathscr{L}$.*

PROOF. From the hypothesis it follows that we can realize an algorithm γ of the type $(\text{н} \to \text{н})$ such that

$$\forall jln(l, n \geqslant \gamma(j) \supset |\underline{H(\varphi_l)}(x_i) - \underline{H(\varphi_n)}(x_i)| < 2^{-j})$$

$(i = 0, \cdots, k-1)$. We construct an algorithm γ_1 as follows:

$$\forall mi(\gamma_1(m\square j) \simeq \gamma(j + \alpha(m\square k))).$$

Then we have

$$\forall mjln(l, n \geqslant \gamma_1(m\square j) \supset \mathscr{N}(m\square H(\varphi_l) - H(\varphi_n)) < 2^{-j}).$$

This means that φ converges in itself in each space C_m, as was required to be proved.

We construct an algorithm $\mathfrak{A}_1$ of the type $(\text{н}\,\text{ш} \to \text{ю})$ as follows. Suppose that P is a word of type ш and that k is a natural number. If $k = 0$, we put $\mathfrak{A}_1(k\square P) \overline{\underline{\circ}}\, 0$. Suppose that $k \geqq 1$. We solve the following system of equations for the unknowns $\alpha_0, \cdots, \alpha_{k-1}$:

$$\alpha_0 + \alpha_1 \cdot x_i + \ldots + \alpha_{k-1} \cdot x_i^{k-1} + \underline{P}(x_i) = 0 \quad (i = 0, \ldots, k-1).$$

This system always has a unique solution. Suppose that $(\bar{\alpha}_0, \cdots, \bar{\alpha}_{k-1})$ is its solution. Then we put $\mathfrak{A}_1(k\square P) \overline{\underline{\circ}}\, \bar{\alpha}_0 \sigma \cdots \sigma \bar{\alpha}_{k-1}$. Obviously

$$\forall k\text{ш}_1(\mathrm{dg}(\mathfrak{A}_1(k\square \text{ш}_1)) \prec \cdot\, k).$$

Moreover, we note that, for any k, ш_1 and $i < k$,

$$\underline{H(\mathfrak{A}_1(k\square \text{ш}_1))}(x_i) = -\underline{\text{ш}_1}(x_i).$$

Now we define an algorithm $\mathfrak{A}$ of the type $(\text{н}\,\text{ш} \to \text{ш})$ as follows:

$$\forall k\text{ш}_1(\mathfrak{A}(k\square \text{ш}_1) \simeq \text{ш}_1 \underset{\text{ш}}{+} H(\mathfrak{A}_1(k\square \text{ш}_1))).$$

2.2.2. a) *For any natural number k, $(\mathfrak{A} \uparrow k\,\square)$ is a linear operator of the type* $(\text{ш} \to \text{ш})$;

b) $\forall k ш_1 i (0 \leqslant i \leqslant k-1 \supset \underline{\mathfrak{A}(k \square ш_1)(x_i)} = 0);$

c) $\forall k ш_1 ш_2 ю_1 (\mathrm{dg}(ю_1) \prec \cdot k \,\&\, ш_1 \underset{ш}{=} ш_2 + H(ю_1)$

$\supset \mathfrak{A}(k \square ш_1) \underset{ш}{=} \mathfrak{A}(k \square ш_2));$

d) $\forall k l ш_1 (l \leqslant k \supset \mathfrak{A}(k \square \mathfrak{A}(l \square ш_1)) \underset{ш}{=} \mathfrak{A}(k \square ш_1)).$

Properties a) and b) follow without difficulty from the definitions of the algorithms $\mathfrak{A}_1$ and $\mathfrak{A}$. Property d) is a direct consequence of property c). Property c) is proved as follows. Suppose that $\mathrm{dg}(ю_1) \prec \cdot k$ and $ш_1 = ш_2 + H(ю_1)$. Then a word $ю_2$ of type ю is realizable such that $\mathrm{dg}(ю_2) \prec \cdot k$ and $\mathfrak{A}(k \square ш_1) = \mathfrak{A}(k \square ш_2) + H(ю_2)$ (and in fact $ю_2 \overline{\underline{\circ}}\, ю_1 + \mathfrak{A}_1(k \square H(ю_1))$. In view of property b) we have $\underline{H(ю_2)(x_i)} = 0$ $(0 < i \leqq k-1)$. Hence we find that $H(ю_2) = \mathfrak{O}^ш$, so that $\mathfrak{A}(k \square ш_1) \underset{ш}{=} \mathfrak{A}(k \square ш_2)$.

2.2.3. *Suppose that* λ *is a sequence of words of type* ш, k *and* m *natural numbers with* $m \geqq 1$. *If the sequence* λ *converges in itself* (*converges to a word* P *of type* ш) *in the space* C_m, *then the sequence* χ *defined by the formula*

$$\forall n (\chi(n) \simeq \mathfrak{A}(k \square \lambda(n))),$$

also converges in itself (*converges to* $\mathfrak{A}(k \square P)$) *in the space* C_m.

Proof. We give the proof for the case of convergence in itself. The proof for the other case is quite analogous.

Suppose that γ is the regulator of convergence in itself of the sequence λ in C_m. Then

(1) $$\forall jln (l, n \geqslant \gamma(j) \supset \mathscr{N}(m \square \lambda(l) - \lambda(n)) \leqslant 2^{-j}).$$

We construct an algorithm γ_1 as follows:

$$\forall j (\gamma_1(j) \simeq \gamma(j + 1 + \alpha(m \square k))).$$

We denote by φ a sequence of words of the type ю, defined by the formula

$$\forall n (\varphi(n) \simeq H(\mathfrak{A}_1(k \square \lambda(n)))).$$

By definition of the sequence χ we have

(2) $$\forall ln (\chi(l) - \chi(n) = \lambda(l) - \lambda(n) + H(\varphi_l) - H(\varphi_n)).$$

Obviously it follows from (1) that

$$\forall jln (l, n \geqslant \gamma(j) \supset |\underline{H(\varphi_l)}(x_i) - \underline{H(\varphi_n)}(x_i)| \leqslant 2^{-j}).$$

Hence from the proof of Lemma 2.2.1 it follows that

(3) $$\forall jln (l, n \geqslant \gamma_1(j) \supset \mathscr{N}(m \square H(\varphi_l) - H(\varphi_n)) \leqslant 2^{-(j+1)}).$$

It follows from (1), (2) and (3) that

$$\forall jln\,(l, n \geqslant \gamma_1(j) \supset \mathcal{N}(m \square \chi(l) - \chi(n)) \leqslant 2^{-j}),$$

as we were required to prove.

REMARK 1. It is possible to construct an algorithm Γ such that, for any natural numbers k and $m \geqq 1$, and any algorithm γ of type (н → н), Γ transforms the word $k \square m \square \mathcal{E}\gamma 3$ into the transcription of an algorithm of the type (н → н), and

$$\forall j\,\langle \Gamma(k \square m \square \mathcal{E}\gamma 3)\rangle_0(j) \simeq \gamma(j + 1 + \alpha(m \square k))).$$

It is clear from the proof of Lemma 2.2.3 that the algorithm Γ has the following property: suppose that λ is a sequence of words of the type ш, k and m natural numbers with $m \geqq 1$, and γ in algorithm of the type (н → н). Then, if

$$\forall jln\,(l, n \geqslant \gamma(j) \supset \mathcal{N}(m \square \lambda(l) - \lambda(n)) \leqslant 2^{-j}),$$

it follows that

$$\forall jln\,(l, n \geqslant \langle \Gamma(k \square m \square \mathcal{E}\gamma 3)\rangle_0(j) \supset \mathcal{N}(m \square \chi(l) - \chi(n)) \leqslant 2^{-j}),$$

where χ is a sequence of words of the type ш, defined by the formula

$$\forall n\,(\chi(n) \simeq \mathfrak{A}(k \square \lambda(n))).$$

REMARK 2. For any word $ш_1$ of type ш and any natural numbers k, m and j $(m \geqq 1)$, if

$$\mathcal{N}(m \square ш_1) \leqslant 2^{-(j+1+\alpha(m \square k))},$$

then

$$\mathcal{N}(m \square \mathfrak{A}(k \square ш_1)) \leqslant 2^{-j}.$$

The following is a consequence of 2.2.3 and property c) of 2.2.2.

2.2.4. *Suppose that λ is a sequence of words of type* ш, *and k and m natural numbers with $m \geqq 1$. Then the condition "one can realize a sequence φ of words of type* ю *having the properties* a) $\forall n(\mathrm{dg}\,\varphi(n)) \prec \cdot\, k$ *and* b) *the sequence ψ is such that $\psi(n) \simeq \lambda(n) + H(\varphi(n))$ converges in itself in C_m" is satisfied if and only if the sequence χ is such that $\chi(n) \simeq \mathfrak{A}(k \square \lambda(n))$ converges in itself in C_m.*

2.3. We shall make use of the set A introduced in §3.4 of Chapter II. We recall that the elements of the set A (and also the words of type e) are "indices" of algorithms of the type (н → н). Suppose that k is a natural number. We denote by $\overline{k}$ a word of type e for which $\forall m(\{\overline{k}\}(m) = k)$.

We construct an algorithm Φ of the type (eн ш → ш), as follows:

$$\forall e_1 m ш_1\,(\Phi(e_1 \square m \square ш_1) \simeq \mathfrak{A}(\{e_1\}(m) \square J(\{e_1\}(m) \square ш_1))).$$

The following follows from the definition and from 2.2.2.

2.3.1. *For any word* e_1 *of type* e *and any number $m \geqq 1$:*

a) $(\Phi \uparrow \text{e}_1 \square m \square)$ *is a linear operator of the type* $(\text{ш} \to \text{ш})$;

b) *for any word* ш_1 *of type* ш, $\Phi(\text{e}_1 \square m \square \text{ш}_1)$ *is the transform of order* $\{\text{e}_1\}(m)$ *of the word* ш_1, *and*

$$\forall i\,(0 \leqslant i \leqslant \{\text{e}_1\}(m) - 1 \supset \underline{\Phi(\text{e}_1 \square m \square \text{ш}_1)}(x_i) = 0).$$

Now we construct an algorithm $\mathfrak{N}$ of the type (еHш → ш) as follows:

$$\forall e_1 m \text{ш}_1 (\mathfrak{N}(e_1 \square m \square \text{ш}_1) \simeq \mathscr{N}(m \square \Phi(e_1 \square m \square \text{ш}_1))),$$

where $\mathscr{N}$ is the multinorm of the space $\mathscr{C}$.

Suppose that U is any word of type e. We shall denote the algorithm $(\mathfrak{N} \uparrow U\square)$ by $\mathfrak{N}_U$. It is not hard to verify that $\mathfrak{N}_U$ defines an enumerable form in the linear space

$$\text{Ч}_4, \mathfrak{P}^{\text{ш}}, \mathfrak{E}^{\text{ш}}, \underset{\text{ш}}{+}, \underset{\text{ш}}{\cdot}, \mathfrak{O}^{\text{ш}}.$$

In other words, *the list*

$$\text{Ч}_4, \mathfrak{P}^{\text{ш}}, \mathfrak{E}^{\text{ш}}, \underset{\text{ш}}{+}, \underset{\text{ш}}{\cdot}, \mathfrak{O}^{\text{ш}}, \mathfrak{N}_U \tag{4}$$

defines an enumerably normed space. We shall denote this space by $\mathscr{C}^U$.

We note the following property of convergence in $\mathscr{C}^U$.

2.3.2. *Suppose that* λ *is a sequence of words of type* ш. *The sequence* λ *converges to the point* P *(converges in itself) in the space* $\mathscr{C}^U$ *if and only if for each* $m \geqq 1$ *the sequence* ψ *defined by the formula*

$$\forall n\,(\psi(n) \simeq \Phi(U \square m \square \lambda(n)))$$

converges to the point $\Phi(U \square m \square P)$ *(converges in itself) in the space* C_m.

This property follows directly from the definition of $\mathfrak{N}$.

For $U \,\underline{\underline{\circ}}\, \bar{0}$ the space $\mathscr{C}^{\bar{0}}$ coincides with the space $\mathscr{C}$, so that $\mathscr{C}^{\bar{0}}$ is complete. Now we shall prove the following theorem:

2.3.3. *Suppose that* U *is a word of type* e *such that* $\exists l \forall m (m \geqq l \supset \{U\}(m) \neq 0)$. *Then the space* $\mathscr{C}^U$ *is not complete (not even almost complete).*

Proof. For simplicity we consider the case when $\forall m(\{U\}(m) \neq 0)$. For the general case the proof is carried through analogously.

We construct a sequence λ of words of type ш such that

$$\forall nx\,((x \leqslant 0 \supset \underline{\lambda(n)}(x) = 0) \,\&\, (0 \leqslant x \leqslant 2^{-n} \supset \underline{\lambda(n)}(x) = 2^n \cdot x) \\ \&\, (x \geqslant 2^{-n} \supset \underline{\lambda(n)}(x) = 1))).$$

It is easy to see that

$$\forall mn\,(\Phi(U \square m \square \lambda(n)) = J(\{U\}(m) \square \lambda(n))). \tag{5}$$

We consider a sequence φ defined by the formula

$$\forall n\,(\varphi(n) \simeq J(1 \square \lambda(n))). \tag{6}$$

Obviously

$$\forall nx\,((x \leqslant 0 \supset \underbrace{\varphi(n)}(x) = 0) \,\&\, \left(0 \leqslant x \leqslant 2^{-n} \supset \underbrace{\varphi(n)}(x) = 2^{n} \cdot \frac{x^2}{2}\right) \\ \&\, (x \geqslant 2^{-n} \supset \underbrace{\varphi(n)}(x) = x - 2^{-(n+1)})).$$

Therefore we find that

$$\forall mjln\,(l, n \geqslant j \supset \mathcal{N}(m \,\square\, \varphi(l) - \varphi(n)) < 2^{-j}).$$

This means that for each $m \geqq 1$ the sequence φ converges in itself in C_m. It is possible to verify that φ converges in C_m to a point Q having the property

$$(7) \qquad \forall x\,((x \leqslant 0 \supset \underbrace{Q}(x) = 0) \,\&\, (x \geqslant 0 \supset \underbrace{Q}(x) = x)).$$

From the fact that φ converges in itself in C_m, and from the hypothesis $\forall m(\{U\}(m) \geqq 1)$, it follows that for each $m \geqq 1$ the sequence ψ_m defined by the formula

$$\forall n\,(\psi_m(n) \simeq \Phi(U \,\square\, m \,\square\, \lambda(n))),$$

converges in itself in C_m. Hence it follows that the sequence λ converges in itself in the space $\mathscr{C}^U$.

Now we suppose that $\mathscr{C}^U$ is complete (or almost complete). Then one can realize a word P of type ш for which λ converges to P in $\mathscr{C}^U$. Hence it follows that the sequence ψ_1 ($\psi_1(n) \simeq \Phi(U \,\square\, 1 \,\square\, \lambda(n))$) converges to $\Phi(U \,\square\, 1 \,\square\, P)$ in $\mathscr{C}^U$. On the other hand, turning our attention to (5) and (6), from the fact that φ converges to Q in C_1 we deduce that the sequence ψ_1 converges to $J(k-1 \,\square\, Q)$ in C_1 (here $k \leftrightharpoons \{U\}(1)$). Thus we find that $\Phi(U \,\square\, 1 \,\square\, P) \underset{1}{=} J(k \,\square\, P) \underset{1}{=} J(k-1 \,\square\, Q)$. Hence it follows that P is the derivative of Q in the segment $-1\triangle 1$. But it is clear from (7) that Q cannot have a derivative which is a word of type ш in the segment $-1\triangle 1$. We have obtained a contradiction. Thus the space $\mathscr{C}^U$ cannot be complete (almost complete), as we were required to prove.

REMARK 1. Suppose that U is a word of type e. By definition the enumerable norm $\mathfrak{N}_U$ depends on the choice of the sequence of points X. In spite of this, we show that the topology in $\mathscr{C}^U$ does not in essence depend on that choice. Suppose that Y is some other sequence of pairwise disjoint duplexes lying in the segment $-1\triangle 1$. We write $y_i \leftrightharpoons Y(i)$. In the same way as in the construction of the algorithm Φ, we may construct an algorithm Φ^1 having the following property: for any natural number m and any word $ш_1$ of type ш, $\Phi^1(U \,\square\, m \,\square\, ш_1)$ is the antiderivative of order $\{U\}(m)$ of the word $ш_1$; and

$$\forall i\,(0 \leqslant i \leqslant \{U\}(m) - 1 \supset \underbrace{\Phi^1(U \,\square\, m \,\square\, ш_1)}(y_i) = 0).$$

Then we construct an algorithm $\mathfrak{N}_U^1$ as follows:

$$\forall m ш_1\,(\mathfrak{N}_U^1(m \,\square\, ш_1) \simeq \mathcal{N}(m \,\square\, \Phi^1(U \,\square\, m \,\square\, ш_1))).$$

Our objective is to prove that *the locally convex topologies defined by the enumerable norms* $\mathfrak{N}_U$ *and* $\mathfrak{N}_U^1$ *are equivalent to one another.* To this end we prove the following formulas:

$$\forall m i \exists j \forall \text{ш}_1 (\mathfrak{N}_U^1 (m \square \text{ш}_1) < 2^{-j} \supset \mathfrak{N}_U (m \square \text{ш}_1) < 2^{-i}). \tag{8}$$

$$\forall m i \exists j \forall \text{ш}_1 (\mathfrak{N}_U (m \square \text{ш}_1) < 2^{-j} \supset \mathfrak{N}_U^1 (m \square \text{ш}_1) < 2^{-i}). \tag{9}$$

These formulas are proved analogously. We shall prove for example formula (8). We note that, for any natural number $m \geqq 1$ and any word ш_1 of type ш, $\Phi(U \square m \square \text{ш}_1)$ and $\Phi^1(U \square m \square \text{ш}_1)$ are antiderivatives of order $k \leftrightharpoons \{U\}(m)$ of the word ш_1. Accordingly, from property c) of 2.2.2 it follows that

$$\Phi (U \square m \square \text{ш}_1) = \mathfrak{A} (k \square \Phi^1 (U \square m \square \text{ш}_1))$$
$$= \Phi^1 (U \square m \square \text{ш}_1) + H(\mathfrak{A}_1 (k \square \Phi^1 (U \square m \square \text{ш}_1))).$$

Suppose given m and i. We choose $j = i + 1 + \alpha(m \square k)$, where α is the algorithm constructed in §2.2. Then, if $\mathfrak{N}_U^1(m \square \text{ш}_1) < 2^{-j}$, we have

$$\mathcal{N} (m \square \Phi^1 (U \square m \square \text{ш}_1)) < 2^{-(i+1)}$$

and

$$\mathcal{N} (m \square H(\mathfrak{A}_1 (k \square \Phi^1 (U \square m \square \text{ш}_1)))) < 2^{-j} \cdot 2^{\alpha(m \square k)} = 2^{-(i+1)}.$$

Thus we find that $\mathfrak{N}_U(m \square \text{ш}_1) < 2^{-i}$, and formula (8) is proved.

REMARK 2. Suppose that F is a Mikusiński sequence. Then one can realize a word U of type e and, for any $m \geqq 1$, a sequence φ of words of type ю, satisfying the following conditions:

a) $\forall n (\mathrm{dg}(\varphi(n)) \prec \cdot k)$, where $k \leftrightharpoons \{U\}(m)$;

b) the sequence G defined by the formula

$$\forall n (G (n) \simeq J (k \square F (n)) + H (\varphi (n)))$$

converges in itself in the space C_m.

In view of Lemma 2.2.4 we can realize a sequence φ which satisfies conditions a) and b) if and only if the sequence χ defined by the formula

$$\forall n (\chi (n) \simeq \mathfrak{A} (k \square J (k \square F (n))))$$

converges in itself in C_m.

Thus, if F is a Mikusiński sequence, we can realize a word U of type e such that for each $m \geqq 1$ the sequence χ defined by the formula

$$\forall n (\chi (n) \simeq \Phi (U \square m \square F (n)))$$

converges in itself in the space C_m (i.e. such that F is a fundamental sequence in the space $\mathscr{C}^U$).

Hence we conclude that *the sequence* F *of words of type* ш *is a Mikusiński sequence if and only if one can realize a word* U *of type* e *for which* F *is a fundamental sequence in the space* $\mathscr{C}^U$.

§3. Constructive generalized functions

3.1. Consider the following list:

$$(1) \qquad Ч_0,\ A,\ Ч_4,\ \mathfrak{P}^{\text{ш}},\ \mathfrak{E}^{\text{ш}},\ \underset{\text{ш}}{+},\ \underset{\text{ш}}{\cdot},\ \mathfrak{D}^{\text{ш}},\ \mathfrak{N}$$

Here the notation is the same as in §2.3. This list defines an A-family of enumerably normed spaces. The spaces of this family differ from one another only in the enumerable norms defined in them. We note that for any word U of type e the space $\mathscr{L}^U$ is the space with index U of this family.

We now consider the alphabets $Ч_4 \cup \{\odot\}$ and $Ч_4 \cup \{\odot, \tau\}$. We denote their indices by ψ and d respectively:

$$A_\psi = Ч_+ \cup \{\odot\}, \quad A_d = A_\psi \cup \{\tau\}.$$

The alphabet A_ψ is intended for the construction of F-constructs and FR-constructs in the spaces of the family (1), and A_d for the formation of constructive generalized functions.

Suppose that $\mathfrak{J}$ is the set of all natural numbers and $\mathfrak{Q}$ is a three-parameter formula defined as follows:

$$\mathfrak{Q} \leftrightharpoons \forall m\, (1 \leqslant m \leqslant i \supset \mathfrak{N}(e_1 \square m \square \text{ш}_1) \leqslant 2^{-i}).$$

The parameters of $\mathfrak{Q}$ are the variables e_1, i and ш_1. We write

$$\mathfrak{Q}^U \leftrightharpoons F^{e_1}_{U} \llcorner \mathfrak{Q} \lrcorner,$$

where U is any object term. Obviously, for any word U of type e the list $(Ч_0, \mathfrak{J}, \mathfrak{Q}^U)$ defines a locally convex topology for $\mathscr{L}^U$.

On the basis of the results of §4.4, Chapter I, we construct the family of F-completions and the family of FR-completions of the spaces of the family (1), as follows.

First we construct formulas $\mathfrak{P}_\nabla$ and $\mathfrak{P}_D$ satisfying the following: a) they are two-parameter formulas, whose parameters are variables e_1 of type e and $t_{\psi,1}$ of type t_ψ; b) for any word U of type e, $\mathfrak{P}^U_\nabla$ defines the set of all F-constructs,[8] and $\mathfrak{P}^U_D$ the set of all FR-constructs in the space $\mathscr{L}^U$. We adjoin these formulas to the list of formulas Δ of the language Я^Δ_Σ and denote the pregeneric letters assigned to the formulas $\mathfrak{P}_\nabla$ and $\mathfrak{P}_D$ by ∂ and d respectively. Suppose that U is a word of type e and P a word in A_ψ. By definition P is a word of type ∂^U if and only if it is a word of type ш or a word of type $Q\odot$, where Q is the transcription of some fundamental sequence in the space $\mathscr{L}^U$; P is a word of type d^U if and only if it is a word of type ш or a word of type $Q \odot R$, where Q is the transcription of some fundamental sequence in $\mathscr{L}^U$ and R is the transcription of the regulator of convergence in itself of the sequence $\langle Q \rangle_4$ in $\mathscr{L}^U$.

---

[8] $\mathfrak{P}^U_\nabla \leftrightharpoons F^{e_1}_U \llcorner \mathfrak{P}_\nabla \lrcorner$; $\mathfrak{P}^U_D \leftrightharpoons F^{e_1}_U \llcorner \mathfrak{P}_D \lrcorner$.

Now we may construct algorithms $\mathfrak{N}^\partial$ and $\mathfrak{N}^d$ satisfying the following conditions:

a) For any word U of type e, $(\mathfrak{N}^\partial \uparrow U\square)$ is an algorithm of type $(\text{н}\,\partial^U \rightarrow \text{в})$, and $(\mathfrak{N}^d \uparrow U\square)$ is an algorithm of the type $(\text{н}\,d^U \rightarrow \text{д})$.

b) Suppose that U is a word of type e, m a natural number $\geqq 1$, P a word of type ∂^U and Q a word of type d^U. Then the F-number $\mathfrak{N}^\partial(U\square m\square P)$ is the limit in the space of F-numbers of the sequence χ defined by the formula

$$\forall n (\chi(n) \simeq \overline{\text{prin}} (\mathfrak{N}(U\square m\square P_{\llcorner\psi}(n)))),$$

and the FR-number $\mathfrak{N}^d(U\square m\square Q)$ is the limit in the space of FR-numbers of the sequence ζ defined by the formula

$$\forall n (\zeta(n) \simeq \mathfrak{N}(U\square m\square Q_{\llcorner\psi}(n)))$$

(see §4.4 of Chapter I). In what follows we shall denote $(\mathfrak{N}^\partial \uparrow U\square)$ and $(\mathfrak{N}^d \uparrow U\square)$ by $\mathfrak{N}^\partial_U$ and $\mathfrak{N}^d_U$ respectively.

Using the general method of construction of addition and multiplication operators in the F-completion and the FR-completion (see §3.2, Chapter I), it is possible to construct algorithms $\underset{\nabla}{+}$, $\underset{\nabla}{\cdot}$, $\underset{D}{+}$ and $\underset{D}{\cdot}$ such that for any word U of type e:

a) $(\underset{\nabla}{+} \uparrow U\square)$ is the algorithm for the addition of words of type ∂^U;

b) $(\underset{\nabla}{\cdot} \uparrow U\square)$ is the algorithm for multiplication of words of type ∂^U by real duplexes;

c) $(\underset{D}{+} \uparrow U\square)$ is the algorithm for adding words of type d^U;

d) $(\underset{D}{\cdot} \uparrow U\square)$ is the algorithm of multiplication of words of type d^U by real duplexes.

In what follows, in place of the notation $(\underset{\nabla}{+} \uparrow U\square)$, $(\underset{\nabla}{\cdot} \uparrow U\square)$, $(\underset{D}{+} \uparrow U\square)$, and $(\underset{\nabla}{\cdot} \uparrow U\square)$, we shall write respectively $\underset{\nabla,U}{+}$, $\underset{\nabla,U}{\cdot}$, $\underset{D,U}{+}$ and $\underset{D,U}{\cdot}$.

We also introduce the formulas

$$\mathfrak{E}_\nabla \leftrightharpoons \forall m (\mathfrak{N}^\partial (\text{e}_1\square m\square t_{\psi,1} - t_{\psi,2}) = 0);$$

$$\mathfrak{E}_D \leftrightharpoons \forall m (\mathfrak{N}^d (\text{e}_1\square m\square t_{\psi,1} - t_{\psi,2}) = 0)$$

and the notation

$$\mathfrak{E}^U_\nabla \leftrightharpoons F^{\text{e}_1}_{U} \llcorner \mathfrak{E}_\nabla \lrcorner; \qquad \mathfrak{E}^U_D \leftrightharpoons F^{\text{e}_1}_{U} \llcorner \mathfrak{E}_D \lrcorner;$$

$$(P \underset{\nabla,U}{=} Q) \leftrightharpoons F^{\text{e}_1,\, t_{\psi,1};\, t_{\psi,2}}_{U,\, P,\, Q} \llcorner \mathfrak{E}_\nabla \lrcorner; \qquad (P \underset{D,U}{=} Q) \leftrightharpoons F^{\text{e}_1,\, t_{\psi,1},\, t_{\psi,2}}_{U,\, P,\, Q} \llcorner \mathfrak{E}_D \lrcorner,$$

where U, P, and Q are object terms. For each word U of type e the formula $\mathfrak{E}^U_\nabla$ defines the equality relation between words of type ∂^U, and the formula $\mathfrak{E}^U_D$ defines the equality relation between words of type d^U.

Thus, by definition, we have

$$\forall \text{e}_1 \partial^{\text{e}_1}_1 \partial^{\text{e}_1}_2 (\partial^{\text{e}_1}_1 \underset{\nabla,\text{e}_1}{=} \partial^{\text{e}_1}_2 \equiv \forall m (\mathfrak{N}^\partial (\text{e}_1\square m\square \partial^{\text{e}_1}_1 - \partial^{\text{e}_1}_2) = 0));$$

$$\forall e_1 d_1^{e_1} d_2^{e_1} (d_1^{e_1} \underset{D,\, e_1}{=} d_2^{e_1} \equiv \forall m\ (\mathfrak{N}^d\ (e_1 \square m \square d_1^{e_1} - d_2^{e_1}) = 0)).$$

Suppose that U is some word of type e. Then the list

$$\text{(2)} \qquad A_\psi,\ \mathfrak{P}_\nabla^U,\ \mathfrak{E}_\nabla^U,\ \underset{\nabla,\, U}{+},\ \underset{\nabla,\, U}{\cdot},\ \mathfrak{O}^{ш},\ \mathfrak{N}_U^\partial$$

is the standard F-completion of the space $\mathscr{L}^U$, and the list

$$\text{(3)} \qquad A_\psi,\ \mathfrak{P}_D^U,\ \mathfrak{E}_D^U,\ \underset{D,\, U}{+},\ \underset{D,\, U}{\cdot},\mathfrak{O}^{ш},\ \mathfrak{N}_U^d$$

is the standard FR-completion of $\mathscr{L}^U$. We denote the space (2) by $\mathscr{D}^U$, and the space (3) by $\mathfrak{D}^U$. $\mathscr{D}^U$ is an almost complete enumerably seminormed space, and $\mathfrak{D}^U$ is an enumerably normed space.

We shall call the points of $\mathscr{D}^U$ *F-generalized functions of order* U, and the points of $\mathfrak{D}^U$ *FR-generalized functions of order* U.

We construct formulas $\mathfrak{Q}_\nabla$ and $\mathfrak{Q}_D$ as follows:

$$\mathfrak{Q}_\nabla \leftrightharpoons \forall m\, (1 \leqslant m \leqslant i \supset \mathfrak{N}^\partial\, (e_1 \square m \square t_{\psi,1}) \leqslant 2^{-i});$$

$$\mathfrak{Q}_D \leftrightharpoons \forall m\, (1 \leqslant m \leqslant i \supset \mathfrak{N}^d\, (e_1 \square m \square t_{\psi,1}) \leqslant 2^{-i}).$$

The parameters of these formulas are the variables e_1, i, and $t_{\psi,1}$. We write

$$\mathfrak{Q}_\nabla^U \leftrightharpoons F_U^{e_1} \llcorner \mathfrak{Q}_\nabla \lrcorner; \qquad \mathfrak{Q}_D^U \leftrightharpoons F_U^{e_1} \llcorner \mathfrak{Q}_D \lrcorner,$$

where U is any object term. For each word U of type e the list $(\mathrm{Ч}_0, \mathfrak{J}, \mathfrak{Q}_\nabla^U)$ defines a locally convex topology for the space $\mathscr{D}^U$, and the list $(\mathrm{Ч}_0, \mathfrak{J}, \mathfrak{Q}_D^U)$ defines a locally convex topology for $\mathfrak{D}^U$ (here $\mathfrak{J}$ is the set of all natural numbers).

It follows from Theorem 4.2.1 of Chapter I that $\mathscr{L}^U$ is a metrizable space. Accordingly, in view of Theorem 4.1.2 of [16] we get

3.1.1. *For any word U of type* ш, *it is not possible to have an algorithm which would transform each point of the space $\mathscr{D}^U$ into a point of the space $\mathfrak{D}^U$ which is equal to it.*

Remark 1. For any word U of type e, each word of type ш is an F-generalized function, and also an FR-generalized function, of order U.

Remark 2 (continuation of Remark 2 in §2.3). A sequence F of words of type ш is a Mikusiński sequence if and only if a word U of type e is realizable such that $\mathcal{E}F$, $\mathrm{Ч}_4^{\text{ша}}\mathfrak{Z} \odot$ is an F-generalized function of order U.

Now we construct algorithms φ and χ as follows:

$$\forall e_1 m d_1^{e_1} n\, (\varphi\, (e_1 \square m \square d_1^{e_1} \square n) \simeq \Phi\, (e_1 \square m \square \underline{d_1^{e_1}}_{\psi}\, (n)));$$

$$\forall e_1 m d_1^{e_1} i\, (\chi\, (e_1 \square m \square d_1^{e_1} \square i) \simeq \overset{\longrightarrow\psi}{d_1^{e_1}}\, (\max\, (m \square i))),$$

Φ being the algorithm defined in §2.3. For any word U of type e, any natural number $m \geqq 1$ and any word P of type d^U, $(\varphi \uparrow U \square m \square P \square)$ is a fundamental sequence in the space C_m and $(\chi \uparrow U \square m \square P \square)$ is its regulator of convergence in itself (in C_m) (see 2.3.2 and the Remark in §4.4 of Chapter I). Suppose that $\mathscr{L}$ is an algorithm such that, for each m, $(\mathscr{L} \uparrow m \,\square)$ is an algorithm for passing to the limit in C_m. We construct an algorithm $\mathfrak{F}$ such that

$$\forall e_1 m d_1^{e_1} (\mathfrak{F}(e_1 \square m \square d_1^{e_1}) \simeq \mathscr{L}(m \square \mathcal{E}(\varphi \uparrow e_1 \square m \square d_1^{e_1} \square) 3 \odot \mathcal{E}(\chi \uparrow e_1 \square m \square d_1^{e_1} \square) 3)).$$

For any word U of type e, any natural number $m \geqq 1$, and any word P of type d^U, $\mathfrak{F}(U \square m \square P)$ is the limit in G_m of the sequence $(\varphi \uparrow U \square m \square P \square)$.

3.1.2. *The algorithm* $\mathfrak{F}$ *has the following properties*:

$\mathfrak{F}1$. $\forall e_1 m d_1^{e_1} (\mathfrak{F}(e_1 \square m \square d_1^{e_1}) \in \text{ш})$.

$\mathfrak{F}2$. $\forall e_1 m d_1^{e_1} i (0 \leqslant i < \{e_1\}(m) \supset \mathfrak{F}(e_1 \square m \square d_1^{e_1})(x_i) = 0)$.

$\mathfrak{F}3$. *For any word U of type* e *and any natural number $m \geqq 1$, $(\mathfrak{F} \uparrow U \square m \square)$ is a linear mapping of the space $\mathfrak{D}^U$ into the space C_m.*

$\mathfrak{F}4$. $\forall e_1 m d_1^{e_1} (\mathfrak{N}^d (e_1 \square m \square d_1^{e_1}) \simeq \mathscr{N}(m \square \mathfrak{F}(e_1 \square m \square d_1^{e_1})))$.

$\mathfrak{F}5$. $\forall e_1 d_1^{e_1} m n (\{e_1\}(m) \leqslant \{e_1\}(n) \supset \mathfrak{F}(e_1 \square n \square d_1^{e_1}) \underset{l}{=} \mathfrak{A}(\{e_1\}(n) \square J(\{e_1\}(n) - \{e_1\}(m) \square \mathfrak{F}(e_1 \square m \square d_1^{e_1}))))$,

where $l \leftrightharpoons \min(m \square n)$. (*In* $\mathfrak{F}5$, *n denotes a variable for positive natural numbers.*)

Properties $\mathfrak{F}1-\mathfrak{F}4$ follow from the definition of $\mathfrak{F}$ and from 2.3.1. Property $\mathfrak{F}5$ is proved as follows. Suppose given a word U of type e, a word P of type d^U, and natural numbers $m, n \geqq 1$. Put $p \leftrightharpoons \{U\}(m)$, $q \leftrightharpoons \{U\}(n)$. Suppose that $p \leqq q$. Put $k = q - p$. Write $Q \leftrightharpoons \mathfrak{F}(U \square m \square P)$, $R \leftrightharpoons \mathfrak{F}(U \square n \square P)$. We need to prove that $R \underset{l}{=} \mathfrak{A}(q \square J(k \square Q))$, where $l \leftrightharpoons \min(m \square n)$. Suppose that φ and ψ are sequences of words of type ш, defined as follows:

$$\forall i (\varphi(i) \simeq \Phi(U \square m \square \underline{P}_{\psi}(i)));$$
$$\forall i (\psi(i) \simeq \Phi(U \square n \square \underline{P}_{\psi}(i))).$$

We shall prove that

$$\forall i (\psi(i) \underset{\text{ш}}{=} \mathfrak{A}(q \square J(k \square \varphi(i)))). \tag{4}$$

Indeed, suppose given a number i. By definition we have

$$\varphi(i) \underset{\text{ш}}{=} J(p \square \underline{P}_{\psi}(i)) + H(\mathfrak{A}_1(p \square J(p \square \underline{P}_{\psi}(i)))),$$

where $\mathfrak{A}_1(p \square J(p \square \underline{P}_{\psi}(i)))$ is a word of type ю and

$$\mathrm{dg}(\mathfrak{A}_1(p \square J(p \square \underline{P}_{\psi}(i))) \prec \cdot p.$$

Therefore it is possible to construct a word $ю_1$ of type ю such that $\mathrm{dg}(ю_1) \prec \cdot q$ and

$$J(k \square \varphi(i)) \underset{\text{ш}}{=} J(q \square \underline{P}_{\psi}(i)) + H(ю_1). \tag{5}$$

In view of property c) of 2.2.2, from (5) we get

$$\mathfrak{A}(p\,\square\,J(k\,\square\,\varphi(i))) \underset{\text{ш}}{=} \mathfrak{A}(q\,\square\,J(q\,\square\,\underline{P}_{\psi}(i))) \underset{\text{ш}}{=} \psi(i).$$

Formula (4) is proved. By the definition of the algorithm $\mathfrak{F}$, the sequence φ converges to Q in C_m, and so in C_l. Therefore the sequence λ defined by the formula

$$\forall i\,(\lambda(i) \simeq J(k\,\square\,\varphi(i))),$$

converges in C_l to $J(k\,\square\,Q)$. Now in view of Lemma 2.2.3 of (4) we deduce that the sequence ψ converges in C_l to $\mathfrak{A}(q\,\square\,J(k\,\square\,Q))$. On the other hand, by definition, the sequence ψ converges in C_n (and so in C_l) to R. Therefore we find that $R \underset{l}{=} \mathfrak{A}(q\,\square\,J(k\,\square\,Q))$, which was required to be proved.

3.2. We introduce an order relation $\mathfrak{X}$ among words of type e as follows:

$$\mathfrak{X}(\mathrm{e}_1, \mathrm{e}_2) \leftrightharpoons \forall m\,(\{\mathrm{e}_1\}(m) \leqslant \{\mathrm{e}_2\}(m)).$$

In place of the notation $\mathfrak{X}(\mathrm{e}_1, \mathrm{e}_2)$, we shall use the notation $\mathrm{e}_1 \prec \mathrm{e}_2$. It is easy to see that the list $(\mathrm{A}, \mathfrak{X})$ defines a directed set.

It is possible to construct algorithms $\underset{\mathrm{e}}{+}$ and $\underset{\mathrm{e}}{\dot{-}}$ of type (ee $\to$ e) such that

$$\forall \mathrm{e}_1\mathrm{e}_2 m\,(\{\underset{\mathrm{e}}{+}(\mathrm{e}_1\,\square\,\mathrm{e}_2)\}(m) = \{\mathrm{e}_1\}(m) + \{\mathrm{e}_2\}(m));$$

$$\forall \mathrm{e}_1\mathrm{e}_2 m\,(\{\underset{\mathrm{e}}{\dot{-}}(\mathrm{e}_1\,\square\,\mathrm{e}_2)\}(m) = \{\mathrm{e}_1\}(m) \dot{-} \{\mathrm{e}_2\}(m)).$$

Here we recall that for any natural numbers k and l we have $k \dot{-} l = \max(k - l\,\square\,0)$.

In what follows, instead of the notation $\underset{\mathrm{e}}{+}(U\,\square\,V)$ and $\underset{\mathrm{e}}{\dot{-}}(U\,\square\,V)$, we shall use the short notation $(U + V)$ and $(U \dot{-} V)$.

Now we shall construct a continuous linear mapping of the space $\mathfrak{D}^U$ into $\mathfrak{D}^V$ for arbitrary words U and V of type e such that $U \prec V$. First of all we note that

(6) $$\forall mk\text{ш}_1\,(\mathscr{N}(m\,\square\,J(k\,\square\,\text{ш}_1)) \underset{\cdot}{\leqslant} m^k \cdot \mathscr{N}(m\,\square\,\text{ш}_1)),$$

where $\mathscr{N}$ is an enumerable norm in $\mathscr{L}$. Denote by η an algorithm of the type (HH $\to$ H) defined by the formula

(7) $$\forall mk\,(\eta(m\,\square\,k) \simeq \mu j\,(m^k \underset{\cdot}{\leqslant} 2^j)).$$

$$\gamma(U\,\square\,V\,\square\,P\,\square\,m) \overset{+}{\in} c_0; \qquad \delta(U\,\square\,V\,\square\,P) \overset{+}{\in} c_0;$$

(8) $$\forall i\,(\langle\gamma(U\,\square\,V\,\square\,P\,\square\,m)\rangle_0(i)$$

$$\simeq \overline{P}^{\psi}(\max(m\,\square\,i + \eta(m\,\square\,\{V\}(m) - \{U\}(m)))));$$

(9) $$\forall i\,(\langle\delta(U\,\square\,V\,\square\,P)\rangle_0(i)$$

$$\simeq \max_{1 \leqslant m \leqslant i} \langle\Gamma(\{V\}(m)\,\square\,m\,\square\,\gamma(U\,\square\,V\,\square\,P\,\square\,m))\rangle_0(i)),$$

where Γ is the algorithm defined in Remark 1 to Lemma 2.2.3.

Now we shall construct an algorithm ω such that, for any words U and V of type e with $U \prec V$ and any word P of type d^U,

$$\omega(U\square V\square P) \simeq \begin{cases} P, \text{ if the sign } \odot \text{ does not enter into } P \\ \quad (\text{i.e. } P \text{ is a word of type ш}), \\ \overline{\mathrm{lef}}_{\downarrow}(P) \odot \delta(U\square V\square P), \text{ if the sign } \odot \text{ does enter into } P. \end{cases}$$

We shall prove that

$$(10) \qquad \forall e_1 e_2 d_1^{e_1}(e_1 \prec e_2 \supset \omega(e_1\square e_2\square d_1^{e_1}) \in d^{e_2}).$$

Suppose that U and V are words of type e such that $U \prec V$, and that P is a word of type d^U. If the sign $\odot$ does not enter into P, i.e. P is a word of type ш, then evidently $P \in d^V$. Now we suppose that the sign $\odot$ does enter into P. We write $\lambda \leftrightharpoons \overline{P_{\downarrow}}$ and $\mu \leftrightharpoons \overline{P^{\downarrow}}$. Suppose that $m \geqq 1$ is a natural number. Put $h \leftrightharpoons \{U\}(m)$ and $l \leftrightharpoons \{V\}(m)$. By hypothesis $h \leq l$. Denote by ζ, ψ and χ sequences of words of type ш, defined as follows:

$$\forall n(\zeta(n) \simeq \Phi(U\square m\square\lambda(n)));$$

$$\forall n(\psi(n) \simeq J(l-h\square\zeta(n)));$$

$$\forall n(\chi(n) \simeq \Phi(V\square m\square\lambda(n))).$$

Since P is a word of type d^U, we have

$$\forall inp(n, p \geqslant \mu(i) \supset \mathfrak{Q}^U(i, \lambda(n) - \lambda(p))),$$

i.e.

$$\forall inpj(n, p \geqslant \mu(i) \,\&\, j \leqslant i \supset \mathcal{N}(j\square\zeta(n) - \zeta(p)) \leqslant 2^{-i}).$$

Hence from (7) and (8) we find that

$$(11)\ \forall inp(n, p \geqslant \langle\gamma(U\square V\square P\square m)\rangle_0(i) \supset \mathcal{N}(m\square\psi(n) - \psi(p)) \leqslant 2^{-i}).$$

Suppose that n is a natural number. By the definition of the algorithm Φ we have

$$\zeta(n) \underset{\text{ш}}{=} J(h\square\lambda(n)) + H(\mathfrak{A}_1(h\square J(h\square\lambda(n)))),$$

while $\mathrm{dg}(\mathfrak{A}_1(h\square J(h\square\lambda(n)))) \prec h$. Therefore we can realize a word ю$_1$ of type ю such that dg(ю$_1$) $\prec \cdot\, l$ and

$$J(l-h\square\zeta(n)) \underset{\text{ш}}{=} J(l\square\lambda(n)) + H(\text{ю}_1).$$

Hence, in view of property c) in 2.2.2, we get

$$\mathfrak{A}(l\square J(l-h\square\zeta(n))) \underset{\text{ш}}{=} \mathfrak{A}(l\square J(l\square\lambda(n))) \underset{\text{ш}}{=} \chi(n).$$

Thus we have proved that

$$(12) \qquad \forall n(\chi(n) = \mathfrak{A}(l\square\psi(n))).$$

Now from (11), (12), (9) and the definition of the algorithm Γ (see Remark 1 to Lemma 2.2.3) we find that

$$\forall inpm\,(n, p \geqslant \langle \delta(U \square V \square P)\rangle_0(i) \,\&\, m \leqslant i \supset \mathcal{N}(m \square \chi(n) - \chi(p)) \leqslant 2^{-}),$$

i.e.

$$\forall inp\,(n, p \geqslant \langle \delta(U \square V \square P)\rangle_0(i) \supset \mathfrak{Q}^V(i, \lambda(n) - \lambda(p))).$$

This shows that $\omega(U \square V \square P)$ belongs to $\mathfrak{Q}^V$ (recall that $\omega(U \square V \square P) = \overline{\mathrm{lef}}_{\downarrow}(P) \odot \delta(U \square V \square P)$). Formula (10) is proved.

In connection with the algorithm ω we note further the following property of the operator $\mathfrak{F}$:

3.2.1. $$\forall e_1 e_2 m d_1^{e_1}(e_1 \prec e_2 \supset \mathfrak{F}(e_2 \square m \square \omega(e_1 \square e_2 \square d_1^{e_1})) \underset{m}{=} \mathfrak{A}(\{e_2\}(m) \square J(\{e_2\}(m) - \{e_1\}(m) \square \mathfrak{F}(e_1 \square m \square d_1^{e_1})))).$$

This property is proved by the same method which we used to prove property $\mathfrak{F}4$ in 3.1.2.

Now we prove the following assertion:

(13) $$\forall e_1 e_2 d_1^{e_1} d_2^{e_1}(e_1 \prec e_2 \supset (d_1^{e_1} \underset{D, e_1}{=} d_2^{e_1} \supset \omega(e_1 \square e_2 \square d_1^{e_1}) \underset{D, e_2}{=} \omega(e_1 \square e_2 \square d_2^{e_1}))).$$

Suppose that U and V are words of type e such that $U \prec V$, and that P and Q are words of type d^U. We suppose that $P \underset{D,U}{=} Q$. Then

$$\forall m\,(\mathfrak{N}^d\,(U \square m \square P \underset{D,U}{-} Q) = 0),$$

so that from $\mathfrak{F}3$ and $\mathfrak{F}4$ it follows that $\forall m(\mathfrak{F}(U \square m \square P) \underset{m}{=} \mathfrak{F}(U \square m \square Q))$, and thus, in view of 3.2.1,

(14) $$\forall m\,(\mathfrak{F}(V \square m \square \omega(U \square V \square P)) \underset{m}{=} \mathfrak{F}(V \square m \square \omega(U \square V \square Q))).$$

Using properties $\mathfrak{F}3$ and $\mathfrak{F}4$ once again, we get

$$\forall m\,(\mathfrak{N}^d\,(V \square m \square \omega(U \square V \square P) \underset{D,V}{-} \omega(U \square V \square Q)) = 0),$$

i.e. $\omega(U \square V \square P) \underset{D,V}{=} \omega(U \square V \square Q)$, as was required to be proved.

From the definition of the algorithm ω it is obvious that

(15) $$\forall e_1 e_2 d_1^{e_1} n\,(e_1 \prec e_2 \supset \underline{\omega(e_1 \square e_2 \square d_1^{e_1})}_{\psi}\,(n) \underset{\text{III} \to \psi}{=} d_1^{e_1}(n)).$$

Using formula (15), it is not difficult to prove the following properties:

(16) $$\forall e_1 e_2 d_1^{e_1} d_2^{e_1}(e_1 \prec e_2 \supset \omega(e_1 \square e_2 \square d_1^{e_1} \underset{D, e_1}{+} d_2^{e_1}) \underset{D, e_2}{=} \omega(e_1 \square e_2 \square d_1^{e_1}) \underset{D, e_2}{+} \omega(e_1 \square e_2 \square d_2^{e_1}));$$

(17) $$\forall e_1 e_2 x\, d_1^{e_1}(e_1 \prec e_2 \supset \omega(e_1 \square e_2 \square x \underset{D, e_1}{\cdot} d_1^{e_1}) \underset{D, e_2}{=} x \underset{D, e_2}{\cdot} \omega(e_1 \square e_2 \square d_1^{e_1}));$$

$$\forall e_1 e_2 e_3 d_1^{e_1}(e_1 \prec e_2 \,\&\, e_2 \prec e_3 \supset \omega(e_1 \square e_3 \square d_1^{e_1}) \underset{D, e_3}{=} \omega(e_2 \square e_3 \square \omega(e_1 \square e_2 \square d_1^{e_1}))).$$

(18)

Properties (10), (13), (16) and (17) show that, for any words U and V of type e with $U \prec V$, $(\omega \uparrow U\Box\, V\Box)$ *is a linear operator from the space* $\mathfrak{D}^U$ *into the space* $\mathfrak{D}^V$.

We prove further the following property:

(19) $$\forall \mathrm{e}_1\mathrm{e}_2 d_1^{\mathrm{e}_1}(\mathrm{e}_1 \prec \mathrm{e}_2 \,\&\, \omega(\mathrm{e}_1 \Box\, \mathrm{e}_2 \Box\, d_1^{\mathrm{e}_1}) \underset{D,\,\mathrm{e}_2}{=} \mathfrak{O}^{\text{ш}} \supset d_1^{\mathrm{e}_1} \underset{D,\,\mathrm{e}_1}{=} \mathfrak{O}^{\text{ш}}).$$

Suppose that U and V are words of type e such that $U \prec V$, and P is a word of type d^U. We write $\lambda \leftrightharpoons \underline{P}_\psi$. Suppose that $m \geqq 1$ is a natural number, and put $h \leftrightharpoons \{U\}(m)$ and $l \leftrightharpoons \{V\}(m)$. By definition, $h \leqq l$. Suppose that ζ, ψ and χ are sequences of words of type ш, defined the same way as in the proof of formula (10). We further introduce the notation

$$Q \leftrightharpoons \mathfrak{F}(U\Box\, m\,\Box\, P);$$

$$R \leftrightharpoons \mathfrak{F}(V\Box\, m\,\Box\, \omega(U\Box\, V\Box\, P)).$$

It is obvious from (15) that

$$\forall n\,(\underline{\omega(U\Box\, V\Box\, P)}_\psi\,(n) \underset{\text{ш}}{=} \lambda(n)),$$

so that

$$\forall n\,(\chi(n) \underset{\text{ш}}{=} \Phi(V\Box\, m\,\Box\, \underline{\omega(U\Box\, V\Box\, P)}_\psi\,(n))).$$

By the definition of the algorithm $\mathfrak{F}$ the sequence ζ converges to Q in C_m, and the sequence χ converges to R in C_m. From the definition of the sequence ψ it follows that ψ converges to $J(l-h\,\Box\, Q)$ in C_m. Hence in view of Lemma 2.2.3 and formula (12) we deduce that ψ converges to $\mathfrak{A}(l\,\Box\, J(l-h\,\Box\, Q))$ in C_m. Thus we find that

(20) $$\mathfrak{A}(l\,\Box\, J(l-h\,\Box\, Q)) \underset{m}{=} R.$$

Now we suppose that $\omega(U\Box\, V\Box\, P) \underset{D,V}{=} \mathfrak{O}^{\text{ш}}$. Then $R \underset{m}{=} \mathfrak{O}^{\text{ш}}$. It is clear from (20) that a word ю_1 of type ю is realizable such that

$$\mathrm{dg}(\text{ю}_1) \prec\cdot\, l \,\&\, J(l-h\,\Box\, Q) \underset{m}{=} H(\text{ю}_1).$$

Hence a word ю_2 of type ю is realizable such that

(21) $$\mathrm{dg}(\text{ю}_2) \prec\cdot\, h \,\&\, Q \underset{m}{=} H(\text{ю}_2).$$

From the definition of the word Q and from property $\mathfrak{F}2$ it follows that

(22) $$\forall i\,(0 \leqslant i \leqslant h-1 \supset \underline{Q}\,(x_i) = 0).$$

From (21) and (22) we get $Q \underset{m}{=} \mathfrak{O}^{\text{ш}}$ (since in the segment $-m\triangle m$ the function $\underline{Q}$ is identically equal to a polynomial of degree less than h, and vanishes at h distinct points). Thus we have proved that if $\omega(U\Box\, V\Box\, P) \underset{D,V}{=} \mathfrak{O}^{\text{ш}}$, then

$$\forall m\,(\mathfrak{F}(U\square m\square P) \underset{m}{=} \mathfrak{O}^{\text{III}}),$$

i.e. $P \underset{D,U}{=} \mathfrak{O}^{\text{III}}$. Formula (19) is proved.

The following property follows from (19) and (16):

(23) $\forall \mathrm{e}_1\mathrm{e}_2 d_1^{\mathrm{e}_1} d_2^{\mathrm{e}_2} (\mathrm{e}_1 \prec \mathrm{e}_2 \supset (\omega(\mathrm{e}_1\square \mathrm{e}_2\square d_1^{\mathrm{e}_1}) \underset{D,\mathrm{e}_2}{=} \omega(\mathrm{e}_1\square \mathrm{e}_2\square d_2^{\mathrm{e}_1}) \supset d_1^{\mathrm{e}_1} \underset{D,\mathrm{e}_1}{=} d_2^{\mathrm{e}_1}))$.

Now we prove that, for any words U and V of type e, if $U \prec V$, then $(\omega \upharpoonright V\square U\square)$ is a continuous mapping of the space $\mathfrak{D}^U$ into the space $\mathfrak{D}^V$. In other words, we shall prove the following formula:

(24) $\forall \mathrm{e}_1\mathrm{e}_2 (\mathrm{e}_1 \prec \mathrm{e}_2 \supset \forall i \exists j \forall d_1^{\mathrm{e}_1} (\mathfrak{Q}_D^{\mathrm{e}_1}(j, d_1^{\mathrm{e}_1}) \supset \mathfrak{Q}_D^{\mathrm{e}_2}(i, \omega(\mathrm{e}_1\square \mathrm{e}_2\square d_1^{\mathrm{e}_1}))))$.

Suppose that U and V are words of type e such that $U \prec V$. For the proof of (24) it suffices for us to prove that

$$\forall m i \exists j \forall d_1^U (\mathfrak{N}^d(U\square m\square d_1^U) \leqslant 2^{-j} \supset \mathfrak{N}^d(V\square m\square \omega(U\square V\square d_1^U)) \leqslant 2^{-i}).$$

(25)

Suppose given numbers m and i with $m \geqq 1$. We write

$$h \leftrightharpoons \{V\}(m);\ k \leftrightharpoons \{V\}(m) - \{U\}(m);\ j \leftrightharpoons i+1+\eta(m\square k)+\alpha(m\square h),$$

where η is an algorithm having property (7), and α the algorithm defined at the beginning of §2.2. Suppose that P is a word of type d^U such that $\mathfrak{N}^d(U\square m\square P) \leqq 2^{-j}$. Then

$$\mathscr{N}(m\square \mathfrak{F}(U\square m\square P)) \leqq 2^{-j}.$$

Hence we find from (6) and (7) that

(26) $\mathscr{N}(m\square J(k\square \mathfrak{F}(U\square m\square P))) \leqslant 2^{-j}\cdot 2^{\eta(m\square k)} = 2^{-(i+1+\alpha(m\square h))}$.

Therefore, in view of the property of the algorithm α, we find that

(27) $\mathscr{N}(m\square H(\mathfrak{A}_1(h\square J(k\square \mathfrak{F}(U\square m\square P))))) \leqslant 2^{-(i+1)}$.

From (26), (27) and the definition of the algorithm $\mathfrak{A}$ it follows that

$$\mathscr{N}(m\square \mathfrak{A}(h\square J(k\square \mathfrak{F}(U\square m\square P)))) \leqslant 2^{-i}.$$

From this formula and from 3.2.1 and 3.1.2 ($\mathfrak{F}$4), we get

$$\mathfrak{N}^d(V\square m\square \omega(U\square V\square P)) \leqslant 2^{-i}.$$

Thus formula (25), and also (24), are proved.

Now consider the following list:

(28) $\qquad \text{Ч}_0,\ A,\ \mathfrak{X},\ \mathrm{A}_\Phi,\ \mathfrak{P}_D,\ \mathfrak{E}_D,\ \underset{D}{+},\ \underset{D}{\cdot},\ \mathfrak{O}^{\text{III}},\ \mathfrak{N}^d,\ \omega.$

From (10), (13), (16)–(18), (23) and (24) we conclude the following:

3.2.2. *The list* (29) *defines an inductive spectrum of enumerably normed spaces.*

For the definition of inductive spectrum see §3.1 in Chapter II.

In the proofs of formulas (10), (13), (16) – (18), (23) and (24), we have simultaneously proved that *for any words U and V of type* e, *if $U \prec V$, then $\mathscr{D}^U$ is a linear subspace of $\mathscr{D}^V$ and the identity mapping of $\mathscr{D}^U$ into $\mathscr{D}^V$ is one-to-one and continuous*. Therefore, if one denotes by $\bar{\omega}$ the algorithm defined by the formula

$$\forall \mathrm{e}_1 \mathrm{e}_2 t_{\Psi,1} (\bar{\omega} (\mathrm{e}_1 \square \mathrm{e}_2 \square t_{\Psi,1}) = t_{\Psi,1}),$$

one finds that the following proposition holds:

3.2.3. *The list*

(29) $$\text{Ч}_0,\ A,\ \mathfrak{X},\ \mathrm{A}_{\Psi},\ \mathfrak{P}_{\nabla},\ \mathfrak{E}_{\nabla},\ \underset{\nabla}{+},\ \underset{\nabla}{\cdot},\ \mathfrak{D}^{\mathrm{III}},\ \mathfrak{N}^{\partial},\ \bar{\omega}$$

defines an expanding inductive spectrum of enumerably seminormed spaces.

3.3. Using the general method of construction of the inductive limit, presented in §3 of Chapter II, we may construct the inductive limit (i.e. the *FR*-inductive limit) of the inductive spectrum (28), and the *F*-inductive limit and the *FR*-inductive limit of the expanding inductive spectrum (29). These inductive limits are locally convex spaces. We suppose that the inductive limit of the spectrum (28) is defined by the list

(30) $$\mathrm{A}_d,\ \mathfrak{P}^d,\ \mathfrak{E}^d,\ \underset{d}{+},\ \underset{d}{\cdot},\ \mathfrak{D}^d,\ \text{Ч}_0,\ \mathfrak{J}^d,\ \mathfrak{Q}^d,$$

the *F*-inductive limit of the spectrum (29) by the list

(31) $$\mathrm{A}_c,\ \mathfrak{P}^{\partial},\ \mathfrak{E}^{\partial},\ \underset{\partial}{+}.\ \underset{\partial}{\cdot},\ \mathfrak{D}^{\partial},\ \text{Ч}_0,\ \mathfrak{J}^{\partial},\ \mathfrak{Q}^{\partial},$$

and the *FR*-inductive limit of the spectrum (29) by the list

(32) $$\mathrm{A}_d,\ \overline{\mathfrak{P}}^{\partial},\ \overline{\mathfrak{E}}^{\partial},\ \overline{\underset{\partial}{+}},\ \overline{\underset{\partial}{\cdot}},\ \overline{\mathfrak{D}}^{\partial},\ \text{Ч}_0,\ \overline{\mathfrak{J}}^{\partial},\ \overline{\mathfrak{Q}}^{\partial}.$$

The spaces (30), (31) and (32) will be denoted respectively by $\mathfrak{D}$, $\mathscr{D}$ and $\overline{\mathscr{D}}$. The objects entering into the lists (30) – (32) are constructed by means of the general method of §3, Chapter II. Here we shall not describe them in detail. We adjoin the formulas $\mathfrak{P}^d$, $\mathfrak{P}^{\partial}$ and $\overline{\mathfrak{P}}^{\partial}$ to the list Σ of formulas of our language $\mathfrak{R}^{\Delta}_{\Sigma}$, and denote the restricted generic letters assigned to these formulas by d, ∂, and $\bar{\partial}$ respectively. Suppose that P is some word in A_d. P is a *word of type d* if it has the form $U \tau Q$, where U is a word of type e and Q a word of type d^U (an *FR*-generalized function of order U); P is a word of type ∂ if we can realize a word U of type e such that P is a word of type ∂^U (an *F*-generalized function of order U); P is a word of type $\bar{\partial}$ if it has the form $U \tau Q$, where U is a word of type e and Q is a word of type ∂^U (an *F*-generalized function of order U). Words of type d will be called *FR-generalized functions*, and words of type ∂ will be called *F-generalized functions*.

The set of words of type d, i.e. the set $\mathfrak{P}^d$, *may be defined by a normal formula.*

In what follows we shall use algorithms Or and Fd defined by the following conditions. For any word P in the alphabet A_d, if the sign τ does not enter into P, then Or is not applicable to P and $\mathrm{Fd}(P) \dot{=} P$. If P has the form $U\tau Q$, where U and Q do not contain the sign τ, then $\mathrm{Or}(P) \dot{=} Q$. If F is an FR-generalized function, then we call $\mathrm{Or}(P)$ the *formal order*, and $\mathrm{Fd}(P)$ the *fundamental FR-generalized function of P.*[9]

Suppose that P and Q are object terms. In place of the relations $\mathfrak{E}^d(P,Q)$, $\mathfrak{E}^\partial(P,Q)$, and $\overline{\mathfrak{E}}^\partial(P,Q)$, denoting the equality relation in $\mathfrak{D}$, $\mathscr{D}$ and $\overline{\mathscr{D}}$, respectively, we will write respectively $(P \underset{d}{=} Q)$, $(P \underset{\partial}{=} Q)$ and $(P \overset{\partial}{=} Q)$. It is possible to construct an algorithm Max of the type (ee → e) such that

$$\forall e_1 e_2 m\, (\{\mathrm{Max}\,(e_1 \square e_2)\}\,(m) \simeq \max\,(\{e_1\}\,(m) \square \{e_2\}\,(m))).$$

By the definition of the equality relation in the F-inductive limit and in the FR-inductive limit we have

$$\forall \partial_1 \partial_2\, ((\partial_1 \underset{\partial}{=} \partial_2) \equiv \exists e_1\, (\partial_1 \in \mathfrak{P}_\nabla^{e_1} \,\&\, \partial_2 \in \mathfrak{P}_\nabla^{e_1} \,\&\, (\partial_1 \underset{\nabla,\, e_1}{=} \partial_2))); \tag{33}$$

$$\forall d_1 d_2\, ((d_1 \underset{d}{=} d_2) \equiv (\omega\,(\mathrm{Or}\,(d_1) \square U \square \mathrm{Fd}\,(d_1)) \underset{D,\,U}{=} \omega\,(\mathrm{Or}\,(d_2) \square U \square \mathrm{Fd}\,(d_2)))), \tag{34}$$

where $U \leftrightharpoons \mathrm{Max}(\mathrm{Or}(d_1) \square \mathrm{Or}(d_2))$. *The equality relation between words of type d is defined by a normal formula.*

Suppose that d_1 and d_2 are words of type d. From formulas (4) and (5) of §3.1 of Chapter II we also obtain the following relations:

$$(d_1 \underset{d}{=} d_2) \equiv \exists e_1\, ((\mathrm{Or}\,(d_1),\ \mathrm{Or}\,(d_2) \prec e_1) \,\&\, \omega\,(\mathrm{Or}\,(d_1) \square e_1 \square \mathrm{Fd}\,(d_1)) \underset{D,e_1}{=} \omega\,(\mathrm{Or}\,(d_2) \square e_1 \square \mathrm{Fd}\,(d_2))); \tag{35}$$

$$(d_1 \underset{d}{=} d_2) \equiv \forall e_1\, ((\mathrm{Or}\,(d_1),\ \mathrm{Or}\,(d_2) \prec e_1) \supset \omega\,(\mathrm{Or}\,(d_1) \square e_1 \square \mathrm{Fd}\,(d_1)) \underset{D,e_1}{=} \omega\,(\mathrm{Or}\,(d_2) \square e_1 \square \mathrm{Fd}\,(d_2))). \tag{36}$$

It is possible to construct an algorithm $\overline{\mathrm{prin}}_\psi$ having the following property:

$$\overline{\mathrm{prin}}_\psi(t_{\psi,1}) = \begin{cases} t_{\psi,1} \text{ if the sign } \odot \text{ does not enter into } t_{\psi,1}; \\ \overline{\mathrm{lef}}(t_{\psi,1})\, \odot, \text{ if } \odot \text{ enters exactly once into } t_{\psi,1}. \end{cases}$$

Suppose that d_1 is a word of type d, $\bar{\partial}_1$ a word of type $\bar{\partial}$, and ∂_1 a word of type ∂. Then $\mathrm{Or}(d_1)\,\tau\,\overline{\mathrm{prin}}_\psi(\mathrm{Fd}(d_1))$ is a word of type $\bar{\partial}$, $\overline{\mathrm{prin}}_\psi(\mathrm{Fd}(d_1))$ is a word of type ∂ and $\mathrm{Fd}(\bar{\partial}_1)$ is a word of type ∂. We write

$$(d_1 \overset{\partial}{=} \bar{\partial}_1) \leftrightharpoons (\mathrm{Or}\,(d_1)\,\tau\,\overline{\mathrm{prin}}_\psi\,(\mathrm{Fd}\,(d_1)) \overset{\partial}{=} \bar{\partial}_1);$$

[9] For the case of words of type d, Or and Fd play the roles of the algorithms $И_1$ and $И_2$ in §3.1, Chapter II.

$$(d_1 \underset{\partial}{=} \partial_1) \leftrightharpoons (\overline{\mathbf{prin}_\Phi}\,(\mathrm{Fd}\,(d_1)) \underset{\partial}{=} \partial_1);$$
$$(\bar{\partial}_1 \underset{\partial}{=} \partial_1) \leftrightharpoons (\mathrm{Fd}\,(\bar{\partial}_1) \underset{\partial}{=} \partial_1).$$

Obviously, for any word ∂_1 of type ∂, we can realize words $\bar{\partial}_1$ of type $\bar{\partial}$ and d_1 of type d such that $\bar{\partial}_1 \underset{\partial}{=} \partial_1$ and $d_1 \underset{\partial}{=} \partial_1$; for any word $\bar{\partial}_1$ of type $\bar{\partial}$ we can realize a word d_1 of type d such that $d_1 \overset{\partial}{=} \partial_1$. On the other hand, we may prove the following assertions:

3.3.1. *There cannot exist an algorithm which constructs relative to each point ∂_1 of the space $\mathscr{D}$ a point d_1 of the space $\mathfrak{D}$ equal to it (i.e. such that $d_1 \underset{\partial}{=} \partial_1$).*

3.3.2. *There cannot exist an algorithm which constructs relative to each point $\bar{\partial}_1$ of the space $\overline{\mathscr{D}}$ a point d_1 of the space $\mathfrak{D}$ equal to it (i.e. such that $d_1 \overset{\partial}{=} \bar{\partial}_1$).*

3.3.3. *There cannot exist an algorithm which constructs relative to each point ∂_1 of the space $\mathscr{D}$ a word U of type e such that $U\tau\,\partial_1$ is a point of $\overline{\mathscr{D}}$.*

Assertion 3.3.1 is a consequence of 3.3.2. At the basis of the proofs of Theorems 3.3.2 and 3.3.3 there lies the theorem on the existence of an algorithm $\mathscr{A}$ of type $(\mathrm{н} \dot{\rightarrow} \mathrm{н})$ whose applicability problem is undecidable. Here we present the proof of 3.3.3.

Proof of 3.3.3. Suppose that $\mathscr{A}$ is an algorithm of type $(\mathrm{н} \dot{\rightarrow} \mathrm{н})$ whose applicability problem is undecidable, and G an algorithm applicable to each pair of natural numbers of the form $k \square n$ and such that $G(k \square n) = \Lambda$ if and only if the process of application of the algorithm $\mathscr{A}$ to the number k terminates in no more than n steps.

It is possible to construct a recursive algorithm φ of the type $(\mathrm{нн} \rightarrow \mathrm{ш})$, satisfying the following conditions:

1) $\forall n(\varphi(0 \square n) \mathrel{\bar{=}} \mathfrak{O}^{\mathrm{ш}})$, where $\mathfrak{O}^{\mathrm{ш}}$ is the zero element of type $\mathrm{ш}$.

2) For any number n, the function $\varphi(1 \square n)$ is defined as follows:

$$\underbrace{\varphi(1 \square n)}(x) = \begin{cases} 0, \text{ if } |x-1| > 2^{-n}; \\ 1, \text{ if } x = 1; \\ \text{is linear in } 1 - 2^{-n}\Delta 1 \text{ and } 1\Delta 1 + 2^{-n}. \end{cases}$$

3) For any natural numbers k and n with $k \geqq 1$, and any duplex x,

$$\underbrace{\varphi(k+1 \square n)}(x) = 2^{\text{Б}}(\underbrace{\varphi(k \square n)}(x + 2^{-\text{Б}}) - \underbrace{\varphi(k \square n)}(x)),$$

where

$$\text{Б} \leftrightharpoons \max(n \square \overbrace{\varphi(k \square n)}(n \square n)).$$

From the definition it easily follows that if $n \geqq k$, then

$$\forall x(x \leqslant 0 \supset \underbrace{\varphi(k \square n)}(x) = 0),$$

so that

$$\forall hknx\,(n \geqslant k \,\&\, x \leqslant 0 \supset J\,(h \,\square\, \varphi\,(k \,\square\, n)).(x) = 0).$$

Hence we find that

(37) $\forall e_1\, knm\,(n \geqslant k \supset \Phi\,(e_1 \,\square\, m \,\square\, \varphi\,(k \,\square\, n)) \underset{m}{=} J\,(\{e_1\}\,(m) \,\square\, \varphi\,(k \,\square\, n))).$

It is possible to construct an algorithm C of the type (НН $\to$ Н) such that

(38) $$\forall hknm\,(n \geqslant m \supset \mathcal{N}\,(m \,\square\, J\,(h + 1 \,\square\, \varphi\,(k + 1 \,\square\, n)) \dot{-} J\,(h \,\square\, \varphi\,(k \,\square\, n))) < 2^{-n+C\,(m\square h)}.$$

This will be proved for a more general case at formula (5) of §4.1.

From (37) and (38) we find that

(39) $$\forall e_1\, knm\,(n \geqslant k,\, m \supset \mathcal{N}\,(m \,\square\, \Phi\,(e_1 + \bar{1} \,\square\, m \,\square\, \varphi\,(k + 1 \,\square\, n)) - \Phi\,(e_1 \,\square\, m \,\square\, \varphi\,(k \,\square\, n))) < 2^{-n+C(m\square\{e_1\}\,(m))}).$$

Now we construct an algorithm Ψ as follows:

$$\forall k\,(\Psi\,(k) \simeq \mathcal{E}\,(\varphi \uparrow k \square),\ Ч_4^{\text{ша}}\, З\, \odot).$$

Let U be a word of type e. The following assertions are evident from (39).

a) If $\Psi(k) \in \mathscr{D}^U$, then $\Psi(k+1) \in \mathscr{D}^{U+\bar{1}}$.

b) If $\Psi(k) \notin \mathscr{D}^U$, then $\Psi(k+1) \notin \mathscr{D}^{U+\bar{1}}$.

Evidently $\Psi(0) \in \mathscr{D}^{\bar{0}}$, $\Psi(1) \in \mathscr{D}^{\bar{1}}$, and for any word U of type e, if $\{U\}\,(2) = 0$, then $\Psi(1) \in \mathscr{D}^U$; hence from properties a) and b), using induction on k, we find that

(i) $\forall k(\Psi(k) \in \mathscr{D}^{\bar{k}})$, so that $\forall k(\Psi(k) \in \mathscr{D})$;

(ii) for any word U of type e, if $\{U\}\,(2) \leq k$, then $\Psi(k+1) \notin \mathscr{D}^U$.

Now we construct an algorithm $\mathfrak{B}$ such that, for any k and n,

$$\mathfrak{B}\,(k \,\square\, n) \simeq \begin{cases} \mathfrak{D}^{\text{ш}}, & \text{if } G\,(k \,\square\, n) \neq \Lambda; \\ \varphi\,(h \,\square\, n), & \text{if } G\,(k \,\square\, n) = \Lambda, \text{ where } h \leftrightharpoons \mu j\,(G\,(k \,\square\, j) = \Lambda). \end{cases}$$

Then we construct an algorithm $\mathfrak{C}$ as follows:

$$\forall k\,(\mathfrak{C}\,(k) \simeq \mathcal{E}\,(\mathfrak{B} \uparrow k \square),\ Ч_4^{\text{ша}}\, З\, \odot).$$

For any natural number k, if $\urcorner\, !\mathscr{A}(k)$, then $\mathfrak{C}\,(k) \underset{\partial}{=} \Psi(0)$, i.e. $\mathfrak{C}\,(k) \in \mathscr{D}^{\bar{0}}$; if $!\mathscr{A}(k)$ and $h = \mu j(G(k \,\square\, j) = \Lambda)$, then $\mathfrak{C}\,(k) \underset{\partial}{=} \Psi(h)$, i.e. $\mathfrak{C}\,(k) \in \mathscr{D}^{\bar{h}}$.

Suppose that P is a word in A_Ψ. We will say that P is a *quasipoint of the space* $\mathscr{D}$ if it has the form $Q\odot$, where Q is the transcription (relation to $Ч_4^{\text{ша}}$) of some algorithm of the type (Н $\to$ Ш), and if it is false that it is not a point of the space $\mathscr{D}$. From what has been said above we obtain the following assertion.

1*) For any natural number k, $\mathfrak{C}\,(k)$ is a quasipoint of the space $\mathscr{D}$.

Now we shall prove the following proposition:

2*) There cannot exist an algorithm which transforms each quasipoint P of the space $\mathscr{D}$ into a word U of type e such that $\urcorner\urcorner\,(P \in \mathscr{D}^U)$.

We suppose that an algorithm $\mathfrak{H}$ with the property indicated in 2*) is realizable. We define an algorithm γ as follows:

$$\forall k\,(\gamma(k) \simeq \{\mathfrak{H}(\mathfrak{C}(k))\}(2)).$$

We shall prove that for any natural number k, if $!\,\mathscr{A}(k)$, then the process of application of $\mathscr{A}$ to the number k terminates after no more than $\gamma(k)$ steps. Indeed, suppose that k is a natural number such that $!\,\mathscr{A}(k)$ and the process of application of $\mathscr{A}$ to the number k stops after exactly h steps (i.e. $h = \mu j(G(k \square j) \doteq \Lambda)$). Then $\mathfrak{C}(k) = \Psi(h)$. Write $U \leftrightharpoons \mathfrak{H}(\mathfrak{C}(k))$ and $l \leftrightharpoons \{U\}(2)$. From the hypothesis on $\mathfrak{H}$ we get $\urcorner\urcorner(\mathfrak{C}(k) \in \mathscr{D}^U)$, so that $\urcorner\urcorner(\Psi(h) \in \mathscr{D}^U)$. In view of property (ii) it is not possible that $l < h$, so that $h \leqq l$, i.e. $h \leqq \gamma(k)$, which we wished to prove. From the fact that an algorithm γ with the property indicated above is realizable, it follows that the problem of application of the algorithm $\mathscr{A}$ is decidable. We have obtained a contradiction. Thus proposition 2*) is proved.

Now we prove Theorem 3.3.3. We suppose that an algorithm $\mathfrak{H}$ with the property indicated in 3.3.3 is realizable. Then for any point Q of the space $\mathscr{D}$, we have $!\,\mathfrak{H}(Q)$, $\mathfrak{H}(Q)$ is a word of type e, and $Q \in \mathscr{D}^U$, where $U \leftrightharpoons \mathfrak{H}(Q)$. Therefore $\urcorner\urcorner\,!\,\mathfrak{H}(P)$ for any quasipoint P of the space $\mathscr{D}$, so that $!\,\mathfrak{H}(P)$; it is false that $\mathfrak{H}(P)$ must be a word of type e (since the set of words of type e is defined by a normal formula); finally, $\urcorner\urcorner(P \in \mathscr{D}^U)$, where $U \leftrightharpoons \mathfrak{H}(P)$. Thus $\mathfrak{H}$ has the property indicated in proposition 2*). This is impossible. Theorem 3.3.3 is proved.

Remark 1. We recall that $\mathscr{D}$ is the F-inductive limit, and $\overline{\mathscr{D}}$ the FR-inductive limit of the expanding inductive spectrum (29). Hence in proving Theorem 3.3.3 we have also proved Theorem 3.2.1 of Chapter II.

Remark 2 (Continuation of Remark 2 in §3.1). A sequence λ of words of type ш is a Mikusiński sequence if and only if $\varepsilon\lambda$, $Ч_4^{\text{sa}}3 \odot$ is an F-generalized function.

Remark 3. In what follows we shall be particularly interested in FR-generalized functions, which we shall call *constructive generalized functions* or simply *generalized functions*. The definitions of the concept of constructive generalized function in the present monograph and in [32] are essentially equivalent to one another. However, in a certain sense the analysis of the topological structure of a Mikusiński sequence provides us, in the present work, with the possibility of making certain contractions in the formulation of the definitions of constructive generalized functions and the relations among them. Thus, for example, in the make-up of a constructive generalized function, we do not have the transcription of a "double" sequence of words of type ю here, as was done in [32]. In the present work the equality relation between constructive generalized functions is defined by the equality relation in some appropriate enumerably normed space.

In the majority of concrete cases of the application of generalized functions, we frequently have to do with generalized functions of a definite order. For generalized functions as defined here, this means that in each such case we will be dealing with the points of some definite complete enumerably normed space.

3.4. Now we introduce a new algorithm F, defined as follows:

$$\forall e_1 m d_1 (F(e_1 \Box m \Box d_1) \simeq \mathfrak{F}(e_1 \Box m \Box \omega(\mathrm{Or}(d_1) \Box e_1 \Box \mathrm{Fd}(d_1)))),$$

where $\mathfrak{F}$ is the algorithm defined in §3.1.

It is obvious from the definition that

$$\forall m d_1 (F(\mathrm{Or}(d_1) \Box m \Box d_1) \underset{m}{=} \mathfrak{F}(\mathrm{Or}(d_1) \Box m \Box \mathrm{Fd}(d_1))).$$

3.4.1. *The algorithm F has the following properties*:

F 1. $\forall e_1 m d_1 (\mathrm{Or}(d_1) \prec e_1 \supset F(e_1 \Box m \Box d_1) \overset{+}{\in} \text{ш});$

F 2. $\forall e_1 m d_1 i (\mathrm{Or}(d_1) \prec e_1 \,\&\, 0 \leqslant i < \{e_1\}(m) \supset \underline{F(e_1 \Box m \Box d_1)}(x_i) = 0);$

F 3. $\forall e_1 d_1 d_2 m (\mathrm{Or}(d_1), \mathrm{Or}(d_2) \prec e_1 \supset F(e_1 \Box m \Box d_1 + d_2)$

$$\underset{m}{=} F(e_1 \Box m \Box d_1) + F(e_1 \Box m \Box d_2));$$

$$\forall e_1 d_1 x m (\mathrm{Or}(d_1) \prec e_1 \supset F(e_1 \Box m \Box x \cdot d_1) \underset{m}{=} x \cdot F(e_1 \Box m \Box d_1));$$

F 4. $\forall e_1 d_1 m n (\mathrm{Or}(d_1) \prec e_1 \,\&\, \{e_1\}(m) \leqslant \{e_1\}(n) \supset F(e_1 \Box n \Box d_1)$

$$\underset{l}{=} \mathfrak{A}(\{e_1\}(n) \Box J(\{e_1\}(n) - \{e_1\}(m) \Box F(e_1 \Box m \Box d_1)))),$$

where $l \leftrightharpoons \min(m \Box n)$;

F 5. $\forall e_1 e_2 d_1 m (\mathrm{Or}(d_1) \prec e_1 \prec e_2 \supset F(e_2 \Box m \Box d_1)$

$$\underset{m}{=} \mathfrak{A}(\{e_2\}(m) \Box J(\{e_2\}(m) - \{e_1\}(m) \Box F(e_1 \Box m \Box d_1))));$$

F 6. *Suppose that* d_1 *and* d_2 *are arbitrary words of type* d, *and set* $U \leftrightharpoons \mathrm{Max}(\mathrm{Or}(d_1) \Box \mathrm{Or}(d_2))$. *Then* $d_1 \underset{d}{=} d_2$ *if and only if one of the following conditions is satisfied*:

$$\forall m (F(U \Box m \Box d_1) \underset{m}{=} F(U \Box m \Box d_2));$$

$$\exists e_1 (U \prec e_1 \,\&\, \forall m (F(e_1 \Box m \Box d_1) \underset{m}{=} F(e_1 \Box m \Box d_2)));$$

$$\forall e_1 m (U \prec e_1 \supset F(e_1 \Box m \Box d_1) \underset{m}{=} F(e_1 \Box m \Box d_2)).$$

Properties F 1 – F 5 follow from the properties 3.1.2 and 3.2.1 of the algorithm $\mathfrak{F}$ and the properties of the algorithm ω. Property F 6 follows from formulas (34) – (36). The algorithm F will be used frequently in what follows.

3.5. Suppose that U is a word of type e and that ψ is a sequence of *FR*-generalized functions of order U (i.e. an algorithm of the type $(\text{н} \to d^U)$. A word X of type d^U is the limit of the sequence ψ in the space $\mathfrak{D}^U$ if

(40) $$\forall i \exists l \forall n (n \geqslant l \supset (\mathfrak{Q}_D^U(i, X \underset{D,U}{-} \psi(n)))).$$

$$\forall m j \exists l \forall n (n \geqslant l \supset \mathfrak{N}^d (U \square m \square X \underset{D,\,U}{-} \psi(n)) \leqslant 2^{-j}). \tag{41}$$

If condition (40) (or (41)) is satisfied, we will write

$$(X \ \underline{\lim}_{\mathfrak{D}^U} \psi) \text{ and } (\psi(n) \xrightarrow[n\to\infty]{\mathfrak{D}^U} X).$$

For a sequence of constructive generalized functions (i.e. words of type d), besides the concept of convergence corresponding to the locally convex topology in $\mathfrak{D}$, we introduce further the concept of l-convergence, as follows. Suppose that φ is a sequence of words of type d and Z a word of type d. We will say that the sequence φ *l-converges to Z* (or Z is the *l-limit* of the sequence φ), and write

$$(\varphi(n) \xrightarrow[n\to\infty]{l} Z) \quad (\text{or} \quad Z \ \underline{\lim}_{l} \varphi),$$

if we can realize a word U of type e, a word X of type d^U and a sequence ψ of words of type d^U such that

$$(Z \underset{d}{=} U \tau X) \,\&\, \forall n (\varphi(n) \underset{d}{=} U \tau \psi(n)) \,\&\, (X \ \underline{\lim}_{\mathfrak{D}^U} \psi). \tag{42}$$

Suppose that U is a word of type e, ψ a sequence of words of type d^U, and X a word of type D^U. Let $Z \leftrightharpoons U \tau X$, and suppose that φ is a sequence of words of type d defined by $\forall n(\varphi(n) \simeq U \tau \psi(n))$. Then from 3.1.1, Chapter II, it follows that if the sequence ψ converges to X in $\mathfrak{D}^U$, then φ converges to Z in $\mathfrak{D}$. Therefore we obtain the following.

3.5.1. *For any sequence φ of constructive generalized functions, if φ l-converges to a constructive generalized function Z, then φ converges to Z in the space $\mathfrak{D}$ as well.*

Remark. The concept of l-convergence is the constructive analogue of the concept of convergence of generalized functions introduced by Mikusiński and Sikorski in [48]. Within the limits of classical mathematics,[10] the concept of convergence of generalized functions introduced in [48] may be formulated as convergence in the almost inductive limit of the family of limit spaces (see the definitions of limit space and almost inductive limit of such spaces in, for example, [**40, 46, 56**]). In more detail, within the limits of classical mathematics it is possible to reconstruct the theory of Mikusiński-Sikorski generalized functions [48] successively, according to the following plan:

a) The initial data are the continuous functions.

b) For each sequence U of natural numbers it is possible to construct a space of generalized functions of order U. This space is complete and countably

---

[10] All the concepts mentioned below are to be understood within the limits of classical mathematics. For some of them we will not introduce precise definitions.

normed, and it is the completion of the space of continuous functions relative to an appropriate topology.

c) After this one may construct the almost inductive limit (directed) of the family of all such spaces of generalized functions. This almost inductive limit is a limit space. The points of this limit space are generalized functions in the sense of Mikusiński and Sikorski, and the convergence in this limit space coincides with convergence in their sense. This limit space is complete, as it is equivalent to the almost inductive limit of complete spaces (see [**40, 46**]).

§4. Differentiation of constructive generalized functions

4.1. In the present subsection we define the operation of differentiation over generalized functions. Suppose that k is a natural number and P a word of type ш. We write $P^{(-k)} \leftrightharpoons J(k \square P)$.

Now we construct auxiliary algorithms γ, $\mathfrak{S}$, $\mathfrak{B}$, $\mathfrak{C}_1$ and $\mathfrak{C}$, satisfying the following requirements:

1) $$\forall Xn\,(\gamma\,(X \square n) \simeq \overline{\underbrace{\mathrm{Fd}\,(X)}_{\psi}\ (n)}\,(n \square n)),$$

where X is a variable for words in A_d.

2) $$\forall d_1\,(\mathfrak{S}\,(d_1) \simeq \mathrm{Or}\,(d_1) + \bar{1});$$

3) For any word X in A_d, if X is a word of type ∂ or d, then $!\mathfrak{B}(X)$ and $\mathfrak{B}(X)$ is the transcription of a sequence of words of type ш having the following property:

$$\forall nx\,(\underbrace{\langle \mathfrak{B}\,(X)\rangle_4\,(n)}\,(x) \simeq 2^{\gamma(X\square n)}\cdot(\underbrace{\underbrace{\mathrm{Fd}\,(X)}_{\psi}\ (n)}\,(x + 2^{-\gamma(X\square n)}) - \underbrace{\underbrace{\mathrm{Fd}\,(X)}_{\psi}\ (n)}\,(x))).$$

4) $\mathfrak{C}_1$ is applicable to any word of the type $d_1 \square m$, where d_1 is a word of type d and m is a natural number $\geqq 1$, and transforms each word of such a form into the transcription of an algorithm of the type $(\mathrm{H} \to \mathrm{H})$ having the property

$$\forall i\,(\langle \mathfrak{C}_1\,(d_1 \square m)\rangle_0\,(i) \simeq \max\,(i + 3 + \eta\,(m \square \{\mathrm{Or}\,(d_1)\}\,(m)) \square m$$
$$\square\, \overline{\mathrm{Fd}\,(d_1)}^{\psi}\,(\max\,(m \square i + 1)))),$$

where $\eta(m \square k) \simeq \mu j (m^k \leqq 2^j)$.

5) $\mathfrak{C}$ is applicable to any word d_1 of type d and transforms each word d_1 of type d into the transcription of an algorithm of the type $(\mathrm{H} \to \mathrm{H})$ having the property

$$\forall i\,(\langle \mathfrak{C}\,(d_1)\rangle_0\,(i) \simeq \max_{1 \leqslant m \leqslant i} \langle \Gamma\,(\{\mathrm{Or}\,(d_1)\}\,(m) + 1 \square m \square \mathfrak{C}_1\,(d_1 \square m))\rangle_0\,(i)),$$

where Γ is the algorithm described in Remark 1 to Lemma 2.2.3.

Finally, we construct algorithms D, D_1 and D^* as follows:

$$\forall d_1\,(D\,(d_1) \simeq \mathfrak{S}\,(d_1)\,\tau\,\mathfrak{B}\,(d_1) \odot \mathfrak{C}\,(d_1));$$

$$\forall \partial_1 (D_1(\partial_1) \simeq \mathfrak{B}(\partial_1) \odot);$$

$$\forall e_1 d_1^{e_1} (D^*(e_1 \square d_1^{e_1}) \simeq \mathfrak{B}(e_1 \tau d_1^{e_1}) \odot \mathfrak{C}(e_1 \tau d_1^{e_1})).$$

4.1.1. *The algorithms D, D_1 and D^* have the following properties:*

a) *D is an operator of the type $(\mathbf{d} \to \mathbf{d})$.*

b) *D_1 is an operator of the type $(\partial \to \partial)$.*

c) *For any word U of type e, $(D^* \uparrow U \square)$ is an operator of the type $(\mathbf{d}^U \to \mathbf{d}^{U+\bar{1}})$.*

PROOF. We shall prove a). From the proof of a) it is not difficult to obtain the proofs for b) and c).

Suppose that d_1 is a word of type d and $d_1 \,\bar{\underline{\circ}}\, U \tau P$. Write $V \leftrightharpoons \mathfrak{S}(d)$, $Q \leftrightharpoons \mathfrak{B}(d_1)$ and $R \leftrightharpoons \mathfrak{C}(d_1)$. Obviously $V \mp U + \bar{1}$. We have to prove that the word $V \tau Q \odot R$ is a word of type d, i.e. that

(1) $$\forall i l n (l, n \geqslant \langle R \rangle (i) \supset \mathfrak{Q}^V (i, \langle Q \rangle (l) - \langle Q \rangle (n))).$$

Suppose that m is a natural number $\geqq 1$. Write $p \leftrightharpoons \{U\}(m)$, so that $\{V\}(m) = p + 1$. For any natural number n write

$$\theta_n \rightleftharpoons \underline{P}_\psi (n),\ \lambda_n \rightleftharpoons \langle Q \rangle (n),\ h_n \rightleftharpoons 2^{-\gamma(d_1 \square n)},\ B_n \rightleftharpoons \mathfrak{A}_1 (p \square \theta_n^{(-p)}),$$

where $\mathfrak{A}_1$ is an algorithm of type (н ш → ю) as defined in §2.2.

In view of condition 3), for any n and x we have

$$\underline{\lambda_n}(x) = \frac{1}{h_n}(\underline{\theta_n}(x + h_n) - \underline{\theta_n}(x)),$$

so that we get

$$\underline{\lambda_n^{(-1)}}(x) = \frac{1}{h_n}[(\underline{\theta_n^{(-1)}}(x + h_n) - \underline{\theta_n^{(-1)}}(x)) - (\underline{\theta_n^{(-1)}}(h_n) - \underline{\theta_n^{(-1)}}(0))],$$

(2) $$\underline{\theta_n}(x) - \underline{\lambda_n^{(-1)}}(x) - \underline{\theta_n}(0) = \left[\underline{\theta_n}(x) - \frac{\underline{\theta_n^{(-1)}}(x + h_n) - \underline{\theta_n^{(-1)}}(x)}{h_n}\right] - \left[\underline{\theta_n}(0) - \frac{\underline{\theta_n^{(-1)}}(h_n) - \underline{\theta_n^{(-1)}}(0)}{h_n}\right].$$

Using Corollary 2 to Theorem 2, §5 of [28], a carry-over of the theorem of Lagrange into constructive mathematics, we may prove that if $|x| \leqq n$, then

(3) $$\left|\underline{\theta_n}(x) - \frac{1}{h_n}[\underline{\theta_n^{(-1)}}(x + h_n) - \underline{\theta_n^{(-1)}}(x)]\right| < 2^{-n}.$$

From (2) and (3) it follows that for $|x| \leqq m \leqq n$ one has

(4) $$\left|\underline{\theta_n^{(-p)}}(x) - \left[\underline{\lambda_n^{(-p-1)}}(x) + \underline{\theta_n}(0) \cdot \frac{x^p}{p!}\right]\right| < 2^{-n} \cdot 2^{\eta(m \square p)+1}.$$

By the definition of the algorithm Φ we have

$$\forall n (\Phi(U \square m \square \theta_n) \simeq \theta_n^{(-p)} + H(B_n)).$$

Now we construct algorithms φ of type (н → ю) and ψ of type (н → ш) as follows:

(i) if $p>0$, then $\forall n(\varphi(n) \simeq B_n \sigma \underline{\theta_n}(0)/p!)$;

(ii) if $p=0$, then $\forall n(\varphi(n) \simeq \underline{\theta_n}(0))$;

(iii) $\forall n(\psi(n) \simeq \lambda_n^{(-p-1)} + H(\underline{\varphi(n)}))$.

In view of (4) we find for $|x| \leqq m \leqq n$:

(5) $$|\underline{\Phi(U \square m \square \theta_n)(x)} - \underline{\psi(n)}(x)| < 2^{-n} \cdot 2^{\eta(m \square p)+1}.$$

Now we prove the following formula:

(6) $$\forall i l n (l, n \geqslant \langle \mathfrak{C}_1(d_1 \square m)\rangle(i) \supset \mathcal{N}(m \square \psi(l) - \psi(n)) < 2^{-i}).$$

Suppose that $l, n \geqq \langle \mathfrak{C}_1(d_1 \square m)\rangle(i)$. Then from condition 4) and formula (5) it follows that for $|x| \leqq m$ we have

$$|\underline{\Phi(U \square m \square \theta_l)}(x) - \underline{\psi(l)}(x)| < 2^{-(i+2)},$$

$$|\underline{\Phi(U \square m \square \theta_n)}(x) - \underline{\psi(n)}(x)| < 2^{-(i+2)}$$

and

$$|\underline{\Phi(U \square m \square \theta_l)}(x) - \underline{\Phi(U \square m \square \theta_n)}(x)| < 2^{-(i+1)}.$$

From these formulas it follows that

$$\forall x(|x| \leqslant m \supset |\underline{\psi(l)}(x) - \underline{\psi(n)}(x)| < 2^{-i}).$$

Thus formula (6) is proved.

Now we prove (1). We note that $\forall n(\mathrm{dg}(\varphi(n)) \prec \cdot p+1)$, so that, from the definition of the algorithm Φ and property c) of 2.2.2, we get

(7) $$\forall n(\Phi(V \square m \square \lambda_n) \underset{\text{ш}}{=} \mathfrak{A}(p+1 \square \psi(n))).$$

Thus from (6) and from the definition of the algorithm Γ we get

(8) $$\begin{aligned} &\forall i l n (l, n \geqslant \langle \Gamma(p+1 \square m \square \mathfrak{C}_1(d_1 \square m))\rangle(i) \\ &\qquad \supset \mathcal{N}(m \square \Phi(V \square m \square \lambda_l) - \Phi(V \square m \square \lambda_n)) < 2^{-i}). \end{aligned}$$

Formula (8) is true for any $m \geqq 1$. Hence, and from condition 5), one deduces formula (1), as required.

Suppose that m is a natural number $\geqq 1$ and $p \leftrightarrows \{U\}(m)$. Denote by χ an algorithm of the type (н → ш), defined by the following formula:

$$\forall n(\chi(n) \simeq \mathfrak{A}(p+1 \square \psi(n))).$$

It is clear from (5) that the sequence ψ converges to $\mathfrak{F}(U \square m \square P)$ in C_m, so that χ converges to $\mathfrak{A}(p+1 \square \mathfrak{F}(U \square m \square P))$ in C_m (see Lemma 2.2.3). On the other hand, it follows from (7) that χ converges to $\mathfrak{F}(V \square m \square C)$ in C_m. Hence we get

$$\mathfrak{F}(V \square m \square Q) \underset{m}{=} \mathfrak{A}(p+1 \square \mathfrak{F}(U \square m \square P)).$$

Thus we have proved that

$$\forall m d_1 (\mathfrak{F} (\mathrm{Or} (D (d_1)) \square m \square \mathrm{Fd} (D (d_1))) \underset{m}{=} \mathfrak{A} (\{\mathrm{Or} (D (d_1))\} (m) \square \mathfrak{F} (\mathrm{Or} (d_1) \square m \square \mathrm{Fd} (d_1))))$$

or

$$\forall m d_1 (F (\mathrm{Or} (D (d_1)) \square m \square D (d_1)) \underset{m}{=} \mathfrak{A} (\{\mathrm{Or} (D (d_1))\} (m) \square F (\mathrm{Or} (d_1) \square m \square d_1))). \tag{9}$$

In order to prove that D is an operator of the type $(d \to d)$, we have to prove further the following formula:

$$\forall d_1 d_2 (d_1 \underset{d}{=} d_2 \supset D (d_1) \underset{d}{=} D (d_2)). \tag{10}$$

Suppose that d_1 and d_2 are words of type d. Write

$$U_1 \leftrightharpoons \mathrm{Or}(d_1), \quad U_2 \leftrightharpoons \mathrm{Or}(d_2), \quad U \leftrightharpoons \mathrm{Max}(U_1 \square U_2),$$

$$V_1 \leftrightharpoons U_1 + \bar{1}, \quad V_2 \leftrightharpoons U_2 + \bar{1}, \quad V \leftrightharpoons \mathrm{Max}(V_1 \square V_2).$$

Suppose that $d_1 \underset{d}{=} d_2$. Then in view of property F 6 (3.4.1) we have

$$\forall m (F (U \square m \square d_1) \underset{m}{=} F (U \square m \square d_2)). \tag{11}$$

Suppose that m is a natural number $\geqq 1$. Write $k \leftrightharpoons \{V\} (m)$, $k_1 \leftrightharpoons \{V_1\} (m)$ and $h \leftrightharpoons k - k_1$. From 2.2.2, 3.4.1, and (9) we get

$$\begin{aligned}
&[\mathrm{F}\,5] & F (V \square m \square D (d_1)) &\underset{m}{=} \mathfrak{A} (k \square J (h \square F (V_1 \square m \square D (d_1)))),\\
&[(9)] & &\underset{m}{=} \mathfrak{A} (k \square J (h \square \mathfrak{A} (k_1 \square F (U_1 \square m \square d_1)))),\\
&[2.2.2, \mathrm{c}] & &\underset{m}{=} \mathfrak{A} (k \square J (h \square F (U_1 \square m \square d_1))),\\
&[2.2.2, \mathrm{c}] & &\underset{m}{=} \mathfrak{A} (k \square \mathfrak{A} (k - 1 \square J (h \square F (U_1 \square m \square d_1)))),\\
&[\mathrm{F}\,5] & &\underset{m}{=} \mathfrak{A} (k \square F (U \square m \square d_1))
\end{aligned}$$

(note that $k - 1 = \{U\} (m)$ and $h = \{U\} (m) - \{U_1\} (m)$).

Analogously we may prove that

$$F (V \square m \square D (d_2)) \underset{m}{=} \mathfrak{A} (k \square F (U \square m \square d_2)).$$

From the formulas obtained above and from (11) it follows that

$$\forall m (F (V \square m \square D (d_1)) \underset{m}{=} F (V \square m \square D (d_2))),$$

i.e. $D(d_1) \underset{d}{=} D(d_2)$. Formula (10) is proved. Thus part a) of Theorem 4.1.1 is proved.

We note that from formula (9), with the aid of properties F 5 (3.4.1) and 2.2.2 c), we may prove the following formula:

$$\forall \mathrm{e}_1 m d_1 (\mathrm{Or}(d_1) \prec \mathrm{e}_1 \supset F(\mathrm{e}_1 + \bar{1} \square m \square D(d_1)) \underset{m}{=} \mathfrak{A}(\{\mathrm{e} + \bar{1}\}(m) \square F(\mathrm{e}_1 \square m \square d_1))). \tag{12}$$

From formula (12) and properties F 3 (3.4.1) and 2.2.2 a) it is easy to see that D is a linear mapping of the space $\mathfrak{D}$ into itself. One may prove also that D_1 is a linear mapping of $\mathscr{D}$ into itself and that, for each word U of type e, $(D^* \uparrow U\square)$ is a linear mapping of $\mathfrak{D}^U$ into $\mathfrak{D}^{U+\bar{1}}$.

Suppose that U is a word of type e and $V \leftrightharpoons U+\bar{1}$. We shall prove that $(D^* \uparrow U\square)$ is a continuous mapping of $\mathfrak{D}^U$ into $\mathfrak{D}^V$. To this end it suffices for us to prove the following formula:

$$\forall m i \exists j \forall d_1^U (\mathfrak{N}^d(U \square m \square d_1^U) \leqslant 2^{-j} \supset \mathfrak{N}^d(V \square m \square D^*(U \square d_1^U)) \leqslant 2^{-i}). \tag{13}$$

Suppose given m and i, with $m \geqq 1$. Choose $j = i+1+\alpha(m \square \{V\}(m))$, where α is the algorithm defined at the beginning of §2.2. Suppose that d_1^U is a word of type d^U. In view of (12) we have

$$\begin{aligned} \mathfrak{F}(V \square m \square D^*(U \square d_1^U)) &\underset{m}{=} F(V \square m\, D(U \tau d_1^U)) \\ &\underset{m}{=} \mathfrak{A}(\{V\}(m) \square F(U \square m \square U \tau d_1^U)) \\ &\underset{m}{=} \mathfrak{A}(\{V\}(m) \square \mathfrak{F}(U \square m \square d_1^U)). \end{aligned} \tag{14}$$

Suppose that $\mathfrak{N}^d(U \square m \square d_1^U) \leqq 2^{-j}$. Then

$$\mathscr{N}(m \square \mathfrak{F}(U \square m \square d_1^U)) \leqq 2^{-j}.$$

Therefore we find from (14) that

$$\mathscr{N}(m \square \mathfrak{F}(V \square m \square D^*(U \square d_1^U))) \leqq 2^{-j}$$

(see Remark 2 to 2.2.3), so that

$$\mathfrak{N}^d(V \square m \square D^*(U \square d_1^U)) \leqq 2^{-i},$$

as was required to be proved.

From what has been proved and from 3.1.3 of Chapter II it follows that D is a continuous mapping of $\mathfrak{D}$ into itself. Hence it follows also that D_1 is a continuous mapping of $\mathscr{D}$ into itself. Thus we have proved the following assertion.

4.1.2. *The operator D is a continuous linear mapping of the space $\mathfrak{D}$ into itself; the operator D_1 is a continuous linear mapping of the space $\mathscr{D}$ into itself; and for each word U of type* e *the operator $(D^* \uparrow U\square)$ is a continuous linear mapping of $\mathfrak{D}^U$ into $\mathfrak{D}^{U+\bar{1}}$.*

Suppose that d_1 is an FR-generalized function, ∂_1 an F-generalized function, U a word of type e and d_1^U an FR-generalized function of order U. Then we will call $D(d_1)$ the *derivative of the FR-generalized function* d_1, $D_1(\partial_1)$ the *derivative of the F-generalized function* ∂_1, and $D^*(U \Box d_1^U)$ the *derivative of the FR-generalized function* d_1^U *of order* U.

REMARK 1. As we know, in the paper [48] of Mikusiński and Sikorski, for the definition of the derivative of the generalized function it was necessary to replace each fundamental sequence of continuous functions by a sequence equivalent to it, whose terms are differentiable. In our definition the "fundamental sequence" defining the derivative is constructed directly from the "fundamental sequence" defining the given constructive generalized function, without any kind of replacement.

REMARK 2. In the definition of the algorithm $\mathfrak{B}$ one may replace γ by any strictly increasing sequence of natural numbers which does not depend on the given FR (or F)-generalized function X. After such a replacement, for any FR (or F)-generalized function X the word $\mathfrak{B}(X) \odot$ nevertheless remains an F-generalized function. However, the definition of the corresponding algorithm $\mathfrak{C}$ under such a substitution will be considerably more complicated.

REMARK 3. From the last part of Assertion 4.1.2 and from the definition of $\dot{l}$-convergence (§3.5) one deduces the following assertion. *Suppose that* φ *is a sequence of FR-generalized functions and* ψ *a sequence of derivatives defined by the formula*

$$\forall n\, (\psi\,(n) \simeq D\,(\varphi\,(n))).$$

If the sequence φ $\dot{l}$*-converges to the FR-generalized function* X, *then* ψ $\dot{l}$*-converges to* $D(X)$.

4.2. Now we shall construct algorithms Dr and Dr_1 satisfying the following recursive schemas:

$$\begin{cases} \mathrm{Dr}\,(0 \Box d_1) \mathrel{\dot{=}} d_1; \\ \mathrm{Dr}\,(n+1 \Box d_1) \simeq D\,(\mathrm{Dr}\,(n \Box d_1)), \end{cases} \qquad \begin{cases} \mathrm{Dr}_1\,(0 \Box \partial_1) \mathrel{\dot{=}} \partial_1; \\ \mathrm{Dr}_1\,(n+1 \Box \partial_1) \simeq D_1\,(\mathrm{Dr}_1\,(n \Box \partial_1)). \end{cases}$$

Here n is a natural number, d_1 a word of type d and ∂_1 a word of type ∂. $\mathrm{Dr}(n \Box d_1)$ will be called the *derivative of order* n *of* d_1, and $\mathrm{Dr}_1(n \Box \partial_1)$ the *derivative of order* n *of* ∂_1.

The following assertion is obvious.

4.2.1. *Every FR-generalized function* (*F-generalized function*) *has a derivative of any order, which is an FR-generalized function* (*F-generalized function*).

From the linearity and the continuity of the operators D and D_1 it follows that, for each natural number n, $(\mathrm{Dr} \uparrow n)$ is a continuous linear mapping of $\mathfrak{D}$ into itself, and $(\mathrm{Dr}_1 \uparrow n \Box)$ is a continuous linear mapping of $\mathscr{D}$ into itself.

For the operator Dr we have the following formulas:

$$\forall n d_1\, (\mathrm{Or}\,(\mathrm{Dr}\,(n \Box d_1)) = \mathrm{Or}\,(d_1) + \bar{n}); \tag{15}$$

$$(16)\qquad \forall \mathrm{e}_1 m n d_1 (\mathrm{Or}(d_1) \prec \mathrm{e}_1 \supset F(\mathrm{e}_1 + \bar{n} \square m \square \mathrm{Dr}(n \square d_1)) \underset{m}{=} \mathfrak{A}(\{\mathrm{e}_1 + \bar{n}\}(m) \square F(\mathrm{e}_1 \square m \square d_1))).$$

Formula (15) follows directly from the definition of the algorithm Dr, and (16) is proved by induction on n, on the basis of (12).

4.2.2. *Suppose that P and Q are words of type d (type ∂). If*

$$(17)\qquad \forall n Д(\underbrace{\mathrm{Fd}(P)}_{\psi}(n), \underbrace{\mathrm{Fd}(Q)}_{\psi}(n)),$$

then $P \underset{d}{=} D(Q)$ $(P \underset{\partial}{=} D_1(Q))$.

(Recall that if $ш_1$ and $ш_2$ are words of type ш, then $Д(ш_1, ш_2)$ denotes that $ш_1$ is the derivative of $ш_2$ (see §1.2).)

Proof. Suppose that P and Q are words of type d satisfying condition (17). Write $U_1 \leftrightharpoons \mathrm{Or}(P)$, $U_2 \leftrightharpoons \mathrm{Or}(Q)$ and $U \leftrightharpoons \mathrm{Max}(U_1 \square U_2 + \bar{1})$. In order to prove that $P \underset{d}{=} D(Q)$, it suffices for us to prove the following formula:

$$(18)\qquad \forall m (F(U \square m \square P) \underset{m}{=} F(U \square m \square D(Q))).$$

Suppose that m is a natural number not less than unity. Write $k \leftrightharpoons \{U_1\}(m)$, $h \leftrightharpoons \{U_2\}(m)$ and $l \leftrightharpoons \{U\}(m)$. In view of property F 5 (3.4.1) and formula (12) we have

$$(19)\qquad F(U \square m \square P) \underset{m}{=} \mathfrak{A}(l \square J(l - k \square F(U_1 \square m \square P)))$$

$$(20)\qquad F(U \square m \square D(Q)) \underset{m}{=} \mathfrak{A}(l \square F(U \dot{-} \bar{1} \square m \square Q));$$

$$(21)\qquad F(U - \bar{1} \square m \square Q) \underset{m}{=} \mathfrak{A}(l - 1 \square J(l - 1 - h \square F(U_2 \square m \square Q))).$$

By the definition of the algorithm F we have

$$(22)\qquad \Phi(U_1 \square m \square \underbrace{\mathrm{Fd}(P)}_{\psi}(n)) \xrightarrow[n \to \infty]{C_m} F(U_1 \square m \square P):$$

$$(23)\qquad \Phi(U_2 \square m \square \underbrace{\mathrm{Fd}(Q)}_{\psi}(n)) \xrightarrow[n \to \infty]{C_m} F(U_2 \square m \square Q).$$

On the basis of the definition of the algorithm Φ, we deduce from property 2.2.2 c) that, for any natural number n,

$$(24)\qquad \mathfrak{A}(l \square J(l \square \underbrace{\mathrm{Fd}(P)}_{\psi}(n))) \underset{ш}{=} \mathfrak{A}(l \square J(l - k \square \Phi(U_1 \square m \square \underbrace{\mathrm{Fd}(P)}_{\psi}(n))));$$

$$(25)\qquad \mathfrak{A}(l - 1 \square J(l - 1 \square \underbrace{\mathrm{Fd}(Q)}_{\psi}(n))) \underset{ш}{=} \mathfrak{A}(l - 1 \square J(l - 1 - h \square \Phi(U_2 \square m \square \underbrace{\mathrm{Fd}(Q)}_{\psi}(n)))).$$

It follows from (22) and (23) that

(26) $J(l - k\square\Phi(U_1\square m\square \underbrace{\mathrm{Fd}(P)}_{\psi}\ (n))) \xrightarrow[n\to\infty]{C_m} J(l-k\square F(U_1\square m\square P)),$

$$J(l-1-h\square\Phi(U_2\square m\square \underbrace{\mathrm{Fd}(Q)}_{\psi}\ (n))) \xrightarrow[n\to\infty]{C_m} J(l-1-h\square F(U_2\square m\square Q)).$$
(27)

Now from (19), (21) and (24) – (27), in view of Lemma 2.2.3, it follows that

(28) $$\mathfrak{A}(l\square J(l\square \underbrace{\mathrm{Fd}(P)}_{\psi}\ (n))) \xrightarrow[n\to\infty]{C_m} F(U\square m\square P);$$

(29) $$\mathfrak{A}(l-1\square J(l-1\square \underbrace{\mathrm{Fd}(Q)}_{\psi}\ (n))) \xrightarrow[n\to\infty]{C_m} F(U \dot{-} \bar{1}\square m\square Q).$$

It follows from (17) that for any n we can realize a word ю$_1$ of type ю such that

$$\mathrm{dg}(\text{ю}_1) \prec \cdot\, l \text{ and } J(l\square \underbrace{\mathrm{Fd}(P)}_{\psi}\ (n)) \underset{\text{ш}}{=} J(l-1\square \underbrace{\mathrm{Fd}(Q)}_{\psi}\ (n)) + H(\text{ю}_1).$$

Therefore in view of property 2.2.2 c) we have

$$\mathfrak{A}(l\square J(l\square \underbrace{\mathrm{Fd}(P)}_{\psi}\ (n))) \underset{\text{ш}}{=} \mathfrak{A}(l\square J(l-1\square \underbrace{\mathrm{Fd}(Q)}_{\psi}\ (n))),$$

so that, again in view of property 2.2.2 c),

(30) $\mathfrak{A}(l\square J(l\square \underbrace{\mathrm{Fd}(P)}_{\psi}\ (n))) \underset{\text{ш}}{=} \mathfrak{A}(l\square \mathfrak{A}(l-1\square J(l-1\square \underbrace{\mathrm{Fd}(Q)}_{\psi}\ (n)))).$

Now, using Lemma 2.2.3, from (29), (30) and (20) we find that

(31) $$\mathfrak{A}(l\square J(l\square \underbrace{\mathrm{Fd}(P)}_{\psi}\ (n))) \xrightarrow[n\to\infty]{C_m} F(U\square m\square D(Q)).$$

From (28) and (31) it follows that

$$F(U\square m\square P) \underset{m}{=} F(U\square m\square D(Q)),$$

so that formula (18) is proved.

For the case of words of type ∂ the proof is carried out analogously.

4.2.3. *Suppose that P and Q are words of type* ш *such that* Д(P,Q). *Then* $P \underset{d}{=} D_1(Q)$ *and* $\bar{0}_T P \underset{d}{=} D(\bar{0}_T Q)$.

This is a direct consequence of Theorem 4.2.2. Roughly speaking, Assertion 4.2.3 shows that, in the case when the generalized functions are almost uniformly continuous functions, the derivative in the sense of generalized functions "coincides" with the derivative in the ordinary sense.

4.2.4. *Suppose that k is a natural number and d_1 an FR-generalized function such that* $\mathrm{Or}(d_1) = \bar{k}$. *Then one can realize a word* ш$_1$ *of type* ш *such that* $d_1 \underset{d}{=} \mathrm{Dr}(k\square \bar{0}_T\, \text{ш}_1)$.

Roughly speaking, this means that every generalized function "of order

$\bar{k}$" is the derivative of order k of (the complete cipher of) some almost uniformly continuous function.

Proof. Suppose that k and d_1 satisfy the hypothesis of the theorem. Then in view of property F 4 (3.4.1) we have

$$\forall mn\,(m \leqslant n \supset F(\bar{k}\square n \square d_1) \underset{m}{=} F(\bar{k}\square m \square d_1)).$$

Hence it is possible to construct a word ш_1 of type ш such that

$$\forall m\,(\text{ш}_1 \underset{m}{=} F(\bar{k}\square m \square d_1)). \tag{32}$$

Suppose that $m \geqq 1$ is a natural number. It follows from (32) that

$$F(\bar{0}\square m \square \bar{0}\tau\text{ш}_1) \underset{m}{=} F(\bar{k}\square m \square d_1).$$

Therefore from (16) we find that

$$\begin{aligned} F(\bar{k}\square m \square \mathrm{Dr}(k\square \bar{0}\tau\text{ш}_1)) &\underset{m}{=} \mathfrak{A}(k\square F(\bar{0}\square m \square \bar{0}\tau\text{ш}_1)), \\ &\underset{m}{=} \mathfrak{A}(k\square F(\bar{k}\square m \square d_1)), \\ &\underset{m}{=} F(\bar{k}\square m \square d_1) \end{aligned} \tag{33}$$

Here the last equality follows from F 2 (3.4.1) and 2.2.2 b). From (33) and from property F 6 (3.4.1) we get $d_1 \underset{d}{=} \mathrm{Dr}(k \square \bar{0}\,\tau\, \text{ш}_1)$, as we were required to prove.

In the simplest case of differential equations with constructive generalized functions the following assertions hold. We will not give their proofs here.

4.2.5. *It is possible to formulate an algorithm which constructs, relative to each pair of the type* $k \square d_1$, *where* k *is a natural number and* d_1 *is a constructive generalized function, a constructive generalized function* d_2 *such that* $\mathrm{Dr}(k \square d_2) \underset{d}{=} d_1$.

4.2.6. *Suppose that* k *is a natural number. The constructive generalized function* d_1 *satisfies the equation* $\mathrm{Dr}(k \square d_1) \underset{d}{=} \mathfrak{D}^d$ *if and only if a word* ю_1 *of type* ю *is realizable such that* $\mathrm{dg}(\text{ю}_1) \prec \cdot\, k$ *and* $d_1 \underset{d}{=} \bar{0}\,\tau\, H(\text{ю}_1)$ *(see the definition of* H *in* §1.2).

4.3. For constructive generalized functions, besides the operations of addition, multiplication by duplexes and differentiation, one may construct other operations as well: for example, the operation of multiplication of generalized functions by complete transcriptions of infinitely differentiable functions,[11] the operation of convolution of generalized functions (in certain restricted cases), and the operation of linear substitution of a variable for

[11] The definition of this concept will be given in Chapter VI.

generalized functions. The majority of the results of the classical elementary theory of generalized functions [48] may be carried over to the constructive theory without particular difficulties. We will not dwell here in detail on these questions. In what follows we consider only the operation of linear substitution of a variable for generalized constructive functions.

Suppose that ζ is an algorithm of the type (ддн → ш), with the property

$$\forall\alpha\beta m\,(\zeta(\alpha\Box\beta\Box m) \geqslant |\alpha|\cdot m + |\beta|),$$

where α and β are variables for duplexes.

Obviously, if $|x| \leqq m$, then $|\alpha x + \beta| \leqq \zeta(\alpha\Box b\Box m)$.

Now we construct algorithms $\mathfrak{S}$, $\mathfrak{B}$, $\mathfrak{C}_1$ and $\mathfrak{C}$ such that, for any duplexes α and β with $\alpha \neq 0$, any word d_1 of type d, and any natural numbers m and n with $m \geqq 1$, one has

$$\mathfrak{S}(\alpha\Box\beta\Box d_1) \overset{+}{\in} c_0;\quad \mathfrak{B}(\alpha\Box\beta\Box d_1) \overset{+}{\in} c_0;$$

$$\mathfrak{C}_1(\alpha\Box\beta\Box d_1\Box m) \overset{+}{\in} c_0.\ \mathfrak{C}(\alpha\Box\beta\Box d_1) \overset{+}{\in} c_0;$$

$$\{\mathfrak{S}(\alpha\Box\beta\Box d_1)\}(m) = \{\mathrm{Or}(d_1)\}(\zeta(\alpha\Box\beta\Box m));$$

$$\langle\mathfrak{B}(\alpha\Box\beta\Box d_1)\rangle_4(n) \simeq L^{\text{ш}}(\alpha\Box\beta\Box \underbrace{\mathrm{Fd}(d_1)}_{\psi}(n));$$

$$\langle\mathfrak{C}_1(\alpha\Box\beta\Box d_1\Box m)\rangle_0(i) \simeq \overbrace{\mathrm{Fd}(d_1)}^{\psi}(\max(\zeta(\alpha\Box\beta\Box n)\Box i + Z(\alpha\Box\{\mathrm{Or}(d_1)\}(\zeta(\alpha\Box\beta\Box m)))));$$

$$\langle\mathfrak{C}(\alpha\Box\beta\Box d_1)\rangle_0(i) \simeq \max_{1\leqslant m\leqslant i}\langle\Gamma(\{\mathfrak{S}(\alpha\Box\beta\Box d_1)\}(m)\Box m\ \mathfrak{C}_1(\alpha\Box\beta\Box d_1\Box m))\rangle_0(i),$$

where $L^{\text{ш}}$ is the operator of linear substitution for words of type ш defined in §1.1, Γ is the algorithm defined in Remark 1 to 2.2.3, and Z is an algorithm of the type (дн → н) with the property $\forall xk(2^{-Z(x\Box k)} < |x|^k)$.

Then we construct an algorithm L^d such that, for any duplexes α and β with $\alpha \neq 0$, and any word d_1 of type d,

$$L^d(\alpha\Box\beta\Box d_1) \simeq \begin{cases} \mathfrak{S}(\alpha\Box\beta\Box d_1)\,\tau\,L^{\text{ш}}(\alpha\Box\beta\Box\mathrm{Fd}(d_1)), \text{ if } \mathrm{Fd}(d_1) \text{ is a word of type ш, i.e. the sign } \odot \text{ does not appear in } d_1; \\ \mathfrak{S}(\alpha\Box\beta\Box d_1)\,\tau\,\mathfrak{B}(\alpha\Box\beta\Box d_1)\odot\mathfrak{C}(\alpha\Box\beta\Box d_1), \text{ if the sign } \odot \text{ appears in } d_1. \end{cases}$$

It is possible to prove that, for any real duplexes α and β ($\alpha \neq 0$), the operator $(L^d \uparrow \alpha\Box\beta\Box)$ is a linear operator of the type $(d \to d)$. Moreover, the following formulas hold:

$$\forall \alpha\beta d_1\,(\mathrm{Or}\,(L^d\,(\alpha\,\square\,\beta\,\square\,d_1)) = \mathfrak{S}\,(\alpha\,\square\,\beta\,\square\,d_1)); \tag{34}$$

$$\forall \alpha\beta d_1 m\,(F\,(\mathfrak{S}\,(\alpha\,\square\,\beta\,\square\,d_1)\,\square\,m\,\square\,L^d\,(\alpha\,\square\,\beta\,\square\,d_1)) \underset{m}{=} \mathfrak{A}\,(k\,\square\,\frac{1}{\alpha^k}\; L^{\text{Ш}}\,(\alpha\,\square\,\beta\,\square\,F\,(\mathrm{Or}\,(d_1)\,\square\,m_1\,\square\,d_1))))); \tag{35}$$

$$\forall \alpha\beta d_1\,(D\,(L^d\,(\alpha\,\square\,\beta\,\square\,d_1)) \underset{d}{=} \alpha\cdot L^d\,(\alpha\,\square\,\beta\,\square\,D\,(d_1))), \tag{36}$$

where

$$k \rightleftharpoons \{\mathfrak{S}\,(\alpha\,\square\,\beta\,\square\,d_1)\}\,(m),\quad m_1 \rightleftharpoons \zeta\,(\alpha\,\square\,\beta\,\square\,m).$$

We will call the algorithm L^d the *operator of linear substitution of a variable for generalized functions*. In the special case $\alpha = 1$ we obtain the shift operator for generalized functions, which we denote by C, defined as follows:

$$\forall \beta d_1\,(C\,(\beta\,\square\,d_1) \simeq L^d\,(1\,\square\,\beta\,\square\,d_1)).$$

In concluding this chapter we present some examples of constructive generalized functions.

1. Suppose that λ is a sequence of words of type Ш, as constructed in the proof of 2.2.3, and γ an algorithm of the type (Н → Н) such that $\forall i(\gamma(i) = i)$. Write

$$\theta \leftrightharpoons \overline{1}\,\tau\,\mathcal{E}\lambda,\ \text{Ч}_4^{\text{sa}}\,\text{З}\odot\mathcal{E}\gamma,\ \text{Ч}_0^{\text{sa}}\,\text{З}.$$

Then θ is a constructive generalized function, which we will call the *generalized Heaviside function*.

2. Suppose that φ is a sequence of words of type Ш, defined in such a way that for any natural number n and an arbitrary duplex x

$$\underline{\varphi(n)}\,(x) = \begin{cases} 0, \text{ if } x \leqq -2 \text{ or } x \geqq 2^{-n}; \\ 2^{2n}(x + 2^{-n}), \text{ if } -2^{-n} \leqq x \leqq 0; \\ -2^{2n}(x - 2^{-n}), \text{ if } 0 \leqq x \leqq 2^{-n}. \end{cases}$$

Write

$$\delta \leftrightharpoons \overline{2}\,\tau\,\mathcal{E}\varphi,\ \text{Ч}_4^{\text{sa}}\,\text{З}\odot\mathcal{E}\gamma,\ \text{Ч}_0^{\text{sa}}\,\text{З},$$

where γ is the algorithm defined in Example 1. Then δ is a constructive generalized function, which we shall called a *generalized Dirac function*.

It is not hard to prove that $D(\theta) \underset{d}{=} \delta$ and

$$\theta \underset{d}{=} D\,(\overline{0}\tau X),\quad \delta \underset{d}{=} \mathrm{Dr}\,(2\,\square\,\overline{0}\,\tau X),$$

where X is a word of type Ш such that

$$\forall x\left(\underline{X}\,(x) = \frac{x + |x|}{2}\right).$$

CHAPTER V

ORDER, SUPPORT, AND VALUE AT A POINT, FOR A CONSTRUCTIVE GENERALIZED FUNCTION

§1. Constructive generalized functions given in an interval

1.1. In the preceding chapter we introduced the concept of constructive generalized function "given" in the entire space of real duplexes. In order to consider the local properties of generalized functions, we define in this section the concept of constructive generalized function in an arbitrary interval. For simplicity we shall consider only rational intervals.[1] We shall denote the subordinate generic letter assigned to the set of all possible rational intervals by ν.

First of all we construct (and fix) an algorithm S such that for each interval Ω the algorithm $(S \uparrow \Omega \square)$ defines a sequence of expanding segments such that each of them is contained in Ω and the system of all the segments of this sequence covers Ω. Then the algorithm S may be constructed, for example, as follows:

$$S(P \nabla Q \square m) \simeq \begin{cases} P + \dfrac{Q-P}{m+2} \Delta Q - \dfrac{Q-P}{m+2}, & \text{if } P \text{ and } Q \text{ are rational numbers;} \\ -m \Delta m, \quad \text{if} \quad P \nabla Q \eqcirc -\infty \nabla \infty; & \\ Q - m - 1 \Delta Q - \dfrac{1}{m+1}, & \text{if } P \eqcirc -\infty,\ Q \text{ is a rational number} \\ P + \dfrac{1}{m+1} \Delta P + m + 1, & \text{if } P \text{ is a rational number and } Q \eqcirc \infty. \end{cases}$$

Here $P \nabla Q$ is an interval and m is a natural number $\geqq 1$.

Suppose that Ω is some interval. We will write $\Omega_m \leftrightarrows S(\Omega \square m)$. Concerning a function f defined in the interval Ω, we shall say that it is *almost uniformly continuous in* Ω if it is uniformly continuous in each segment Ω_m. Analogously to the way in which the concept of the complete cipher of an almost uniformly continuous function (i.e. a word of type ш) was defined, it is possible to define the concept of the "complete cipher of an almost uniformly continuous function in Ω".[2] Suppose that $\mathfrak{P}^0$ is a formula such that, for any interval Ω and any word P in the alphabet Ч_4, $\mathfrak{P}^0(\Omega, P)$ expresses the condition "P is a complete cipher of an almost uniformly continuous function in the interval Ω". We adjoin this formula to the list Δ of the language Я_Σ^Δ and

[1] See the definitions and notation in [6].

[2] See the definitions and notation in [6].

denote the corresponding generic letter by ц. Then, for any interval Ω, words of type $ц^{\Omega}$ are the complete ciphers of almost uniformly continuous functions in Ω.

Starting from words of type $ц^{\Omega}$, we may define constructive generalized functions in the interval Ω in quite the same way as we defined constructive generalized functions from words of type ш. Here we will not repeat this construction in detail. Rather we describe briefly certain main points.

Suppose that $\mathscr{N}^0$ is an algorithm such that, for any interval Ω, any natural number $m \geqq 1$ and any word $ц_1^{\Omega}$ of type $ц^{\Omega}$, $\mathscr{N}^0(\Omega \square m \square ц_1^{\Omega})$ is the least upper bound of the function $\underline{ц_1^{\Omega}}$ on the segment Ω_m. Here Ω is an interval. The space of words of type $\overline{ц^{\Omega}}$ with the enumerable norm $(\mathscr{N}^0 \uparrow \Omega \square)$ will be denoted by $\mathscr{C}_{\Omega}$, and the space of words of type $ц^{\Omega}$ with the norm $(\mathscr{N}^0 \uparrow \Omega \square m \square)$ will be denoted by C_{Ω_m}. We note that in the special case when $\Omega \,\underline{\underline{\circ}}\, - \infty \triangle \infty$, the space $\mathscr{C}_{\Omega}$ coincides with the space $\mathscr{C}$.

We fix an algorithm X^0 of the type $(\nu н \to д)$ such that for each interval Ω the algorithm $(X^0 \uparrow \Omega \square)$ defines a sequence of distinct points in the segment Ω_1 (and thus in any segment Ω_m with $m \geqq 1$). We write $X_i^{\Omega} \leftrightharpoons X^0(\Omega \square i)$, where i is a natural number.

It is possible to construct an algorithm H^0 which transforms each word of the type $\Omega \square P$, where Ω is an interval and P a word of type ю, into a representative of the polynomial frame P in the space $\mathscr{C}_{\Omega}$; if $P \,\underline{\underline{\circ}}\, \alpha_0 \sigma \cdots \sigma \alpha_k$, then $H^0(\Omega \square P)$ is a word of type $ц^{\Omega}$ and

$$\forall x\, (x \in \Omega \supset \underline{H^\circ(\Omega \square P)}\, (x) = \alpha_0 + \ldots + \alpha_k \cdot x^k).$$

Then we construct an algorithm $\mathfrak{A}^0$ such that, for any interval Ω, any natural number k and any word $ц_1^{\Omega}$ of type $ц^{\Omega}$:

a) $\mathfrak{A}^0(\Omega \square k \square ц_1^{\Omega})$ is a word of type $ц^{\Omega}$;

b) $\forall i\, (0 \leqslant i < k \supset \underline{\mathfrak{A}^\circ(\Omega \square k \square ц_1^{\Omega})}\, (x_i^{\Omega}) = 0)$;

c) a word $ю_1$ of type ю is realizable such that

$$\mathrm{dg}(ю_1) \prec \cdot\, k \quad \text{and} \quad \mathfrak{A}^0(\Omega \square k \square ц_1^{\Omega}) \underset{ц,\Omega}{=} ц_1^{\Omega} \underset{ц,\Omega}{+} H^0(\Omega \square ю_1),$$

where $\underset{ц,\Omega}{=}$ denotes equality in $\mathscr{C}_{\Omega}$ and $\underset{ц,\Omega}{+}$ denotes addition in $\mathscr{C}_{\Omega}$.

It is possible to construct an algorithm Γ^0 with the following property: for any segment Ω, any natural numbers m and k with $m \geqq 1$, any algorithm γ of the type $(н \to н)$ and any sequence λ of words of type $ц^{\Omega}$, if γ is a regulator of convergence in itself of the sequence λ in C_{Ω_m}, then $\Gamma^0(\Omega \square k \square m \square \mathcal{E}\gamma\mathcal{3})$ is the transcription of the regulator of convergence in itself in the space C_{Ω_m} of the sequence χ defined by the formula

$$\forall n\, (\chi(n) \simeq \mathfrak{A}^\circ(\Omega \square k \square \lambda(n))).$$

We construct, and then fix, an algorithm J^0 which transforms each word of the form $\Omega \square k \square P$, where Ω is an interval, k a natural number and P a word of type ц^{Ω}, into a word $J^0(\Omega \square k \square P)$ of type ц^{Ω} which is the antiderivative of order k of the word P in the interval Ω.

On the basis of the algorithms J^0 and $\mathfrak{A}^0$ it is possible to construct an algorithm Φ^0 such that, for any interval Ω, any word U of type e, any natural number $m \geqq 1$ and any word P of type ц^{Ω}, $\Phi^0(\Omega \square U \square m \square P)$ *is a word of type* ц^{Ω} *which is the antiderivative of order* $\{U\}(m)$ *of the word* P *(in the interval* Ω*) and satisfies the condition*

$$\forall i\,(0 \leqslant i \leqslant \{U\}(m) - 1 \supset \underbrace{\Phi^0(\Omega \square U \square m \square P)}(x_i^{\Omega}) = 0)$$

(see 2.3.1, Chapter III). Then we construct an algorithm $\mathfrak{N}^0$ such that

$$\forall \nu_1 e_1 m \text{ц}_1^{\nu_1} (\mathfrak{N}^0 (\nu_1 \square e_1 \square m \square \text{ц}_1^{\nu_1}) \simeq \mathcal{N}^0 (\nu_1 \square m \square \Phi^0 (\nu_1 \square e_1 \square m \square \text{ц}_1^{\nu_1}))),$$

ν_1 being the variable for rational intervals.

Suppose that Ω is an interval and U a word of type e. Then the algorithm $(\mathfrak{N}^0 \uparrow \Omega \square U \square)$ defines an enumerable norm in the linear space of words of type ц^{Ω}. The linear space of words of type ц^{Ω}, equipped with the enumerable norm $(\mathfrak{N}^0 \uparrow \Omega \square U \square)$, becomes an enumerably normed space. We shall denote this space by $\mathscr{L}_{\Omega}^{U}$. The *FR*-standard completion of this space will be denoted by $\mathfrak{D}_{\Omega}^{U}$. By the same method by which we constructed the algorithm ω in §3.2, Chapter IV, we may construct an algorithm ω^0 such that, for any interval Ω and any words U and V of type e, if $U \prec V$, then $(\omega^0 \uparrow \Omega \square U \square V \square)$ is a continuous linear mapping of $\mathfrak{D}_{\Omega}^{U}$ into $\mathfrak{D}_{\Omega}^{V}$ satisfying the condition

$$\forall \text{ц}_1^{\Omega}\, n((\underbrace{\omega^0(\Omega \square U \square V \square \text{ц}_1^{\Omega})}(n) \underset{\text{ц}.\Omega}{=} \text{ц}_1^{\Omega}(n)).$$

Suppose that Ω is an interval. The family of spaces $\mathfrak{D}_{\Omega}^{U}$, considered together with an algorithm of mappings $(\omega^0 \uparrow \Omega \square)$, forms an inductive spectrum of enumerably normed spaces. The inductive limit of this inductive spectrum will be denoted by $\mathfrak{D}_{\Omega}$. This is the *space of constructive generalized functions in the interval* Ω.

We note that a word P in the alphabet A_d is a constructive generalized function in the interval Ω if and only if it has one of the following forms:

a) $P \,\overline{\underline{\circ}}\, U \tau Q$, where U is a word of type e and Q a word of type ц^{Ω};

b) $P \,\overline{\underline{\circ}}\, U \tau Q \odot R$, where U is a word of type e, Q is the transcription of a sequence of words of type ц^{Ω} and R is the transcription of the regulator of convergence in itself of the sequence $\langle Q \rangle_4$ in the space $\mathscr{L}_{\Omega}^{U}$ (i.e. $Q \odot R$ is a point of $\mathfrak{D}_{\Omega}^{U}$).

It is possible to construct algorithms $\mathfrak{F}^0$ and F^0 such that:

1) $\mathfrak{F}^0$ is applicable to any word of the form $\Omega \square U \square m \square Q$, where Ω is an interval, U a word of type e, $m \geqq 1$ and Q is a point of $\mathfrak{D}_{\Omega}^{U}$, and $\mathfrak{F}^0(\Omega \square U \square m \square Q)$ is the limit in C_{Ω_m} of the sequence λ of words of type ц^{Ω} defined by the formula

$$\forall n (\lambda (n) \simeq \Phi^0 (\Omega \square U \square m \square \underline{Q}_{\psi} (n))).$$

2) For any interval Ω, any constructive generalized function P in the interval Ω, and any word U of type e, if $\mathrm{Or}(P) \prec U$, then

$$\forall m (F^0(\Omega \square U \square m \square P) \simeq \mathfrak{F}^0(\Omega \square U \square m \square \omega^0(\Omega \square \mathrm{Or}(P) \square U \square \mathrm{Fd}(P)))).$$

For each fixed interval Ω the algorithms $(\mathfrak{F}^0 \uparrow \Omega \square)$ and $(F^0 \uparrow \Omega \square)$ have properties analogous to the properties of the corresponding algorithms $\mathfrak{F}$ and F (see §§3.1 and 3.4 in Chapter IV).

1.2. Suppose that Ω is some interval. In order to be able to suppose that every constructive generalized function (i.e. word of type d) is also a constructive generalized function in the interval Ω, we now construct the "canonical mapping" of the space $\mathfrak{D}$ into the space $\mathfrak{D}_\Omega$, as follows.

First of all we construct algorithms ζ and w with the following properties: if Ω is an interval and $m \geqq 1$, then

$$\zeta (\Omega \square m) \simeq \underline{\mu j} (\Omega_m \subset \underline{-} \, j \Delta j);$$

if Ω is an interval and ш_1 is a word of type ш, then $w(\Omega \square \text{ш}_1)$ is a word of type ц^Ω, and

$$\forall x (x \in \Omega \supset \underline{w(\Delta \square \text{ш}_1)} (x) = \underline{\text{ш}_1} (x));$$

$$\forall mi \; (\overline{w(\Omega \square \text{ш}_1)} (m \square i) \simeq \overline{\text{ш}_1} (\zeta (\Omega \square m) \square i)).$$

$w(\Omega \square \text{ш}_1)$ will be regarded as the representative of the word ш_1 of type ш in $\mathscr{C}_\Omega$.

It is possible to construct an algorithm $\mathfrak{S}$ which transforms each word of the type $\Omega \square U$, where Ω is an interval and U a word of type e, into a word of type e such that

$$\forall m (\{\mathfrak{S} (\Omega \square U)\} (m) = \{U\} (\zeta (\Omega \square m))).$$

We now construct algorithms $\mathfrak{B}$ and $\mathfrak{C}$ which are applicable to every word of the form $\Omega \square d_1$, where Ω is an interval and d_1 is a word of type d, and satisfies the following conditions for any Ω and d_1:

$$\forall n (\langle \mathfrak{B} (\Omega \square d_1) \rangle_4 (n) \simeq \omega (\Omega \square \underline{\mathrm{Fd} (d_1)}_{\psi} \; (n)));$$

$$\forall i (\langle \mathfrak{C} (\Omega \square d_1) \rangle_0 (i) \simeq \max_{1 \leqslant m \leqslant i} \langle \Gamma^0 (\Omega \square \{\mathfrak{S} (\Omega \square \mathrm{Or} (d_1))\} (m) \square m$$

$$\square \, \mathcal{E} (\overline{\mathrm{Fd} (d_1)}^{\psi} \uparrow \zeta (\Omega \square m) \square) \, 3) \rangle_0 (i)).$$

Finally, we construct an algorithm W such that, for any interval Ω and any word d_1 of type d,

$$W(\Omega \square d_1) \simeq \begin{cases} \mathfrak{S}(\Omega \square \mathrm{Or}(d_1)) \, \tau \, w(\Omega \square \mathrm{Fd}(d_1)), \text{ if } \mathrm{Fd}(d_1) \text{ is a word of type} \\ \quad \text{ш, i.e. if } \odot \text{ does not enter into } d_1; \\ \mathfrak{S}(\Omega \square \mathrm{Or}(d_1)) \, \tau \, \mathfrak{B}(\Omega \square d_1) \odot \mathfrak{C} (\Omega \square d_1), \text{ if } \odot \text{ enters into } d_1. \end{cases}$$

Using the properties of the algorithms $\mathfrak{S}$, $\mathfrak{B}$ and $\mathfrak{C}$, it is not hard to prove that $W(\Omega \square d_1)$ is a constructive generalized function in the interval Ω. $W(\Omega \square d_1)$ will be regarded as the *representative of the word* d_1 *of type* d *in the space* $\mathfrak{D}_\Omega$. It is possible to show further that for each interval Ω the algorithm $(W \uparrow \Omega \square)$ is a continuous linear mapping of $\mathfrak{D}$ into $\mathfrak{D}_\Omega$.

We note some relations: for any interval Ω, any words U of type e and ш$_1$ of type ш, and any natural number $m \geq 1$ we have the equation

$$\Phi^0(\Omega \square \mathfrak{S}(\Omega \square U) \square m \square w(\Omega \square \text{ш}_1)) \underset{\Omega_m}{=} \mathfrak{A}^0(\Omega \square \{U\}(\zeta(\Omega \square m)) \square w(\Omega \square \Phi(U \square \zeta(\Omega \square m) \square \text{ш}_1)), \tag{1}$$

where $\underset{\Omega_m}{=}$ denotes equality in the space C_{Ω_m}. We denote the left and right sides of equation (1) by P and Q respectively. For the proof of (1) it suffices to note that the functions $\underline{P}$ and $\underline{Q}$ are antiderivatives of order $\{U\}(\zeta(\Omega \square m))$ of one and the same function ш$_1$ in the interval Ω and vanish simultaneously at $\{U\}(\zeta(\Omega \square m))$ distinct points in Ω_m.

On the basis of formula (1) we may prove the following assertion.

1.2.1. *For any interval* Ω, *any words* d_1 *of type* d *and* U *of type* e, *and any natural number* $m \geq 1$, *if* $\mathrm{Or}(d_1) \prec \cdot\, U$, *then one can realize a word* ю$_1$ *of type* ю *such that*

$$\mathrm{dg}(\text{ю}_1 \prec \cdot \{U\}(\zeta(\Omega \square m));$$

$$\forall x (x \in \Omega_m \supset \underline{F^0(\Omega \square \gamma(\Omega \square U) \square m \square W(\Omega \square d_1))(x)} \underset{\text{B}}{=} \underline{F(U \square \zeta(\Omega \square m) \square d_1)(x)} + \underline{H(\text{ю}_1)}(x)). \tag{2}$$

1.3. Suppose that Ω is an interval, and that P and Q are constructive generalized functions in Ω. We will write

$$(P \underset{d,\Omega}{=} Q)$$

if P and Q are equal to one another in the space $\mathfrak{D}_\Omega$. Set

$$U \leftrightharpoons \mathrm{Max}(\mathrm{Or}(P) \square \mathrm{Or}(Q)).$$

We note that $(P \underset{d,\Omega}{=} Q)$ if and only if one of the following conditions is satisfied:

$$\forall m (F^0(\Omega \square U \square m \square P) \underset{\Omega_m}{=} F^0(\Omega \square U \square m \square Q));$$

$$\exists \text{e}_1 (U \prec \text{e}_1 \,\&\, \forall m (F^0(\Omega \square \text{e}_1 \square m \square P) \underset{\Omega_m}{=} F^0(\Omega \square \text{e}_1 \square m \square Q)));$$

$$\forall \text{e}_1 (U \prec \text{e}_1 \supset \forall m (F^0(\Omega \square \text{e}_1 \square m \square P) \underset{\Omega_m}{=} F^0(\Omega \square \text{e}_1 \square m \square Q))),$$

where $\underset{\Omega_m}{=}$ denotes the equality relation in C_{Ω_m} (see property F6, 3.4.1, Chapter IV).

Suppose that d_1 and d_2 are constructive generalized functions. We will say that d_1 *is equal to* d_2 *in* Ω, and write

$$(d_1 \underset{d,\Omega}{=} d_2),$$

if

$$W(\Omega \,\square\, d_1) \underset{d,\Omega}{=} W(\Omega \,\square\, d_2),$$

where W is the algorithm defined in §1.2.

On the basis of Assertion 1.2.1, properties F4–F6 of the algorithm F (see 3.4.1, Chapter IV), and the remark made above on the equality relation in the space $\mathfrak{D}_\Omega$, it is not hard to prove the following assertion:

1.3.1. *Suppose that d_1 and d_2 are constructive generalized functions, U a word of type* e, *Ω an interval and m a natural number such that $\Omega \subset -m \triangle m$ and* $\mathrm{Max}(\mathrm{Or}(d_1) \,\square\, \mathrm{Or}(d_2)) \prec U$. *For $(d_1 \underset{d,\Omega}{=} d_2)$ it is necessary and sufficient that a word ю_1 of type* ю *be realizable for which the following conditions are satisfied:*

$$\mathrm{dg}\,(\text{ю}_1) \prec\cdot \{U\}(m);$$

(3) $$\forall x\,(x \in \Omega \supset \underline{F\,(U \,\square\, m \,\square\, d_1)}\,(x) \underset{\text{B}}{=} \underline{F\,(U \,\square\, m \,\square\, d_2)(x)} + \underline{H\,(\text{ю}_1)}\,(x)).$$

1.3.2. *Suppose that d_1 is a constructive generalized function and Ω a finite interval.*[3] *Then a natural number k and a word P of type* ш *are realizable such that*

$$d_1 \underset{d,\Omega}{=} \mathrm{Dr}\,(k \,\square\, \bar{0}\tau P).$$

In other words, in each finite interval each constructive generalized function is the derivative of some order of (the complete cipher of) an almost uniformly continuous function.

Proof. Suppose that d_1, Ω and m satisfy the hypotheses of the theorem. We write $U \leftrightharpoons \mathrm{Or}(d_1)$, $k \leftrightharpoons \{U\}(m)$ and $P \leftrightharpoons F(U \,\square\, m \,\square\, d_1)$. Then we write $d_2 \leftrightharpoons \bar{0}\tau P$. Obviously $\mathrm{Or}(d_2) = \bar{0}$ and $\mathrm{Or}(\mathrm{Dr}(k \,\square\, d_2)) = \bar{k}$. Suppose that $V \leftrightharpoons \mathrm{Max}(U \,\square\, \bar{k})$. Then $\bar{0} \prec V \dot{-} \bar{k}$ and $\{V \dot{-} \bar{k}\}(m) = 0$. We note that $F(V \dot{-} \bar{k} \,\square\, m \,\square\, d_2) \underset{m}{=} P$.[4] Hence it follows from formula (16) in §4.2, Chapter IV, that

$$F\,(V \,\square\, m \,\square\, \mathrm{Dr}\,(k \,\square\, d_2)) \underset{m}{=} \mathfrak{A}\,(k \,\square\, P).$$

Hence we can realize a word ю_1 of type ю such that

(4) $$\forall x\,(x \in -m \triangle m \supset \underline{F\,(V \,\square\, m \,\square\, \mathrm{Dr}\,(k \,\square\, d_2))}\,(x) \underset{\text{B}}{=} \underline{P\,(x)} + \underline{H\,(\text{ю}_1)}\,(x)).$$

In view of properties F5 and F2 (3.4.1, Chapter IV) we have

(5) $$P \underset{\circ}{=} F\,(U \,\square\, m \,\square\, d_1) \underset{m}{=} F\,(V \,\square\, m \,\square\, d_1).$$

In view of 1.3.1 it follows from (4) and (5) that $d_1 \underset{d,\Omega}{=} \mathrm{Dr}(k \,\square\, d_2)$, as we were required to prove.

---

[3] The interval Ω is said to be finite if a natural number m is realizable such that $\Omega \subseteq -m \triangle m$.

[4] It follows from the properties of the algorithms F, $\mathfrak{F}$, and Φ that if P is a word of type ш, U a word of type e and m a natural number such that $\{U\}(m) = 0$, then $F(U \,\square\, m \,\square\, \bar{0}\tau P) \underset{m}{=} P$.

REMARK. From the proof of 1.3.2 it is clear that in each interval $-m \nabla m$ the constructive generalized function d_1 is the derivative of order $\{\text{Or}(d_1)\}(m)$ of the constructive generalized function $\overline{0}\,\tau\, F(\text{Or}(d_1) \square m \square d_1)$.

1.3.3. *Suppose that* d_1 *and* d_2 *are constructive generalized functions, and* Ω_1 *and* Ω_2 *rational intervals.*

a) *If* $\Omega_2 \subseteq \Omega_1$ *and* $d_1 \underset{d,\Omega_1}{=} d_2$, *then* $d_1 \underset{d,\Omega_2}{=} d_2$.

b) *If* Ω_1 *and* Ω_2 *intersect, i.e.* $\exists x (x \in \Omega_1 \,\&\, x \in \Omega_2)$, *and if* $(d_1 \underset{d,\Omega_1}{=} d_2) \,\&\, (d_1 \underset{d,\Omega_2}{=} d_2)$, *then* $d_1 \underset{d,\Omega}{=} d_2$, *where* $\Omega \leftrightharpoons \Omega_1 \cup \Omega_2$.[5)]

If Ω_1 and Ω_2 are finite intervals, this assertion is a direct consequence of 1.3.1. In the remaining cases it is proved using 1.3.1 and the properties of the algorithms F and F^0.

§2. Order and support of a constructive generalized function

2.1. Suppose that d_1 is any constructive generalized function. The word $\text{Or}(d_1)$ of type e will be called the *formal order* of the generalized function d_1, and for each natural number $m \geqq 1$ the number $\{\text{Or}(d_1)\}(m)$ will be called the *formal order of* d_1 *in the segment* $-m \triangle m$.

Now suppose that k is a natural number. We will say that d_1 is a *generalized function of order* $\leqq k$, and write

$$(d_1 \in \underline{\text{order} \leqq} \, k),$$

if

$$\exists d_2 (d_2 \underset{d}{=} d_1 \,\&\, \forall m (\{\text{Or}(d_2)\}(m) \leqslant k)). \tag{1}$$

We will say that d_1 is a *generalized function of finite order*, if

$$\exists k (d_1 \in \underline{\text{order} \leqq} \, k). \tag{2}$$

We will say that d_1 is a *generalized function of order* k, if

$$(d_1 \in \underline{\text{order} \leqslant} \, k) \,\&\, (k \geqslant 1 \supset \neg (d_1 \in \underline{\text{order} \leqslant} \, k-1)). \tag{3}$$

Finally, d_1 is a *regular generalized function*, if it is a generalized function of order 0.

2.1.1. *Every constructive generalized function is the l-limit* (§3.5, *Chapter* IV) *of a sequence of regular generalized functions.*

PROOF. Suppose that d_1 is a constructive generalized function, and write $U \leftrightharpoons \text{Or}(d_1)$ and $P \leftrightharpoons \text{Fd}(d_1)$. Obviously P is the limit in the space $\mathfrak{D}^U$ of a sequence $\underline{P}_\downarrow$ of words of type III. Therefore d_1 is the l-limit of a sequence λ defined by the formula

$$\forall n (\lambda (n \simeq U \tau \underline{P}_\downarrow (n)).$$

---

5) If $\Omega_1 \overset{\circ}{=} a \nabla b$, $\Omega_2 \overset{\circ}{=} c \nabla d$, and Ω_1 and Ω_2 intersect, then

$$\Omega_1 \cup \Omega_2 \leftrightharpoons \min(a \square c) \nabla \max(b \square d).$$

Since $\underset{\smile}{P}_\psi(n)$ is a word of type ш,

$$U\tau \underset{\smile\psi}{P}(n) \underset{d}{=} \bar{0}\tau \underset{\smile\psi}{P}(n).$$

Thus λ is a sequence of regular generalized functions. The theorem is proved.

The following is a direct consequence of 3.5.1 of Chapter IV, and of 2.1.1.

2.1.2. *Every constructive generalized function is the limit in the space* $\mathfrak{D}$ *of a sequence of regular generalized functions.*

We note that if the generalized function d_1 satisfies the condition

$$\forall m\,(\{\mathrm{Or}\,(d_1)\}\,(m) \leqslant k),$$

then we can realize a generalized function d_2 such that $(d_2 = d_1 \,\&\, \mathrm{Or}(d_2) = \bar{k})$. For this it suffices for us to choose

$$d_2 \rightleftharpoons \bar{k}\,\tau\omega\,(\mathrm{Or}\,(d_1)\,\square\,\bar{k}\,\square\,\mathrm{Fd}\,(d_1)).$$

Using Theorem 4.2.4, Chapter IV, we therefore obtain

2.1.3. *For any natural number k and constructive generalized function d_1 of order $\leqq k$, one can realize a word Q of type* ш *such that* $d_1 \underset{d}{=} \mathrm{Dr}(k \,\square\, \bar{0}\,\tau\, Q)$.

Roughly speaking, this means that every generalized function of order $\leqq k$ "is" the derivative of order k of (the complete cipher of) some almost uniformly continuous function.

It is clear from 2.1.3 that *for any regular generalized function d_1, one can realize a word Q of type* ш *such that* $d_1 \underset{d}{=} \bar{0}\,\tau\, Q$.

On the basis of the definition of the derivative of a constructive function it is easy to see that if d_1 is a generalized function of order $\leqq k$, then $D(d_1)$ is a generalized function of order $\leqq k+n$, where n is any natural number.[6)]

Now we consider certain examples of constructive generalized functions of finite order. Suppose that X is a word of type ш such that

$$\forall x\left(\underset{\smile}{X}(x) = \frac{x + |x|}{2}\right).$$

Suppose that k is a natural number. Write $X_k \rightleftharpoons \mathrm{Dr}(k \,\square\, \bar{0}\,\tau\, X)$. It is not difficult to show that *for any natural number k, X_k is a generalized function of order k.* Thus the Heaviside generalized function θ is a generalized function of order 1, and the generalized Dirac δ-function is a generalized function of order 2.

Write $Y_k \rightleftharpoons \mathrm{Fd}(X_k)$ and $Y_{k,n} \rightleftharpoons \underset{\smile}{Y}_{k_\psi}(n)$, where k and n are natural numbers. Obviously $Y_{k,n}$ is a word of type ш. By the definition of the derivative (§4.1, Chapter IV) we have

---

6) It is possible to prove further that if $k \geqq 1$ and if d_1 is a generalized function of order k, then $D(d_1)$ is a generalized function of order $k+1$ and $\mathrm{Dr}(n \,\square\, d_1)$ is a generalized function of order $k+n$.

(4) $$\forall x(\underline{Y_{k+1,n}}(x) = 2^{\gamma_{k,n}}\cdot[\underline{Y_{k,n}}(x+2^{-\gamma_{k,n}}) - \underline{Y_{k,n}}(x)]),$$

where k and n are natural numbers, and $\gamma_{k,n} \leftrightarrows \overline{Y_{k,n}}(n \square n)$. We note that $Y_{0,n} \underset{ш}{=} X$ for any n, so that

$$\gamma_{0,n} = \overline{Y_{0,n}}(n \square n) \geqslant n.$$

It is possible to prove that

(5) $$\forall kn(\gamma_{k,n} \geqslant n).$$

From (4), (5) and the obvious property

$$\forall nx(x \leqslant 0 \supset \underline{Y_{0,n}}(x) = 0)$$

it follows that

$$\forall nkx(n \geqslant k \,\&\, x \leqslant -k\cdot 2^{-k} \supset \underline{Y_{k,n}}(x) = 0),$$

so that

(6) $$\forall nkx\left(n \geqslant k \,\&\, x \leqslant \frac{-1}{2} \supset \underline{Y_{k,n}}(x) = 0\right).$$

This property will be used in the proof of Theorem 2.1.5.

We now note the following assertion.

2.1.4. *It is not true that for any constructive generalized function d_1 of finite order one can realize a natural number k such that d_1 is a generalized function of order k.*

This assertion is a direct consequence of Lemma 2 in §2 of **[28]**. Indeed, we construct an algorithm ψ of type $(д \to d)$ as follows:

$$\forall x(\psi(x) \simeq x \underset{d}{\cdot} \delta),$$

where δ is the generalized Dirac function. If $x = 0$, then $\psi(x) \underset{d}{=} \mathfrak{O}^d$, i.e. $\psi(x)$ is a generalized function of order 0. If $x \neq 0$, then $\sigma(x)$ is a generalized function of order 2. Therefore, if for each generalized function d_1 of finite order we could realize a word k such that d_1 is a generalized function of order k, then we could also realize an algorithm φ of type $(д \to н)$ having the property

(7) $$\forall x((x \underset{в}{=} 0 \supset \varphi(x) = 0) \,\&\, (x \underset{в}{\neq} 0 \supset \varphi(x) = 2)).$$

It follows from (7) that φ also has the property

$$\forall x((x = \overline{0} \equiv \varphi(x) = 0) \,\&\, (x \neq 0 \equiv \varphi(x) = 2)).$$

In view of Lemma 2 in §2 of **[28]**, an algorithm φ with such a property is not possible. Assertion 2.1.4 is proved.

By definition it is obvious that for any generalized function d_1 of finite order one can realize a natural number k such that ($d_1 \in$ <u>order $\leq k$</u>). On the other hand, the following theorem holds.

2.1.5. *It is not possible to have an algorithm which constructs relative to each generalized function* d_1 *of finite order a natural number such that* d_1 *is a generalized function of order* $\leq k$.

Proof. We shall again use an algorithm $\mathscr{A}$ of type $(\text{н} \dot{\rightarrow} \text{н})$ whose applicability problem is undecidable. We denote by G an algorithm applicable to any word of the form $l \square n$, where l and n are natural numbers such that $G(l \square n) \dot{=} \Lambda$ if and only if the process of application of $\mathscr{A}$ to the number l terminates in $\leq n$ words.

We construct a sequence Z of generalized functions as follows:

$$\forall k \left(Z(k) \simeq C \left(-k - \frac{1}{2} \square X_k \right) \right),$$

where $X_k \leftrightharpoons \mathrm{Dr}(k \square \mathring{0}_T X)$, and C is the shift algorithm for generalized functions defined in §4.3 of Chapter IV.

Write $Z_k \leftrightharpoons Z(k)$ and $P_{k,n} \leftrightharpoons \underline{\mathrm{Fd}(Z_k)}(n)$. For any natural numbers k and n we have

$$\forall x \left(\underline{P_{k,n}}(x) = \underline{Y_{k,n}} \left(x - k - \frac{1}{2} \right) \right).$$

Therefore we find from (6) that

$$\forall nkx\, (n \geqslant k \,\&\, x \leqslant k \supset \underline{P_{k,n}}(x) = 0). \tag{8}$$

From formula (36) of §4.3, Chapter IV, it follows that for any natural number k we have $Z_k \underset{d}{=} \mathrm{Dr}(k \square \bar{0}_T Q_k)$, where Q_k is a word of type ш such that

$$\forall x \left(\underline{Q_k}(x) = \frac{1}{2} \left(x - k - \frac{1}{2} + \left| x - k - \frac{1}{2} \right| \right) \right).$$

Hence we deduce that Z_k is a generalized function of order k for any natural number k, and

$$\forall d_1 m\, (d_1 \underset{d}{=} Z_k \,\&\, m > k \supset \{\mathrm{Or}(d_1)\}(m) \geqslant k). \tag{9}$$

Construct an algorithm γ of type $(\text{н} \rightarrow \text{e})$ such that for any natural numbers l and m with $m \geq 1$

$$\{\gamma(l)\}(m) \simeq \begin{cases} 0, & \text{if } G(l \square m-1) \neq \Lambda; \\ \mu j (j \leqslant m \,\&\, G(l \square j) \dot{=} \Lambda), & \text{if } G(l \square m-1) \dot{=} \Lambda. \end{cases}$$

Now we construct an algorithm $\mathfrak{B}$ of type $(\text{нн} \rightarrow \text{ш})$ and an algorithm $\mathfrak{C}$ of type $(\text{нн} \rightarrow \text{н})$ such that, for any natural numbers l, n, and i,

$$\mathfrak{B}(l \square n) \simeq \begin{cases} \mathfrak{O}^{\text{ш}}, & \text{if } G(l \square n) \dot{\neq} \Lambda; \\ P_{k_0,n}, & \text{if } G(l \square n) \dot{=} \Lambda, \text{ where } k_0 \leftrightharpoons \mu j (j \leqslant n \,\&\, (G(l \square j) \dot{=} \Lambda). \end{cases}$$

$$\mathfrak{C}(l \square i) \simeq \begin{cases} 0, & \text{if } G(l \square i) \dot{\neq} \Lambda; \\ \max(k_0 \square \overline{\mathrm{Fd}(Z_{k_0})}^{\psi}(i)), & \text{if } G(l \square i) \dot{=} \Lambda, \\ \text{where } k_0 \leftrightharpoons \mu j (j \leqslant i \,\&\, G(l \square j) \dot{=} \Lambda). \end{cases}$$

Next we construct an algorithm $\mathfrak{X}$ as follows:

$$\forall l(\mathfrak{X}(l) \simeq \Upsilon(l) \tau \mathcal{E}(\mathfrak{B} \uparrow l\Box), \, Ч_4^{\text{sa}} \, 3 \odot \mathcal{E}(\mathfrak{C} \uparrow l\Box), \, Ч_4^{\text{sa}} \, 3).$$

We shall show that

$$(10) \qquad \forall l(\mathfrak{X}(l) \overset{+}{\in} d).$$

Let l be a natural number. If $\neg ! \mathscr{A}(l)$, then obviously $\mathfrak{X}(l) \underset{d}{=} \mathfrak{D}^d$. Now we suppose that $! \mathscr{A}(l)$. We write

$$k_0 \leftrightharpoons \mu j(G(l\Box j) = \Lambda), \quad U \leftrightharpoons \Upsilon(l), \quad \lambda_n \leftrightharpoons \mathfrak{B}(l\Box n).$$

In order to prove that $\mathfrak{X}(l) \in d$, it suffices for us to prove the following formula:

$$\forall minp(n, p \geqslant \mathfrak{C}(l\Box i) \,\&\, m$$

$$(11) \qquad \leqslant i \supset \mathscr{N}(m\Box \Phi(U\Box m\Box \lambda_n) \underset{\text{III}}{-} \Phi(U\Box m\Box \lambda_p)) \lessdot 2^{-i}).$$

First of all we note that

$$(12) \qquad \forall nx(x \leqslant k_0 \supset \underline{\lambda_n}(x) = 0).$$

Indeed, if $n < k_0$, then $\lambda_n \doteq \mathfrak{D}^{\text{III}}$, and if $n \geqq k_0$, then $\lambda_n \doteq P_{k_0,n}$, so that (12) follows from (8). Obviously the following formula is a consequence of (12):

$$(13) \quad \forall mnp(m \leqslant k_0 \supset \mathscr{N}(m\Box \Phi(U\Box m\Box \lambda_n) - \Phi(U\Box m\Box \lambda_p)) = 0).$$

We shall prove also the following formula:

$$(14) \quad \begin{aligned} &\forall minp(n, p \geqslant \mathfrak{C}(l\Box i) \,\&\, k_0 \lessdot m \leqslant i \supset \mathscr{N}(m\Box \Phi(U\Box m\Box \lambda_n) \\ &\qquad - \Phi(U\Box m\Box \lambda_p)) \lessdot 2^{-i}). \end{aligned}$$

To this end we note that if $n,p \geqq k_0$, then $\lambda_n \doteq P_{k_0,n}$ and $\lambda_p \doteq P_{k_0,p}$. Moreover, if $m > k_0$, then $\{U\}(m) = k_0$. Hence we get

$$(15) \quad \begin{aligned} &\forall mnp(m \geqslant k_0 \,\&\, n, p \geqslant k_0 \supset \Phi(U\Box m\Box \lambda_n) \\ &\qquad - \Phi(U\Box m\Box \lambda_p) \underset{\text{III}}{=} \Phi(\bar{k}_0\Box m\Box P_{k_0,n}) - \Phi(\bar{k}_0\Box m\Box P_{k_0,p})). \end{aligned}$$

On the other hand, it is obvious that

$$(16) \quad \begin{aligned} &\forall minp(n, p \geqslant \overline{\mathrm{Fd}(Z_{k_0})}^{\Phi}(i) \,\&\, m \leqslant i \supset \mathscr{N}(m\Box \Phi(\bar{k}_0\Box m\Box P_{k_0,n}) \\ &\qquad - \Phi(\bar{k}_0\Box m\Box P_{k_0,p})) \lessdot 2^{-i}). \end{aligned}$$

Now from (15) and (16) it follows immediately that (14) holds. It is clear that (13) and (14) imply (11). Thus (11) is proved.

Thus we have proved that

$$(17) \qquad \forall l((\neg ! \mathscr{A}(l) \in \mathfrak{X}(l) \overset{+}{\in} d) \,\&\, (! \mathscr{A}(l) \supset \mathfrak{X}(l) \in d)).$$

From (17) and the obvious formula

$$(18) \qquad \forall l \neg\neg(! \mathscr{A}(l) \vee \neg ! \mathscr{A}(l))$$

it follows that $\forall l \neg\neg (\mathfrak{X}(l) \in d)$. Since the set of words of type d can be defined by a normal formula, (10) follows.

We note that

$$\forall l ((!\mathcal{A}(l) \supset \mathfrak{X}(l) \underset{d}{=} Z_{k_0}) \& (\neg !\mathcal{A}(l) \supset \mathfrak{X}(l) \underset{d}{=} \mathfrak{D}^d)),$$

where

$$\bar{k}_0 \leftrightharpoons \mu j (G(l \square j) \mathrel{\dot{=}} \Lambda.$$

We shall prove further the following auxiliary formula:

$$\forall l d_2 (d_2 \underset{d}{=} \mathfrak{X}(l) \supset \forall m (\{\mathrm{Or}(\mathfrak{X}(l))\}(m) \leqslant \{\mathrm{Or}(d_2)\}(m))). \tag{19}$$

Suppose that l is a natural number and d_2 a word of type d such that $d_2 \underset{d}{=} \mathfrak{X}(l)$. First we show that

$$!\mathcal{A}(l) \vee \neg !\mathcal{A}(l) \supset \forall m (\{\mathrm{Or}(\mathfrak{X}(l))\}(m) \leqslant \{\mathrm{Or}(d_2)\}(m)), \tag{20}$$

If $\neg !\mathscr{A}(l)$, then $\forall m(\{\gamma(l)\}(m) = 0)$, and hence

$$\forall m (\{\mathrm{Or}(\mathfrak{X}(l))\}(m) \leqslant \{\mathrm{Or}(d_2)\}(m)).$$

Now suppose that $!\mathscr{A}(l)$. Then we have

$$\forall m ((m \leqslant k_0 \supset \{\gamma(l)\}(m) = 0) \& (m > k_0 \supset \{\gamma(l)\}(m) = k_0)), \tag{21}$$

where $k_0 \leftrightharpoons \mu j (G(l \square j) \mathrel{\dot{=}} \Lambda)$. On the other hand, $\mathfrak{X}(l) \underset{d}{=} Z_{k_0}$. Therefore it follows from (9) and $d_2 \underset{d}{=} \mathfrak{X}(l)$ that

$$\forall m (m > k_0 \supset \{\mathrm{Or}(d_2)\}(m) \geqslant k_0). \tag{22}$$

It follows from (21) and (22) also that

$$\forall m (\{\mathrm{Or}(\mathfrak{X}(l))\}(m) \leqslant \{\mathrm{Or}(d_2)\}(m)).$$

Thus formula (20) is proved. From (20) and (18) it follows that

$$\neg\neg \forall m (\{\mathrm{Or}(\mathfrak{X}(l))\}(m) \leqslant \{\mathrm{Or}(d_2)\}(m)),$$

so that

$$\forall m (\{\mathrm{Or}(\mathfrak{X}(l))\}(m) \leqslant \{\mathrm{Or}(dt)\}(m)).$$

Thus (19) is proved.

Now we turn to the proof of 2.1.5. First we prove the following assertion:

1) *There cannot exist an algorithm which constructs relative to each generalized function d_1 satisfying the condition*

$$\neg\neg \exists k \forall m (\{\mathrm{Or}(d_1)\}(m) \leqslant k) \& \forall d_2 (d_2 = d_1 \supset \forall m (\{\mathrm{Or}(d_1)\}(m) \leqslant \{\mathrm{Or}(d_2)\}(m))), \tag{23}$$

a natural number k such that $\forall m(\{\mathrm{Or}(d_1)\}(m) \leq k)$.

We suppose that an algorithm $\mathfrak{H}$ having the indicated property is realizable. It is not difficult to verify that

(24) $$\forall l \neg\neg \exists k \forall m (\{\mathrm{Or}(\mathfrak{X}(l))\}(m) \leqslant k).$$

It follows from (24) and (19) that each generalized function of the sequence $\mathfrak{X}$ satisfies condition (23). We construct an algorithm ζ as follows:

$$\forall l (\zeta(l) \simeq \mathfrak{H}(\mathfrak{X}(l))).$$

Obviously ζ is an algorithm of the type $(\text{H} \to \text{H})$. Suppose that l is a natural number such that $!\mathscr{A}(l)$ and $k_0 \leftrightharpoons \mu j(G(l \square j) \doteq \Lambda)$. Then it follows from (21) that

(25) $$\forall m (m \geqslant k_0 \supset \{\mathrm{Or}(\mathfrak{X}(l))\}(m) = k_0).$$

On the other hand, from the property of the algorithm $\mathfrak{H}$ we find that

(26) $$\forall m (\{\mathrm{Or}(\mathfrak{X}(l))\}(m) \leqslant \zeta(l)).$$

It follows directly from (25) and (26) that $\zeta(l) \geqq k_0$. Thus we can realize an algorithm ζ of the type $(\text{H} \to \text{H})$ such that for any natural number l, if $!\mathscr{A}(l)$, then the process of application of $\mathscr{A}$ to the number l terminates after no more than $\zeta(l)$ steps. Hence it follows that the problem of the applicability of $\mathscr{A}$ to the natural numbers is decidable. This contradicts the property of the algorithm $\mathscr{A}$. Thus proposition 1) is proved.

It is not difficult to prove from proposition 1) the following proposition:

2) *There cannot exist an algorithm which constructs for each generalized function d_1 satisfying the condition*

$$\exists k \forall m (\{\mathrm{Or}(d_1)\}(m) \leqslant k) \,\&\, \forall d_2 (d_2 = d_1 \supset \forall m (\{\mathrm{Or}(d_1)\}(m) \leqslant \{\mathrm{Or}(d_2)\}(m))),$$
(27)

a natural number k such that $\forall m(\{\mathrm{Or}(d_1)\}(m) \leqq k)$.

Finally, using proposition 2), we prove 2.1.5 as follows. We suppose that we can realize an algorithm $\mathfrak{G}$ such that if d_1 is a generalized function of finite order, then

$$!\mathfrak{G}(d_1) \,\&\, (\mathfrak{G}(d_1) \in \text{H}) \,\&\, (d_1 \in \underline{\text{order} \leqslant \mathfrak{G}(d_1)}).$$

Such an algorithm $\mathfrak{G}$ is applicable to any generalized function d_1 satisfying the condition (27). We shall show that if d_1 satisfies condition (27), then

(28) $$\forall m (\{\mathrm{Or}(d_1)\}(m) \leqslant \mathfrak{G}(d_1)).$$

Indeed, suppose that d_1 satisfies (27) and $k \leftrightharpoons \mathfrak{G}(d_1)$. Then $(d_1 \in \underline{\text{order} \leqq k})$, i.e. a function d_2 is realizable such that

(29) $$d_2 \underset{d}{=} d_1 \,\&\, \forall m (\{\mathrm{Or}(d_2)\}(m) \leqslant k).$$

It follows from (29) and (27) that $\forall m(\{\mathrm{Or}(d_1)\}(m) \leqq k)$, i.e. formula (28) is proved. Thus the algorithm $\mathfrak{G}$ has the property indicated in proposition 2). But by proposition 2), such an algorithm is not possible. Theorem 2.1.5 is completely proved.

In connection with 2.1.3 we prove the following theorem.

2.1.6. *For any natural number k, it is not possible to have an algorithm which transforms each generalized function d_1 of order $\leqq k$ into a word Q of type* ш *such that* $d_1 \underset{d}{=} \mathrm{Dr}(k \square \bar{0}\, \tau\, Q)$.

Proof. We will use the construction in the proof of Theorem 10.2.2 in [**16**], or rather in the corrected version of that proof given in [**17**].* On the basis of this construction it is possible to construct an algorithm φ of type $(\bar{q} \to \text{ш})$ having the following properties:

1) For any word $\bar{q}_1$ of type $\bar{q}$ we can realize a word of type ш which is the derivative of the word $\varphi(\bar{q}_1)$.

2) There does not exist an algorithm ψ of the type $(\bar{q} \to \text{ш})$ such that, for any word $\bar{q}_1$ of type $\bar{q}$, $\psi(\bar{q}_1)$ is the derivative of $\varphi(\bar{q}_1)$.

Here $\bar{q}$ is the generic letter of the set of F-constructs in the space Π_0 (see [**16**]).

Suppose that k is any fixed natural number. We construct an algorithm $\mathfrak{X}$ of the type $(\bar{q} \to d)$ as follows:

$$\forall \bar{q}_1 (\mathfrak{X}(\bar{q}_1) \simeq \mathrm{Dr}(k+1 \square \bar{0} \tau \varphi(\bar{q}_1))).$$

First we prove that $\mathfrak{X}(\bar{q}_1)$ is a generalized function of order $\leqq k$ for any word $\bar{q}_1$ of type $\bar{q}$. Suppose that $\bar{q}_1$ is a word of type $\bar{q}$. In view of property 1) we can realize a word P of type ш such that $\text{Д}(P, \varphi(\bar{q}_1))$. Hence from 4.2.3, Chapter IV, it follows that $\bar{0}\, \tau\, P \underset{d}{=} D(\bar{0}\, \tau\, \varphi(\bar{q}_1))$, so that

$$\mathfrak{X}(\bar{q}_1) = \mathrm{Dr}(k+1 \square \bar{0} \tau \varphi(\bar{q}_1)) = \mathrm{Dr}(k \square \bar{0} \tau P).$$

Therefore $\mathfrak{X}(\bar{q}_1)$ is a generalized function of order $\leqq k$.

Now we suppose that we can realize an algorithm $\mathfrak{S}$ which transforms each generalized function d_1 of order $\leqq k$ into a word $\mathfrak{S}(d_1)$ of type ш such that $d_1 \underset{d}{=} \mathrm{Dr}(k \square \bar{0}\, \tau\, \mathfrak{S}(d_1))$. We construct an algorithm $\mathfrak{H}$ of type $(\bar{q} \to \text{ш})$ as follows:

$$\forall \bar{q}_1 (\mathfrak{H}(\bar{q}_1) \simeq \mathfrak{S}(\mathfrak{X}(\bar{q}_1))).$$

We have

$$\forall \bar{q}_1 (\mathrm{Dr}(k+1 \square \bar{0} \tau \varphi(\bar{q}_1)) \underset{d}{=} \mathrm{Dr}(k \square \bar{0} \tau \mathfrak{H}(\bar{q}_1))). \tag{30}$$

Suppose that $\bar{q}_1$ is a word of type $\bar{q}$ and m is a natural number $\geqq 1$. In view of formula (16) in §4.2, Chapter IV, we have

$$F(\overline{k+1} \square m \square \mathrm{Dr}(k+1 \square \bar{0} \tau \varphi(\bar{q}_1))) \underset{m}{=} \mathfrak{A}(k+1 \square \varphi(\bar{q}_1)); \tag{31}$$

$$F(\bar{k} \square m \square \mathrm{Dr}(k \square \bar{0} \tau \mathfrak{H}(\bar{q}_1))) \underset{m}{=} \mathfrak{A}(k \square \mathfrak{H}(\bar{q}_1)). \tag{32}$$

From (32) and property F 5 (3.4.1, Chapter IV) we get

---

* *Editor's note.* The corrections given in [**17**] were incorporated into the English translation of [**16**].

$$(33)\qquad F(\overline{k+1}\square m\square \mathrm{Dr}(k\square\bar{0}\tau\mathfrak{H}(\bar{q}_1))) \underset{m}{=} \mathfrak{A}(k+1\square J(1\square\mathfrak{H}(\bar{q}_1))).$$

In view of property F 6 (3.4.1, Chapter IV), it follows from (30), (31) and (33) that

$$(34)\qquad \forall\bar{q}_1(\mathfrak{A}(k+1\square\varphi(\bar{q}_1)) \underset{\text{Ш}}{=} \mathfrak{A}(k+1\square J(1\square\mathfrak{H}(\bar{q}_1))).$$

Therefore we can realize an algorithm ζ of type $(\bar{q}\to \text{Ю})$ such that

$$\forall\bar{q}_1(\varphi(\bar{q}_1) \underset{\text{Ш}}{=} J(1\square\mathfrak{H}(\bar{q}_1)) + H(\zeta(\bar{q}_1))).$$

We construct an algorithm ψ of type $(\bar{q}\to \text{Ш})$ as follows:

$$\forall\bar{q}_1(\psi(\bar{q}_1)\simeq\mathfrak{H}(\bar{q}_1) + H(D^{\text{Ю}}(\zeta(\bar{q}_1)))),$$

where $D^{\text{Ю}}$ is the algorithm of differentiation of words of type Ю (see §1.2, Chapter IV).

Obviously, for any word $\bar{q}_1$ of type $\bar{q}$, $\psi(\bar{q}_1)$ is the derivative of $\varphi(\bar{q}_1)$. Such an algorithm, in view of property 2), is impossible. Hence the algorithm $\mathfrak{S}$ is impossible. Theorem 2.1.6 is proved.

The following corollary is obvious:

2.1.7. *There cannot exist an algorithm which transforms each regular generalized function d_1 into a word Q of the type* Ш *such that* $d_1 \underset{d}{=} \bar{0}_\tau Q$.

2.2. Suppose that d_1 is a constructive generalized function. Concerning the segment $a\triangle b$, we shall say that it is the *support* of the generalized function, and write $(a\triangle b\,\underline{\mathrm{supp}}\,d_1)$, if d_1 is equal to $\mathfrak{O}^d$ in the intervals $-\infty\triangle a$ and $b\triangle\infty$ (see §1.3).

We will say that d_1 is a *generalized function with bounded support* if

$$\exists m(-m\triangle m\,\underline{\mathrm{supp}}\,d_1).$$

2.2.1. *Every generalized function with bounded support is a generalized function of finite order.*

Proof. Suppose that d_1 is a generalized function with bounded support. Then we can realize a natural number m such that $(-m\triangle m\,\underline{\mathrm{supp}}\,d_1)$. Write $U\leftrightharpoons\mathrm{Or}(d_1)$, $k\leftrightharpoons\{U\}(m+1)$ and $Q\leftrightharpoons F(U\square m+1\square d_1)$. Since d_1 is equal to $\mathfrak{O}^d$ in the intervals $-\infty\nabla -m$ and $m\triangle\infty$, it is equal to $\mathfrak{O}^d$ in $-(m+1)\triangle -m$ and $m\triangle m+1$ (see 1.3.3). In view of Theorem 1.3.1 we can realize words ю_1 and ю_2 of type Ю such that

$$(35)\qquad dg(\text{ю}_1)\prec\!\cdot\, k,\ dg(\text{ю}_2)\prec\!\cdot\, k,$$

$$(36)\qquad \forall x((-m-1<x<-m\supset \underline{Q}(x) = \underline{H(\text{ю}_1)}(x))\,\&\,(m<x<m+1\supset\underline{Q}(x) = \underline{H(\text{ю}_2)}(x))).$$

On the basis of formula (36) it is possible to construct a word P of type Ш such that for any duplex x

$$\underset{\smile}{P}(x) = \begin{cases} \underset{\smile}{Q}(x), & \text{if } |x| \leqslant m+1, \\ \underset{\smile}{H(\text{ю}_1)}(x), & \text{if } x \leqslant -m, \\ \underset{\smile}{H(\text{ю}_2)}(x), & \text{if } x \geqslant m. \end{cases}$$

Then we write $d_2 \leftrightarrows \mathrm{Dr}(k \square \bar{0}_T P)$. It is clear from the remark to Theorem 1.3.2 that d_1 is equal to d_2 in $-(m+1)\triangle(m+1)$. Moreover, d_2 is equal to $\mathfrak{O}^d$ in $-\infty\triangle -m$ and $m\triangle\infty$. This last is proved using Theorem 1.3.1. Therefore in these intervals d_1 is equal to d_2. In view of 1.3.3 it follows from what has been said that $d_1 \underset{d}{=} d_2$. On the other hand, it is obvious that $\forall m(\{\mathrm{Or}(d_2)\}(m) \leqq k)$. Thus d_1 is a generalized function of finite order, as was required to be proved.

2.2.2. *Every constructive generalized function is the l-limit (and hence the limit in the space $\mathfrak{D}$) of a sequence of regular generalized functions with bounded support.*

Proof. Suppose given a generalized function d_1. We write

$$U \leftrightarrows \mathrm{Or}(d_1), \quad P \leftrightarrows \mathrm{Fd}(d_1), \quad \lambda_n \leftrightarrows \underset{\smile}{P}_{\psi}(n).$$

Obviously

(37) $$\forall minp\,(n,\, p \geqslant \overline{P}^{\psi}(i) \,\&\, m \leqslant i \supset \mathscr{N}(m \square \Phi(U \square m \square \lambda_n) - \Phi(U \square m \square \lambda_p)) \leqslant 2^{-i}).$$

It is possible to construct a sequence φ of words of type ш such that, for any natural number n and any real duplex x,

$$\underset{\smile}{\varphi(n)}(x) \begin{cases} = \underset{\smile}{\lambda_n}(x), \text{ if } |x| \leqq n, \\ = \bar{0}, \text{ if } |x| \geqq n+1, \\ \text{is linear in the segments } -n-1\triangle -n \text{ and } n\triangle n+1. \end{cases}$$

We have

(38) $$\forall mn\,(n \geqslant m \supset \Phi(U \square m \square \varphi(n)) \underset{m}{=} \Phi(U \square m \square \lambda_n)).$$

It follows from (37) and (38) that

(39) $$\forall minp\,(n,\, p \geqslant \max(i \square \overline{P}^{\psi}(i)) \,\&\, m \leqslant i \supset \mathscr{N}(m \square \Phi(U \square m \square \varphi(n)) - \Phi(U \square m \square \varphi(p))) \leqslant 2^{-i}).$$

Suppose that ζ is an algorithm of the type (н → н), defined by the formula

$$\forall i\,(\zeta(i) \simeq \max(i \square \overline{P}^{\psi}(i))).$$

We write $Q \leftrightarrows \mathfrak{E}\varphi$, Ч$^{\text{ш}}_4$З$\odot\mathfrak{E}\zeta$, Ч$^{\text{ш}}_4$З and $d_2 \leftrightarrows U_T Q$. It follows from (39) that $Q \in \mathfrak{D}^U$, so that d_2 is a generalized function.

It follows from (38) that

$$\forall inm\,(n \geqslant i \,\&\, m \leqslant i \supset \mathscr{N}(m \square \Phi(U \square m \square \varphi(n)) - \Phi(U \square m \square \lambda_n)) \leqslant 2^{-i}),$$

so that $Q \underset{D,U}{=} P$, whence $d_1 \underset{d}{=} d_2$.

Suppose that ψ is a sequence of words of type d, defined by the formula

$$\forall n (\psi(n) \simeq \bar{0}\, \tau \varphi(n)).$$

In view of 2.1.1, d_2 (and hence d_1 as well) is the l-limit of the sequence ψ. We note that if $|x| \geqq n+1$, then $\underline{\varphi(n)}(x) = 0$, so that

$$(-(n+1) \triangle (n+1) \underline{\text{supp }\psi(n)})$$

for any n. Thus d_1 is the l-limit of a sequence of regular generalized functions with bounded support.

In concluding this section we prove the following theorem:

2.2.3. *There cannot exist an algorithm which constructs relative to each generalized function d_1 with bounded support a natural number m such that* $(-m \triangle m \underline{\text{supp }d_1})$.

PROOF. Consider a sequence $\mathfrak{X}$ of generalized functions as constructed in the proof of Theorem 2.1.5. It is easily verified that, for any natural number l, if $\neg ! \mathscr{A}(l)$, then $\mathfrak{X}(l) = \mathfrak{O}^d$, and if $! \mathscr{A}(l)$ and $k \leftrightharpoons \mu j (G(l \square j) \underset{\cdot}{=} \Lambda)$, then

$$(-(k+1) \triangle (k+1) \underline{\text{supp }\mathfrak{X}(l)}).$$

Therefore

$$\forall l \neg\neg \exists m (-m \triangle m \underline{\text{supp }\mathfrak{X}(l)}).$$

Using the same considerations as we used to prove proposition 1) in the proof of Theorem 2.1.5, we can prove the following proposition:

1*) *There cannot exist an algorithm which constructs relative to each generalized function d_1 satisfying the condition*

$$\neg\neg \exists (-m \triangle m \underline{\text{supp }d_1})$$

a natural number m such that $(-m \triangle m \underline{\text{supp }d_1})$.

After this, Theorem 2.2.3 follows from proposition 1*) and from the fact that the condition $(-m \triangle m \underline{\text{supp }d_1})$ may be expressed by a normal formula.

§3. Value of a constructive generalized function at a point

3.1. In order to consider the connection between the concept of a constructive generalized function and the concept of a constructive function, in this subsection we introduce the concept of the value of a constructive generalized function at a point, copying the definition of Łojasiewicz on the value of a generalized function in classical mathematics (see [47], [48]).

First of all we introduce some auxiliary concepts and notation. Suppose that P is a word of type ш and k a natural number. We will write

$$P^{(k)} \leftrightharpoons \text{Dr}(k \square \bar{0}_{\tau} P).$$

Suppose that a is a real duplex, m a natural number $\geqq 1$ and φ an algorithm of type (д $\dot{\rightarrow}$ ш) applicable to any real duplex $\alpha \neq a$. We shall say that the word P of type ш is the *limit* in C_m of the algorithm φ as $\alpha \to a$, and write

$$(\varphi(\alpha) \xrightarrow[\alpha \to a]{c_m} P),$$

if the following condition is satisfied:

$$\forall i \exists l \forall \alpha (0 < |\alpha - a| < 2^{-l} \supset \mathcal{N}(m \square \varphi(\alpha) \underset{\text{ш}}{-} P) < 2^{-i}).$$

Suppose that U is a word of type e, Q a word of type d^U, and ψ an algorithm of type $(\text{д} \dot{\to} d^U)$ applicable to any real duplex $\alpha \neq a$. We will say that Q is the *limit* in the space $\mathfrak{D}^U$ of the algorithm ψ as $\alpha \to a$, and write

$$(\psi(\alpha) \xrightarrow[\alpha \to a]{\mathfrak{D}^U} Q),$$

if the following condition is satisfied:

$$\forall i \exists l \forall \alpha (0 < |\alpha - a| < 2^{-l} \supset \mathfrak{Q}_D^U(i, \psi(\alpha) \underset{D, U}{-} Q)). \tag{1}$$

Condition (1) is equivalent to the following condition:

$$\forall mi \exists l \forall \alpha (0 < |\alpha - a| < 2^{-l} \supset \mathfrak{N}^d(U \square m \square \psi(\alpha) \underset{D, U}{-} Q) < 2^{-i}). \tag{2}$$

Now suppose that a is a real duplex, Z a constructive real function (i.e. a word of type d), and λ an algorithm of the type $(\text{д} \dot{\to} d)$ applicable to any real duplex $\alpha \neq a$. In keeping with the definition of l-limit in §3.5 of Chapter IV, we will say that Z is the *l-limit of the algorithm* λ *as* $\alpha \to a$, and write

$$(\lambda(\alpha) \xrightarrow[\alpha \to a]{l} Z),$$

if we can realize a word U of type e, a word Q of type d^U, an algorithm ψ of type $(\text{д} \dot{\to} d^U)$ applicable to any real duplex $\alpha \neq a$, and a natural number k, such that

$$(Z \underset{d}{=} U \tau Q) \,\&\, \forall \alpha (0 < |\alpha - a| < 2^{-k} \supset \lambda(\alpha) \underset{d}{=} U \tau \psi(\alpha)) \,\&\, (\psi(\alpha) \xrightarrow[\alpha \to a]{\mathfrak{D}^U} Q).$$

It is possible to prove that Z is the l-limit of the algorithm λ as $\alpha \to a$ if and only if the following condition is satisfied:

$$\forall m \exists k \text{ш}_1 \varphi l \Big((Z \underset{d, -m \nabla m}{=\!=\!=} \text{ш}_1^{(k)}) \,\&\, \forall \alpha (0 < |\alpha - a| < 2^{-l} \supset \lambda(\alpha) \underset{d, -m \nabla m}{=\!=\!=} \varphi(\alpha)^{(k)}) \,\&\, \Big(\varphi(\alpha) \xrightarrow[\alpha \to a]{c_m} \text{ш}_1\Big)\Big),$$

where φ is a functorial variable for algorithms of the type $(\text{д} \dot{\to} \text{ш})$ applicable to any duplex $\alpha \neq a$, and the sign $\underset{d, -m \nabla m}{=}$ denotes equality of generalized functions in the interval $-m \nabla m$.

Suppose that a is a real duplex and f a constructive function defined at all points $x \neq a$. We shall say that the real duplex c is the *limit* of the function f as $x \to a$, and write

$$f(x) \xrightarrow[x \to a]{} c,$$

if the following condition is satisfied:

$$\forall i \exists l \forall x (0 < |x-a| < 2^{-l} \supset |f(x) - c| < 2^{-i}).$$

Suppose that c is a real duplex. Then the word $\overline{0}_T H(c)$ is a word of type d. We shall write $\tilde{c} \leftrightharpoons \overline{0}_T H(c)$.

Now we define the concept of the value of a constructive generalized function at a point. Suppose that d_1 is a constructive generalized function and a a real duplex. Concerning the real duplex c, we shall say that it is the *value of the generalized function d_1 at the point a*, and write

$$(c \,\underline{\mathrm{val}}_a\, d_1),$$

if the condition

$$(L^d(\alpha \square a \square d_1) \xrightarrow[\alpha \to 0]{l} \tilde{c})$$

is satisfied, where L^d is the algorithm of linear substitution of a variable for generalized functions, defined in §4.3 of Chapter III.[7]

The majority of the theorems in the paper [47] of Łojasiewicz on the value of a generalized function may be carried over in appropriate form to constructive mathematics. Among these we mention Theorem 2.3 and Corollary 3 of Theorem 5.2 of that paper. They carry over to constructive mathematics in the following form:

3.1.1. *Suppose that d_1 is a constructive generalized function and a a real duplex. The real duplex c is the value of d_1 at the point a if and only if one can realize an interval Ω which contains the point a, a natural number k and a word P of type* ш, *such that*

$$d_1 \underset{d,\Omega}{=} P^{(k)} \quad \text{and} \quad (\underline{P(x) : (x-a)^k}) \underset{x \to a}{\rightarrow} (c : k!).$$

3.1.2. *Suppose that Ω is an interval, and d_1 and d_2 are constructive generalized functions having values at every point of the interval Ω. If*

$$\forall xy (x \in \Omega \supset (y \,\underline{\mathrm{val}}_x\, d_1 \equiv y \,\underline{\mathrm{val}}_x\, d_2)),$$

then d_1 is equal to d_2 in the interval Ω.

We shall not present the proofs of these theorems here in detail. We note only the following peculiarity: Theorem 3.1.1 is the result of carrying over Theorem 2.3 of [47] to constructive mathematics. The proof presented in

[7] We note that within the limits of constructive mathematics the following assertion holds. Suppose that d_1 is a constructive generalized function and α a real duplex. If we can realize a constructive generalized function Z such that

$$(L^d(\alpha \square a \square d_1) \xrightarrow[\alpha \to a]{l} Z),$$

then Z is a constant, i.e. a duplex c is realizable such that $Z \underset{d}{=} \tilde{c}$. This assertion is a result of carrying over a theorem of Zieleźny (see [48, 57]) to constructive mathematics.

[48], §17, meets the requirements of constructive mathematics. We note only that in that same Theorem 2.3 of [48] the generalized functions in question were of finite order, and that the d_1 in the formulation of our Theorem 3.1.1 is a constructive function of any order. However, the concept of the value of a generalized function at a point has a local character, so that we nevertheless have the possibility of copying the proof of Theorem 2.3 of [47] as presented in [48], §17, so as to obtain a proof of Theorem 3.1.1 within the limits of constructive mathematics. This may be done with only minor alterations of a technical character.

Theorem 3.1.2 is the result of carrying over Corollary 3 to Theorem 5.2 of [47] to constructive mathematics. In [47] that corollary was based on Lemmas and Theorems 2.3, 2.8, 5.1 and 5.2 (see [47]). These lemmas and theorems may be carried over into constructive mathematics with only minor alterations, so that Theorem 3.1.2 is proved on the basis of their constructive analogues.

Now we shall prove the following lemma.

3.1.3. *Suppose that d_1 is a generalized function, a and c real duplexes and Ω an interval containing the point a. If for some natural number k one can realize a word P of type* ш *such that*

$$(d_1 \underset{d,\,\Omega}{=} P^{(k)}) \,\&\, \left(\frac{\underset{\sqcup}{P}(x)}{(x-a)^k} \underset{x\to a}{\longrightarrow} \frac{c}{k!} \right),$$

then for any natural number $l \geq k$ one can realize a word Q of type ш *such that*

$$(d_1 \underset{d,\,\Omega}{=} Q^{(l)}) \,\&\, \left(\frac{\underset{\sqcup}{Q}(x)}{(x-a)^l} \underset{x\to a}{\longrightarrow} \frac{c}{l!} \right).$$

PROOF. Suppose that d_1, a, c, Ω, k and P satisfy the hypotheses of the lemma. First we consider the case $l = k+1$. It is possible to construct a word Q of type ш such that

$$\forall x.(\underset{\sqcup}{Q}(x) = \int_a^x (P)).$$

Then P is the derivative of the word Q, so that $\overline{0}_T P \underset{d}{=} D(\overline{0}_T Q)$ (see 4.2.3 of Chapter IV), whence $P^{(k)} \underset{d}{=} Q^{(k+1)}$. Hence, and from the assumption $(d_1 \underset{d,\Omega}{=} P^{(k)})$, it follows that $(d_1 \underset{d,\Omega}{=} Q^{(k+1)})$. Moreover, from the hypothesis that

$$\left(\frac{\underset{\sqcup}{P}(x)}{(x-a)^k} \underset{x\to a}{\longrightarrow} \frac{c}{k!} \right)$$

it follows that

$$\left(\frac{\underset{\sqcup}{Q}(x)}{(x-a)^{k+1}} \underset{x\to a}{\longrightarrow} \frac{c}{(k+1)!} \right).^{8)}$$

8) This is proved using the following assertion. Suppose that Ω is an interval containing

Thus the lemma is proved for the case $l=k+1$. Starting from what has been proved, we may prove the lemma by induction for any $l>k$.

3.1.4. *Suppose that* d_1 *and* d_2 *are constructive generalized functions,* a, c_1 *and* c_2 *real duplexes,* Ω *an interval and* P *a word of type* ш.

1) *If* $d_1 \underset{d,\Omega}{=} \overline{0}_T P$, *then* $\forall x(x \in \Omega \supset \underset{\sqcup}{P}(x)\ \underline{\text{val}}_x d_1)$;
2) *If* $d_1 \underset{d,\Omega}{=} d_2$ *and* $c_1\ \underline{\text{val}}_a d_1$, *then* $c_1\ \underline{\text{val}}_a d_2$;
3) *If* $c_1\ \underline{\text{val}}_a d_1$ *and* $c_2\ \underline{\text{val}}_a d_1$, *then* $c_1 \underset{B}{=} c_2$;
4) *If* $c_1\ \underline{\text{val}}_a d_1$ *and* $c_2\ \underline{\text{val}}_a d_2$, *then* $(c_1+c_2)\ \underline{\text{val}}_a (d_1+d_2)$ *and* $(c_1-c_2)\ \underline{\text{val}}_a (d_1-d_2)$.

Proof. Assertions 1) and 2) follow from Theorem 3.1.1. Assertion 3) follows from the definition of the value of a generalized function d_1 at the point a. Assertion 4) is proved using Theorem 3.1.1 and Lemma 3.1.3.

Suppose that d_1 is a generalized function and a a real duplex. The point a is called a *regular point* for the generalized function d_1 if

$$\exists x(x\ \underline{\text{val}}_a d_1).$$

The point a is said to be a *singular point for* d_1 if

$$\neg \exists x(x\ \underline{\text{val}}_a d_1).$$

3.1.5. *Suppose that* d_1 *and* d_2 *are constructive generalized functions such that* $D(d_2) \underset{d}{=} d_1$. *Then every regular point for* d_1 *is also a regular point for* d_2.

Proof. Suppose that a is a regular point for d_1, and that $(c\ \underline{\text{val}}_a d_1)$. In view of Theorem 3.1.1 we can realize an interval Ω containing the point a, a natural number k and a word P of type ш such that

$$(d_1 \underset{d,\,\Omega}{=} P^k) \& \left(\frac{\underset{\sqcup}{P}(x)}{(x-a)^k} \xrightarrow[x \to a]{} \frac{c}{kl} \right). \tag{3}$$

In view of Lemma 3.1.3 we may suppose that $k \geqq 1$. Obviously $P^{(k)} \underset{d}{=} D(P^{(k+1)})$, so that $d_1 \underset{d,\Omega}{=} D(P^{(k-1)})$. Hence, using the assumption $D(d_2) \underset{d}{=} d_1$, we find that $D(d_2 - P^{(k-1)}) \underset{d,\Omega}{=} \mathfrak{O}^d$. In view of Theorem 4.2.6 of Chapter III,[9] it follows that we can realize a real duplex α such that

$$d_2 \underset{d,\Omega}{=} P^{(k-1)} + \tilde{\alpha}.$$

It follows from (3) that

$$\left(\frac{\underset{\sqcup}{P}(x)}{(x-a)^{k-1}} \xrightarrow[x \to a]{} 0 \right),$$

the point a, and that f, g, f' and g' are constructive functions defined in Ω; here f' is the derivative of f, g' the derivative of g, $f(a)=g(a)=0$ and $g'(x) \neq 0$ for any $x \neq a$. Then for any duplex c we have

$$\left(\frac{f'(x)}{g'(x)} \xrightarrow[x \to a]{} c \right) \supset \left(\frac{f(x)}{g(x)} \xrightarrow[x \to a]{} c \right).$$

This assertion is valid in constructive mathematics.

[9] This theorem was formulated for constructive generalized functions. It is also valid for constructive generalized functions in the interval Ω.

so that $0\,\underline{\text{val}}_a\,P^{(k-1)}$. Moreover, in view of property 1) of 3.1.4 we have $(\alpha\,\underline{\text{val}}_a\,\tilde{\alpha})$. Hence, in view of properties 4) and 2), we find from 3.1.4 that $(\alpha\,\underline{\text{val}}_a\,d_2)$. So a is a regular point of d_2.

Suppose that d_1 is a generalized function, and that a and b are real duplexes. Suppose further that d_2 is the "antiderivative" of the generalized function d_1, i.e. a word of type d such that $D(d_2) = d_1$. If a and b are regular points for d_2 and $(\alpha\,\underline{\text{val}}_a\,d_2)\,\&\,(\beta\,\underline{\text{val}}_b\,d_2)$, then the number $\beta - \alpha$ will be called the *definite integral* of the generalized function d_1 between the limits a and b. It will be denoted by $\int_a^b [d_1]$.

We note that if d_2 and d_3 are antiderivatives of one and the same function d_1, i.e. if $D(d_2) = d_1$ and $D(d_3) = d_1$, then a real duplex c such that $d_2 = d_3 + \tilde{c}$ is realizable. Accordingly, every regular point for d_2 is also a regular point for d_3, and conversely. Moreover, if

$$(\alpha\,\underline{\text{val}}_a\,d_2)\,\&\,(\beta\,\underline{\text{val}}_b\,d_2) \quad \text{and} \quad (\alpha_1\,\underline{\text{val}}_a\,d_3)\,\&\,(\beta_1\,\underline{\text{val}}_b\,d_3),$$

then $\beta - \alpha = \beta_1 - \alpha_1$. Thus the definite integral of a generalized function does not depend on the choice of its antiderivative.

From what has been said it is clear that the integral $\int_a^b [d_1]$ is defined if and only if a and b are regular points of some generalized function d_1 which is an antiderivative.

Basing ourselves on 3.1.4, we can prove the following assertion.

3.1.6. *Suppose that d_1 and d_2 are constructive generalized functions, a and b real duplexes, Ω an interval and P a word of type* ш.

1) *If a and b lie in Ω, and $d_1 \underset{d,\Omega}{=} \overline{0}_T P$, then $\int_a^b [d_1] = \int_a^b (P)$.*

2) *If $\int_a^b [d_1]$ and $\int_a^b [d_2]$ are defined, then $\int_a^b [d_1 + d_2]$ and $\int_a^b [d_1 - d_2]$ are defined, and*

$$\int_a^b [d_1 \pm d_2] = \int_a^b [d_1] \pm \int_a^b [d_2].$$

3.2. We have defined constructive generalized functions on the basis of almost uniformly continuous functions. Each almost uniformly continuous function "is" in a certain sense a constructive generalized function. Thus the concept of constructive generalized function is a "generalization" of the concept of almost uniformly continuous function. However, as is shown by the following theorem, the concept of constructive generalized function is not a "generalization" of the concept of constructive continuous function.

3.2.1. *It is possible to determine a constructive generalized function f, defined on the segment $0\triangle 1$, such that there cannot exist a constructive generalized function d_1 satisfying the condition*

$$\forall x (x \in 0\triangle 1 \supset f(x)\,\underline{\text{val}}_x\,d_1).$$

PROOF. In the paper [6] the concept of "generalized integral on a segment" was introduced. For our purposes we shall make some changes in the definition of this concept and introduce the following concept. Suppose that $\alpha\triangle\beta$ is a segment and Φ some exact disjunct covering by segments of the segment $\alpha\triangle\beta$.[10] The two-place predicate I^* is called the *generalized integral on the segment $\alpha\triangle\beta$ relative to the covering* Φ, if for arbitrary real duplexes u, v and for arbitrary constructive functions f, g, given on $\alpha\triangle\beta$, the following conditions are satisfied:

1) If f is polygonal on each segment Φ_n, if $I^*(\mathcal{E} f \mathcal{3}, u)$, and if

$$\forall x(x \in \alpha\triangle\beta \supset f(x) \geqq 0),$$

then $u \geqq 0$.

2) If $I^*(\mathcal{E} f \mathcal{3}, u)$ and $I^*(\mathcal{E} g \mathcal{3}, v)$, then $I^*(\mathcal{E} f - g \mathcal{3}, u - v)$.

3) If f is polygonal on $\alpha\triangle\beta$, then $I^*(\mathcal{E} f \mathcal{3}, \int_a^b f(x)\,dx)$.

We fix the covering Φ used in the proof of Theorem 3.5 of [6]. This is a singular exact disjunct covering by segments of the segment $0\triangle1$. It is easy to see that the function f constructed in the proof of that theorem is polygonal on each segment Φ_n, where n is any natural number. Therefore that theorem may be formulated as follows.

One can realize a constructive function f given on the segment $0\triangle1$ having the following property. For any generalized integral I^* on the segment $0\triangle1$ relative to the covering Φ there cannot exist a real duplex u satisfying the condition $I^*(\mathcal{E} f \mathcal{3}, u)$.

Now we define a predicate I^* as follows. For any function f given on $0\triangle1$ and any real duplex u, $I^*(\mathcal{E} f \mathcal{3}, u)$ if and only if we can realize a constructive generalized function d_1 satisfying the following conditions:

a) $\forall x(x \in 0\triangle1 \supset f(x)\ \underline{\mathrm{val}}_x\, d_1)$;

b) $u = \int_0^1 [d_1]$ (this integral is defined because 0 and 1 are regular points of d_1).

We shall show that the predicate I^* thus defined is a generalized integral on $0\triangle1$ relative to the covering Φ.

1) Suppose that f is polygonal on each segment Φ_n, and that $I^*(\mathcal{E} f \mathcal{3}, u)$ and

$$\forall x\,(x \in 0\triangle1 \supset f(x) \geqslant 0).$$

Suppose that n is a real number and $\Phi_n \mathrel{\bar{\bar{\circ}}} a_n\triangle b_n$. Write $\Omega \leftrightharpoons a_n \nabla b_n$. The function f is uniformly continuous on Φ_n, so that we can construct a word P of type ш such that

$$\forall x\,(a_n \leqslant x \leqslant b_n \supset \underset{\smile}{P}(x) \geqslant f(x)). \tag{4}$$

In view of Theorem 3.1.2 it follows from (4) and condition a) that $d_1 \underset{d,\Omega}{=} \overline{0}_T P$. Hence, in view of property 1) of 3.1.6, we conclude that, for arbitrary duplexes α and β with $\alpha \leqq \beta$ and $\alpha\triangle\beta \subseteq \Omega$, one has $\int_\alpha^\beta [d_1] = \int_\alpha^\beta (P)$. Since $f(x) \geqq 0$ for $x \in 0\triangle1$, it therefore follows that $\int_\alpha^\beta [d_1] \geqq 0$.

10) Here we apply concepts introduced in [6], §1.

Suppose that d_2 is the antiderivative of d_1. It follows from condition a) that each point in the segment $0\triangle 1$ is a regular point of d_2, i.e. that

$$\forall x(x \in 0\triangle 1 \supset \exists y(y\,\underline{\mathrm{val}}_x d_2)).$$

By the principle of the constructive interpretation of a mathematical inference it therefore follows that we can realize an algorithm g of the type $(\text{д} \dot{\rightarrow} \text{д})$ with the property that

$$\forall x(0 \leqq x \leqq 1 \supset !g(x) \,\&\, (g(x)\,\underline{\mathrm{val}}_x d_2)).$$

It is not hard to verify that if $x =_{\text{в}} y$ and $0 \leqq x,\ y \leqq 1$, then $g(x) =_{\text{в}} g(y)$. Therefore g is a constructive function given on the segment $0\triangle 1$.

By the definition of definite integral we have $\int_\alpha^\beta [d_1] = g(\beta) - g(\alpha)$. Summing up the above results, we get

(5) $$\forall \alpha\beta\,(a_n < \alpha \leqslant \beta < b_n \supset g(\beta) - g(\alpha) \geqslant 0).$$

Being a constructive function, the function g is continuous on the segment $0\triangle 1$, so that one may prove from (5) that

$$\forall \alpha\beta\,(a_n \leqslant \alpha \leqslant \beta \leqslant b_n \supset g(\beta) - g(\alpha) \geqslant 0).$$

Thus the function g does not decrease on each segment Φ_n. Hence it follows that g is a nondecreasing function on the segment $0\triangle 1$ (see the footnote on page 481 (transl. p. 66) of [**6**]). And we get

$$u = \int_0^1 [d_1] = g(1) - g(0) \geqslant 0,$$

as we wished to prove.

2) Suppose that $I^*(\mathcal{E} f \mathfrak{Z}, u)$ and $I^*(\mathcal{E} g \mathfrak{Z}, v)$. Then we can realize constructive generalized functions d_1 and d_2 such that

$$\forall x(x \in 0\triangle 1 \supset (f(x)\,\underline{\mathrm{val}}_x d_1) \,\&\, (g(x)\,\underline{\mathrm{val}}_x d_2)),$$

$$u = \int_0^1 [d_1] \text{ and } v = \int_0^1 [d_2].$$

Hence we get (see 3.1.4 and 3.1.6)

$$\forall x(x \in 0\triangle 1 \supset (f(x) - g(x)\,\underline{\mathrm{val}}_x (d_1 - d_2)) \,\&\, (u - v = \int_0^1 [d_1 - d_2])),$$

i.e. $I^*(\mathcal{E} f - g \mathfrak{Z},\ u - v)$.

3) Suppose that f is a polygonal function given on $0\triangle 1$. It is possible to construct a word P of type ш such that

$$\forall x(x \in 0\triangle 1 \supset \underset{\smile}{P}(x) = f(x)).$$

Write $d_1 \leftrightharpoons \overline{0}\,\tau\, P$. Obviously

$$\forall x(x \in 0\triangle 1 \supset f(x)\,\underline{\mathrm{val}}_x d_1).$$

Moreover, in view of property 1) of 3.1.6 we have

$$\int_0^1 [d_1] = \int_0^1 (P) = \int_0^1 f(x)\,dx.$$

Thus we have found that $I^*(\mathcal{E} f \mathfrak{Z}, \int_0^1 f(x)\,dx)$.

Thus we have proved that the predicate I^* defined above is a generalized integral on $0\triangle 1$ relative to the covering Φ.

We note that, by the definition of the predicate I^*, for any function f given on $0\triangle 1$ we have

$$\exists u I^*(\mathcal{E} f \mathfrak{Z}, u) \equiv \exists d_1 \forall x (x \in 0 \triangle 1 \supset f(x) \underline{\text{ val }}_x d_1).$$

The function f constructed in the proof of Theorem 3.6 of [6] has the property $\neg \exists u I^*(\mathcal{E} f \mathfrak{Z}, u)$, so that for it we have

$$\neg \exists d_1 \forall x (x \in 0\triangle 1 \supset f(x) \underline{\text{val}}_x d_1).$$

Theorem 3.2.1 is proved. (The fact that the function f is continuous on $0\triangle 1$ follows from the well-known theorem of G. S. Ceĭtin in [27]).

We now prove the following further theorem.

3.2.2. *It is possible to define a constructive generalized function d_1 having the property that there cannot exist a constructive function f such that, for any regular point x of the generalized function d_1, f is defined at x and $(f(x) \underline{\text{val}}_x d_1)$.*

In other words, there is a constructive generalized function whose values cannot be computed by means of a constructive function.

Proof. We construct a summable FR-construct X (i.e. a point of the space $\mathfrak{L}_1$; see [26], §15) such that for any natural numbers n and i

$$\underline{X}(n) \underset{\circ}{=} 1 \triangle \frac{-1}{2} \tau \frac{-1}{2} \triangle \frac{-1}{2^2} \tau \ldots \tau \frac{-1}{2^n} \triangle 0 \sigma 1 \tau \frac{1}{2} \tau \ldots \tau \frac{1}{2^n} \tau 0;\ \overline{X}(i) \underset{\circ}{=} i.$$

Then it is possible to construct a word P of type ш such that

$$\forall x \left(\underline{P}(x) \simeq \int_\alpha^x (X) - \int_\alpha^{-1} (X) \right),$$

where $\alpha \leftrightharpoons \min(-1 \,\square\, x)$. We can show that:

if $x \leq -1$, then $P(x) = 0$;

if $-1/2^n \leq x \leq -1/2^{n+1}$, then

$$\underline{P}(x) = \frac{1}{2^n}\left(x + \frac{1}{2^n}\right) + \frac{4^n - 1}{6 \cdot 4^{n-1}},$$

where n is any natural number;

if $x \geq 0$, then $P(x) = 2/3$.

Write $d_1 \leftrightharpoons D(\dot{\overline{0}} \tau P)$. Using Theorem 3.1.1, we can prove that:

a) if $x < -1$ or $x > 0$, then $(0 \underline{\text{val}}_x d_1)$;

b) if $-1/2^n < x < -1/2^{n+1}$, then $(2^{-n} \underline{\text{val}}_x d_1)$;

c) for any natural number n, the point -2^{-n} is not a regular point of d_1.

Moreover, it is not difficult to verify that

$$\frac{\underset{\sqcup}{P}(x) - \frac{2}{3}}{x} \xrightarrow[x \to 0]{} 0,$$

so that the point 0 is a regular point of d_1 and $(0 \, \underline{\text{val}}_0 \, d_1)$ (we note that $d_1 = D(\overline{0}_{\mathcal{T}} P) = D(\overline{0}_{\mathcal{T}} P \underset{d}{+} \tilde{\tfrac{2}{3}})$).

Suppose that a constructive function f is realizable with the property that for any regular point x of the generalized function d_1 we have $!f(x)$ and $f(x) \, \underline{\text{val}}_x \, d_1$. Then:

if $x < -1$ or $x > 0$, then $!f(x)$ and $f(x) = 0$;

if $-1/2^n < x < -1/2^{n+1}$, then $!f(x)$ and $f(x) = 1/2^n$; $!f(0) \,\&\, f(0) = 0$.

Since the function f is continuous at each point where it is defined, f cannot be defined at the points $x = -1/2^n$, n any natural number. Thus f is defined at the point $x = 0$ and is not defined at the points $x = -1/2^n$. Obviously the sequence of points $\{-1/2^n\}$ converges to the point 0. In view of Lemma §1 of Chapter III of [27] (see also the remark to that lemma), such a function f is not possible. Theorem 3.2.2 is proved.

CHAPTER VI

SPACES OF CONSTRUCTIVE INFINITELY DIFFERENTIABLE FUNCTIONS AND CONSTRUCTIVE FUNCTIONALS ON THEM

§1. Spaces of constructive almost uniformly continuous and infinitely differentiable functions

1.1. In this chapter we shall define spaces of constructive infinitely differentiable functions and spaces adjoint to them. At the same time we shall determine the constructive analogue of the concept of generalized function defined by Sobolev and Schwartz in classical mathematics.

By *function* we shall always understand a constructive function defined everywhere, i.e. an operator of the type $(\text{д} \to \text{д})$.

Suppose that f is a function. We shall say that f is *infinitely differentiable* if an algorithm φ is realizable having the following properties: a) for any natural number n the algorithm $(\varphi \uparrow n \, \square)$ is a function, while $\forall x(\varphi(0 \, \square \, x) = f(x))$, i.e. the function $(\varphi \uparrow 0 \, \square)$ "coincides" with the function f; b) for any n the function $(\varphi \uparrow (n+1) \, \square)$ is the derivative of the function $(\varphi \uparrow n \, \square)$.

We note that it is possible to realize constructive functions which are infinitely differentiable but which are not uniformly continuous on each finite segment. For instance, in the paper [8] of B. A. Kušner an example was given of an analytic constructive function on a segment which was unbounded on it. It is also possible to construct an infinitely differentiable function which is not bounded and which is therefore not uniformly continuous on any segment of the form $-m \triangle m$, where m is a positive natural number. At the basis of the construction of such examples lies the theorem on the realizability of a singular disjunct covering by segments of any nondegenerate segment ([6]). At the same time, in classical mathematics, with the classical understanding of terms, each infinitely differentiable function is uniformly continuous on any finite segment. In connection with this divergence, in the further development of the theory we will not choose the concept of infinitely differentiable constructive function as the constructive analogue of the classical concept of infinitely differentiable function. As the constructive analogue of this classical concept we shall choose the constructive concept "complete transcription of an almost uniformly continuous and infinitely differentiable constructive function", which we introduce as follows.

We consider the set $\mathfrak{P}^f$ of words in the alphabet Ч_0, defined by the formula

$$(1) \qquad \mathfrak{P}^f \rightleftharpoons (\langle c_{0,1} \rangle_4 \in (\text{н} \to \text{ш})) \,\&\, \forall n \, \text{Д}(\langle c_{0,1} \rangle_4 (n+1), \ \langle c_{0,1} \rangle_4 (n)).$$

The parameter for this formula is the variable $c_{0,1}$. The word P in the alphabet $Ч_0$ is an element of the set $\mathfrak{P}^f$ if and only if it is the transcription of a sequence of words of the type ш, while, for each natural number n, $\langle P\rangle_4(n+1)$ is the derivative of the word $\langle P\rangle_4(n)$. We recall that the predicate Д may be expressed by a normal formula (see §1.2, Chapter IV). Therefore the set $\mathfrak{P}^f$ may be defined by a normal formula.

We adjoin the formula $\mathfrak{P}^f$ to the list Σ of formulas of the language $Я_\Sigma$ and denote by f the subordinate generic letter set into correspondence with the formula $\mathfrak{P}^f$ in this language. A word of type f, i.e. an element of the set $\mathfrak{P}^f$, will be called the *complete transcription of a constructive almost uniformly continuous and infinitely differentiable function*. In what follows we shall call it for short the *transcription of an infinitely differentiable function*.

We agree to write

$$P^{(n)} \rightleftharpoons \langle P\rangle_4(n),$$

where P is any word of type f and n is any natural number.

We define a two-parameter formula $\mathfrak{E}^f$ such that, for any words f_1 and f_2 of type f,

$$\mathfrak{E}^f(f_1, f_2) \equiv (f_1^{(0)} \underset{ш}{=} f_2^{(0)}).$$

In place of $\mathfrak{E}^f(f_1, f_2)$ we shall use the notation $(f_1 \underset{f}{=} f_2)$. It is easy to verify that, for any words f_1 and f_2 of type f,

$$(f_1 \underset{f}{=} f_2) \equiv \forall n (f_1^{(n)} \underset{ш}{=} f_2^{(n)}).$$

We denote by $\mathfrak{O}^f$ the transcription (relative to the alphabet $Ч_4^{ш}$) of an algorithm which transforms any natural number into the zero element $\mathfrak{O}^ш$ of type ш. It is possible to construct algorithms $\underset{f}{+}$ of type $(ff\to f)$ and $\underset{f}{\cdot}$ of type $(дf\to f)$ such that, for any words f_1 and f_2 of type f and any duplex α,

$$\forall n (\langle \underset{f}{+} (f_1 \square f_2)\rangle_4 (n) \simeq f_1^{(n)} \underset{ш}{+} f_2^{(n)});$$

$$\forall n (\langle \underset{f}{\cdot} (\alpha \square f_1)\rangle_4 (n) \simeq \alpha \underset{ш}{\cdot} f_1^{(n)}).$$

It is not hard to verify that the list

(2) $$Ч_0,\ \mathfrak{P}^f,\ \mathfrak{E}^f,\ \underset{f}{+},\ \underset{f}{\cdot},\ \mathfrak{O}^f$$

defines a constructive linear space.

Now we define some further operations over words of type f.

a) It is possible to construct an algorithm D^f of the type $(\text{н}f\to f)$ such that, for any natural number k and any word f_1 of type f,

$$\forall n (\langle D^f(k \square f_1)\rangle_4 (n) \simeq f_1^{(k+n)}).$$

$D^f(k \square f_1)$ is called the *derivative of order k of f_1*. Evidently

$$\forall k f_1 (D^f(k \square f_1)^{(0)} = f_1^{(k)}).$$

b) It is possible to construct an algorithm $\underset{f}{\times}$ of type $(ff\to f)$ such that, for any words f_1 and f_2 of type f,

$$\forall n(\langle \underset{f}{\times}(f_1\Box f_2)\rangle_4(n)\simeq\sum_{i=0}^{n}C_n^i\cdot(f_1^{(n-i)}\underset{\text{ш}}{\times}f_2^{(i)})).$$

$\underset{f}{\times}(f_1\Box f_2)$ (for short, $(f_1\underset{f}{\times}f_2)$) is called the *product of the words f_1 and f_2*.

c) It is possible to construct an algorithm L^f such that, for any real duplexes α and β with $\alpha\neq 0$, and any word f_1 of type f,

$$\forall n(\langle L^f(\alpha\Box\beta\Box f_1)\rangle_4(n)\simeq\alpha^n\cdot L^{\text{ш}}(\alpha\Box\beta\Box f_1^{(n)})),$$

where $L^{\text{ш}}$ is the algorithm of linear substitution of a variable for words of type ш, defined in §1.1 of Chapter IV. It is easy to verify that for any real duplexes α and β with $\alpha\neq 0$ the algorithm $(L^f\upharpoonright\alpha\Box\beta\Box)$ is an operation of the type $(f\to f)$. Obviously for any duplex x

$$\underline{L^f(\alpha\Box\beta\Box f_1)^{(0)}}(x)=\underline{f_1^{(0)}}(\alpha x+\beta).$$

The algorithm L^f will be called the *algorithm of linear substitution of a variable for words of type f*.

We present some examples of words of type f which will be used in what follows.

EXAMPLE 1. It is possible to construct a function χ such that, for any real duplex x,

$$\chi(x)=\begin{cases}\exp\left(\dfrac{-1}{1-x^2}\right), & \text{if } |x|<1;\\ 0, & \text{if } |x|\geqslant 1.\end{cases}$$

This function may be constructed, for example, using the algorithm for extending uniformly continuous functions given in the set of rational numbers. In the same way it is possible to construct an algorithm $\mathfrak{X}$ such that for any natural number n the algorithm $(\mathfrak{X}\upharpoonright n\Box)$ is a function which is the derivative of order n of the function χ. The functions $(\mathfrak{X}\upharpoonright n\Box)$ are uniformly continuous. Therefore it is possible to construct a word X_1 of type f such that

$$\forall x(\underline{X_1^{(0)}}(x)=\chi(x)).$$

EXAMPLE 2. Write $X\leftrightharpoons\exp(1)\underset{f}{\cdot}X_1$. X is a word of type f. Obviously $\underline{X^{(0)}}(0)=1$ and $\underline{X^{(0)}}(\pm 1)=0$. Hence it is possible to prove that for any natural number $m\geqq 1$ we have

$$\forall n(\mathcal{N}(m\Box X^{(n)})\leqslant\mathcal{N}(m\Box X^{(n+1)})),$$

where $\mathcal{N}$ is an enumerable norm for the space $\mathscr{C}$ (see §1.1, Chapter IV).

For the proof of this property one can use Corollary 3 to Theorem 2 of [28]. This corollary is a constructive variant of the theorem of Lagrange.

EXAMPLE 3. Suppose that α is a positive real duplex. Write $X_a \leftrightharpoons L^f(a \square 0 \square X)$. Obviously

$$\forall x (\underline{X_a^{(0)}}(x) = \underline{X^{(0)}}(ax));$$

$$\forall nx (\underline{X_a^{(n)}}(x) = a^n \cdot \underline{X^{(n)}}(a \cdot x)).$$

Hence if $a \geqq 1$, then for any $m \geqq 1$ we have

$$\forall n (\mathcal{N}(m \square X_a^{(n+1)}) \geqslant a \cdot \mathcal{N}(m \square X_a^{(n)})).$$

EXAMPLE 4. Imitating the construction of "basic functions" by averaging continuous functions (see for example [1], Chapter II, §1.4), we may construct for any segment $a \triangle b$ and for any natural number n a word Z of type f such that $0 \leqq Z^{(0)}(x) \leqq 1$ for any duplex x, while if $x \in a \triangle b$, then $Z^{(0)}(x) = 1$, and if $x \notin a - 2^{-n} \triangle b + 2^{-n}$, then $Z^{(0)}(x) = 0$.

1.2. We introduce the following formula:

$$\mathfrak{P}_K \leftrightharpoons (c_{0,1} \in f) \,\&\, \forall x (|x| \geqslant m \supset \underline{c_{0,1}^{(0)}}(x) = 0),$$

where m is a variable whose admissible values are positive natural numbers. $\mathfrak{P}_K$ is a two-parameter formula whose parameters are the variables m and $c_{0,1}$.

We adjoin this formula $\mathfrak{P}_K$ to the list Δ of formulas of the language $Я_\Sigma^\Delta$ and denote by $Ф$ a pregeneric letter assigned to the formula $\mathfrak{P}_K$ in this language. Write

$$\mathfrak{P}_K^M \leftrightharpoons F_{M \llcorner}^m \mathfrak{P}_{K \lrcorner},$$

where M is any positive natural number. For any positive natural number M, a *word of type* $Ф^M$ (an element of the set $\mathfrak{P}_K^M$) is a word P of type f satisfying the condition

(3) $$\forall x (|x| \geqslant M \supset \underline{P^{(0)}}(x) = 0).$$

Obviously the condition (3) is equivalent to the following:

$$\forall nx (|x| \geqslant M \supset \underline{P^{(n)}}(x) = 0).$$

We construct an algorithm $\mathfrak{N}^f$ of type $(\text{нн}f \to \text{д})$ such that, for any natural numbers m and n with $m \geqq 1$, and any word f_1 of type f,

$$\mathfrak{N}^f (m \square n \square f_1) \simeq \max (\mathcal{N}(m \square f_1^{(0)}) \square \ldots \square \mathcal{N}(m \square f_1^{(n)})).$$

Now we fix any positive natural number M and consider the following list:

(4) $$Ч_0,\ \mathfrak{P}_K^M,\ \mathfrak{S}^f,\ \underset{f}{+},\ \underset{f}{\cdot},\ \mathfrak{D}^f,\ (\mathfrak{N}^f \uparrow M \square).$$

It is not hard to verify that this list defines a constructive enumerably normed space. This space will be denoted by K_M.

We denote by $\mathfrak{Z}$ the set of all natural numbers, and by $\mathfrak{Q}_K^M$ a two-parameter formula such that, for any natural number i and any word f_1 of type f,

$$\mathfrak{Q}_K^M(i,\ f_1) \equiv (\mathfrak{N}^f(M \square i \square f_1) \leqslant 2^{-i}).$$

The list $(\mathrm{Ч}_0, \mathfrak{Z}, \mathfrak{Q}_K^M)$ defines a locally convex topology in the space K_M (see §4.4, Chapter I).

1.2.1. *The space K_M is complete and separable.*

PROOF. The proof of this theorem is carried out in a way analogous to the proof of the corresponding assertion in classical mathematics. Here we will not construct that proof in detail, but rather note some of its peculiarities:

1. *Polynomial*: We note that if λ is a fundamental sequence of points of the space K_M and γ is a regulator of convergence for λ in itself in K_M, then for any natural number k the sequence φ of words of type ш defined by the formula

(5) $$\forall n (\varphi(n) \simeq \lambda(n)^{(k)}),$$

is a fundamental sequence in the space $\mathscr{C}$, while the algorithm ζ of type $(\mathrm{H} \to \mathrm{H})$ defined by the formula

$$\forall i (\zeta(i) \simeq \gamma(\max(k \square i)))$$

is a regulator of convergence in itself of the sequence φ in $\mathscr{C}$. We note that the space $\mathscr{C}$ is complete. On the basis of what was said above it is possible to construct an algorithm $\mathscr{G}$ which transforms each FR-construct of the form $\mathcal{E}\lambda\mathcal{3} \odot \mathcal{E}\gamma\mathcal{3}$ in the space K_M into the transcription of a sequence ξ of words of type ш such that, for any k, $\xi(k)$ is the limit in $\mathscr{C}$ of the sequence φ defined by the formula (5). It is easy to prove that $\xi(k+1)$ is the derivative of $\xi(k)$ for any k, so that $\mathcal{E}\xi\mathcal{3}$ is a point of the space K_M. The algorithm $\mathscr{G}$ is the algorithm of passing to the limit in the space K_M. So, the space K_M is complete.

2. *Separability*: On the basis of the results of §7 in [8], we may construct a "basis" for the space K_M, i.e. a sequence B of words of type Φ^M having the following property: for any point Φ_1^M of the space K_M we can realize a sequence η of real duplexes such that the sequence λ of words of the type Φ^M defined by the formula

$$\forall n \Big(\lambda(n) \simeq \sum_{i=0}^{n} \eta(i) \underset{f}{\cdot} B(i)\Big),$$

converges in Φ_1^M in the space K_M (here the sign $\sum$ denotes the summation of words of type f).[1)]

[1)] In [18] B. S. Mitjagin proved that the functions

$$\varphi_n(x) = h_n\left(\tan \frac{\pi}{2} x\right), \text{ where } h_n(t) = e^{t^2:2} \cdot \frac{d^n}{dt^n}(e^{-t^2}),$$

form a basis for the space of infinitely differentiable functions in the segment $[-1, 1]$, vanishing along with their derivatives at the points ± 1.

Hence for any point Φ_1^M of the space K_M, and any natural number i, we can realize a σ-system of rational numbers $a_1 \sigma \cdots \sigma a_k$ such that

$$\mathfrak{N}^f\left(M \square i \square \Phi_1^M - \sum_{j=0}^{k} a_j \cdot B_j\right) \leqslant 2^{-i}.$$

Obviously the set of all σ-systems of rational numbers is algorithmically enumerable. On the basis of what has been said it is possible to construct algorithms α of the type $(\text{н} \to \Phi^M)$ and β of the type $(\Phi^M \text{н} \to \text{н})$ such that

$$\forall \Phi_1^M i\, (\mathfrak{N}^f (M \square i \square \Phi_1^M - \alpha(\beta(\Phi_1^M \square i))) \leqslant 2^{-i}).$$

Thus the space K_M is separable.

We recall some notation introduced in §1.2 of Chapter III. $\mathfrak{U}$ denotes a universal arithmetic algorithm, A the set of "indices" of all general recursive functions (i.e. the set of all natural numbers m such that $\forall n\,!\,\mathfrak{A}(m \square n)$), and e is a generic letter assigned to the set A.

Denote by R^M a two-parameter formula such that, for any words Φ_1^M of type Φ^M and e_1 of type e,

$$R^M(e_1, \Phi_1^M) \equiv \forall n\, (\mathfrak{N}^f (M \square n \square \Phi_1^M) \leqslant \{e_1\}(n) + 1).$$

The list $(\text{Ч}_0, A, R^M)$ defines a normal fundamental system of bounded sets in the space K_M (see §1.2 of Chapter III).

Now we shall prove the following theorem.

1.2.2 (*Non-bornologicity of the enumerably normed space* K_M). *In the space* K_M *one can realize a convex set which absorbs every bounded set in* K_M *but which does not contain any of its neighborhoods of the origin.*

Proof. Consider the set $\mathfrak{A}$ of points of the space K_M, defined as follows:

$$(\Phi_1^M \in \mathfrak{A}) \rightleftharpoons \forall nk\, (\forall j\, (j \leqslant n \,\&\, !\,\mathfrak{U}(j \square n) \supset \mathfrak{U}(j \square n) \leqslant k) \supset \mathfrak{N}^f (M \square n \square \Phi_1^M) \leqslant k + 1).$$

Obviously $\mathfrak{A}$ is a convex set in K_M.

1) We shall prove that $\mathfrak{A}$ absorbs every bounded set in K_M, i.e. that

(6) $$\forall e_1 \exists a\, (a > 0 \,\&\, \forall \Phi_1^M (R^M(e_1, \Phi_1^M) \supset a \underset{j}{\cdot} \Phi_1^M \in \mathfrak{A})).$$

Suppose that a word e_1 of type e is given. Write

(7) $$b \rightleftharpoons \max(\mathfrak{U}(e_1 \square 0) \square \ldots \square \mathfrak{U}(e_1 \square e_1)) + 1; \quad a \rightleftharpoons 1 : b.$$

Now suppose that Φ_1^M is a point of K_M such that $R^M(e_1, \Phi_1^M)$. Then

(8) $$\forall n\, (\mathfrak{N}^f (M \square n \square \Phi_1^M) \leqslant \mathfrak{U}(e_1 \square n) + 1).$$

We shall prove that $a \cdot \Phi_1^M \in \mathfrak{A}$, i.e. that

$$\forall kn\, (\forall j\, (j \leqslant n \,\&\, !\,\mathfrak{U}(j \square n) \supset \mathfrak{U}(j \square n) \leqslant k) \supset \mathfrak{N}^f (M \square n \square a \cdot \Phi_1^M) \leqslant k + 1).$$

(9)

If $n \leqq e_1$, it follows from (7) and (8) that

$$\mathfrak{N}^f(M \square n \square a \cdot \Phi_1^M) \leqq a \cdot (\mathfrak{U}(e_1 \square n) + 1) \leqq 1,$$

so that $\mathfrak{N}^f(M \square n \square a \cdot \Phi_1^M) \leqq k + 1$ for any k. Now we suppose that $n \geqq e_1$. Then it is obvious that

$$\forall k (\forall j (j \leqslant n \& ! \mathfrak{U}(j \square n) \supset \mathfrak{U}(j \square n) \leqslant k) \supset \mathfrak{U}(e_1 \square n) \leqslant k). \tag{10}$$

It follows from (8) and (10) that for any k, if

$$\forall j (j \leqslant n \& ! \mathfrak{U}(j \square n) \supset \mathfrak{U}(j \square n) \leqslant k),$$

then

$$\mathfrak{N}^f(M \square n \square a \cdot \Phi_1^M) \leqslant a \cdot (\mathfrak{U}(e_1 \square n) + 1) \leqslant \mathfrak{U}(e_1 \square n) + 1 \leqslant k + 1.$$

Thus formula (9), and hence (6), are proved.

2) It remains to prove that $\mathfrak{A}$ does not contain any neighborhood of zero of the space K_M, i.e. that

$$\neg \exists i \forall \Phi_1^M (\mathfrak{Q}_K^M(i, \Phi_1^M) \supset \Phi_1^M \in \mathfrak{A}). \tag{11}$$

First of all we prove the following formula:

$$\forall n \neg\neg \exists k \forall j (j \leqslant n \& ! \mathfrak{U}(j \square n) \supset \mathfrak{U}(j \square n) \leqslant k). \tag{12}$$

Suppose given a natural number n. Write

$$B_j \leftrightharpoons ! \mathfrak{U}(j \square n), \quad C_j \leftrightharpoons \neg ! \mathfrak{U}(j \square n) \quad (j = 0, \ldots, n).$$

From the obvious formulas

$$\neg\neg(B_0 \vee C_0), \ldots, \neg\neg(B_n \vee C_n)$$

it follows that

$$\neg\neg[(B_0 \& \ldots \& B_n) \vee \ldots \vee (B_0 \& \ldots \& C_n) \vee \ldots \vee (C_0 \& \ldots \& C_n)]. \tag{13}$$

We denote the formula to the right of the double negation sign in (13) by $\mathfrak{C}$. It is not hard to show that

$$\mathfrak{C} \subset \exists k \forall j (j \leqslant n \& ! \mathfrak{U}(j \square n) \supset \mathfrak{U}(j \square n) \leqslant k). \tag{14}$$

It follows from (13) and (14) that

$$\neg\neg \exists k \forall j (j \leqslant n \& ! \mathfrak{U}(j \square n) \supset \mathfrak{U}(j \square n) \leqslant k).$$

Formula (12) is proved.

Now we shall prove (11). First we note that the formulas $\mathfrak{Q}_K^M$ and $\mathfrak{A}$ are normal. Hence (11) is equivalent successively to the following formulas:

$$\forall i \neg \forall \Phi_1^M \neg\neg (\mathfrak{Q}_K^M(i, \Phi_1^M) \supset (\Phi_1^M \in \mathfrak{A})),$$

$$\forall i \neg\neg \exists \Phi_1^M \neg (\mathfrak{Q}_K^M(i, \Phi_1^M) \supset (\Phi_1^M \in \mathfrak{A})), \tag{15}$$

$$\forall i \neg\neg \exists \Phi_1^M (\mathfrak{Q}_K^M(i, \Phi_1^M) \& \neg (\Phi_1^M \in \mathfrak{A})).$$

Suppose given a natural number i. Write $n \leftrightharpoons 2i+2$. We shall show that

$$\exists k \forall j (j \leqslant n \,\&\, !\, \mathfrak{U}(j \square n) \supset \mathfrak{U}(j \square n) \leqslant k) \supset \exists \Phi_1^M (\mathfrak{Q}_K^M(i, \Phi_1^M) \,\&\, \neg(\Phi_1^M \in \mathfrak{A})). \tag{16}$$

Now suppose that we can realize a natural number k such that

$$\forall j (j \leqslant n \,\&\, !\, \mathfrak{U}(j \square n) \supset \mathfrak{U}(j \square n) \leqslant k). \tag{17}$$

Write $a \leftrightharpoons \max(k \square 2)$. It is possible to construct a word X of the type Φ^M such that

$$\neg(X = \mathfrak{O}^f) \,\&\, \forall j (\mathcal{N}(M \square X^{(j+1)}) \geqslant a \cdot \mathcal{N}(M \square X^{(j)})) \tag{18}$$

(Example 3, §1.1). Write

$$Z \leftrightharpoons \frac{2^{-i}}{\mathcal{N}(M \square X^{(i)})} \cdot X.$$

We have $\mathcal{N}(M \square Z^{(i)}) = 2^{-i}$, so that $\mathfrak{N}^f(M \square i \square Z) \leq 2^{-i}$, i.e. $\mathfrak{Q}_K^M(i, Z)$. Moreover, from (18) we find that

$$\mathcal{N}(M \square Z^{(n)}) = \frac{2^{-i}}{\mathcal{N}(M \square X^{(i)})} \cdot \mathcal{N}(M \square X^{(n)}) \geqslant \frac{2^{-i}}{\mathcal{N}(M \square X^{(i)})} \cdot a^{n-i}$$
$$\times \mathcal{N}(M \square X^{(i)}) = 2^{-i} \cdot a^{n-i} = 2^{-i} \cdot a^{(2i+2)-i} \geqslant a^2 > a+1 \geqslant k+1,$$

so that

$$\mathfrak{N}^f(M \square n \square Z) > k+1. \tag{19}$$

It follows from (17) and (19) that $\neg(Z \in \mathfrak{A})$. Thus formula (16) is proved. It follows from (16) and (12) that

$$\neg\neg \exists \Phi_1^M (\mathfrak{Q}_K^M(i, \Phi_1^M) \,\&\, \neg(\Phi_1^M \in \mathfrak{A})).$$

Thus formula (15), and hence (11), are proved. Theorem 1.2.2 is completely proved.

Remark. As we know, in classical functional analysis (see for example [7]) every countably normed space is bornological (in such spaces every convex set absorbing every bounded set contains some neighborhood of the origin). This theorem serves as the basis for the proofs of some other theorems on linear operators defined in countably normed spaces. Theorem 1.2.2 shows that the constructive analogue of this theorem is refuted by an example. Thus Theorem 1.2.2 represents a point of essential divergence between the classical theory of countably normed spaces and the constructive theory of enumerably normed spaces.

1.3. We note that for any natural numbers m and l, if $1 \leq m \leq l$, then the space K_m is a linear subspace of K_l. Moreover, it is obvious that

$$\forall \Phi_1^m (\mathfrak{N}^f(m \square n \square \Phi_1^m) = \mathfrak{N}^f(l \square n \square \Phi_1^m)).$$

Therefore K_m is a subspace of K_l.

We shall prove further that *if* $1 \leq m \leq l$, *then the space* K_m (*more precisely, the set of all points of* K_m) *is a closed set in* K_l. To this end we have to prove the following two formulas:

$$\forall \phi_1^l (\neg(\phi_1^l \in K_m) \supset \exists i \forall \phi_2^l (\mathfrak{N}^f (l \square i \square \phi_2^l - \phi_1^l) \leqslant 2^{-i} \supset \neg(\phi_2^l \in K_m))). \tag{20}$$

$$\forall \phi_1^l (\forall i \exists \phi_2^l (\phi_2^l \in K_m \,\&\, \mathfrak{N}^f (l \square i \square \phi_2^l - \phi_1^l) \leqslant 2^{-i}) \supset (\phi_1^l \in K_m)). \tag{21}$$

In this case (21) is a consequence of (20). We shall prove (20). Let Z be a point of K_l not lying in K_m. Then

$$\neg \forall x (m < |x| \leqslant l \supset \underline{Z}^{(0)}(x) = 0). \tag{22}$$

Since $Z^{(0)}$ is a continuous function, it follows from (22) that

$$\neg \forall a (m < |a| \leqslant l \supset \underline{Z}^{(0)}(a) = 0), \tag{23}$$

where a is a variable for rational numbers. Inasmuch as the set of all rational numbers satisfying the condition $m < |z| \leq l$ is algorithmically enumerable, from (23) we find that

$$\exists a (m < |a| \leqslant l \,\&\, \underline{Z}^{(0)}(a) \neq 0)$$

(see Lemma 1 in the Appendix). Thus we can realize a duplex x and a natural number i such that

$$|x| > m \,\&\, |\underline{Z}^{(0)}(x)| > 2^{-i}.$$

It is easy to verify that

$$\forall \phi_2^l (\mathfrak{N}^f (l \square i \square \phi_2^l - Z) \leqslant 2^{-i} \supset \neg(\phi_2^l \in K_m)).$$

Thus formula (20) is proved.

In what follows we denote by N the directed set of all positive natural numbers. Consider the following list:

$$Ч_0,\ N,\ Ч_0,\ \mathfrak{P}_K,\ \mathfrak{C}^f,\ \underset{f}{+},\ \underset{f}{\cdot},\ \mathfrak{D}^f,\ \mathfrak{N}^f, \tag{24}$$

where $\mathfrak{P}_K$ is a two-parameter formula as defined at the beginning of §1.2.

From what has been said above we conclude that *the list* (24) *defines a strictly expanding sequence of enumerably normed spaces* (see §3.2, Chapter II). For each positive natural number M the space K_M is the space with the index M in this sequence.

For the expanding sequence (24) of spaces it is possible to construct both the F-inductive limit and the FR-inductive limit. Denote by r the index of the alphabet $Ч_0 \cup \{\tau\}$, i.e. $\mathrm{A}_r = Ч_0 \cup \{\tau\}$. For the construction of these inductive limits we introduce the following objects:

a) The set $\widetilde{\mathfrak{P}}^\Phi$ of words f_1 of type f defined by the formula

$$\exists m \forall x (|x| \geqslant m \supset \underline{f_1}^{(0)}(x) = 0).$$

b) The set $\mathfrak{P}^\Phi$ of all possible words in the alphabet A_r of the type $m \tau f_1$, where m is a positive natural number and f_1 is a word of type f, satisfying the condition

$$\forall x(|x| \geqslant m \supset \underline{f_1^{(0)}}(x) = 0).$$

We adjoin the formulas $\widetilde{\mathfrak{P}}^{\Phi}$ and $\mathfrak{P}^{\Phi}$ to the list Σ of formulas of the language $Я_{\Sigma}$, and denote by $\widetilde{ф}$ and $ф$ the subordinate generic letters assigned to $\widetilde{\mathfrak{P}}^{\Phi}$ and $\mathfrak{P}^{\Phi}$ respectively. We note that $\mathfrak{P}^{\Phi}$ is a normed space.

c) A two-parameter formula $\mathfrak{E}^{\Phi}$ defined in such a way that for any words $ф_1$ and $ф_2$ of type $ф$, if $ф_1 \overline{\underline{\circ}}\, m_1 \tau f_1$ and $ф_2 \overline{\underline{\circ}}\, m_2 \tau f_2$, then

$$\mathfrak{E}^{\Phi}(ф_1, ф_2) \equiv (f_1 \underset{f}{=} f_2).$$

d) An algorithm $\underset{\Phi}{+}$ of the type $(фф \to ф)$, defined in such a way that, for any words $ф_1$ and $ф_2$ of type $ф$, if $ф_1 \overline{\underline{\circ}}\, m_1 \tau f_1$ and $ф_2 \overline{\underline{\circ}}\, m_2 \tau f_2$, then

$$\underset{\Phi}{+}(ф_1 \square ф_2) \doteq m\tau(f_1 \underset{f}{+} f_2),$$

where $m \leftrightarrows \max(m_1 \square m_2)$.

e) An algorithm $\underset{\Phi}{\cdot}$ of type $(дф \to ф)$, defined in such a way that, for any real duplex α and any word $ф_1$ of type $ф$, if $ф_1 \overline{\underline{\circ}}\, m \tau f_1$, then

$$\underset{\Phi}{\cdot}(\alpha \square ф_1) \doteq m\tau(\alpha \underset{f}{\cdot} f_1).$$

f) $\mathfrak{O}^{\Phi} \leftrightarrows 1 \tau \mathfrak{O}^f$ ($\mathfrak{O}^{\Phi}$ is a word of type $ф$).

g) The set A of "indices" of all general recursive functions (see §1.2 of this chapter and §3.4 of Chapter II).

h) A two-parameter formula $\widetilde{\mathfrak{Q}}^{\Phi}$, defined in such a way that for any element e_1 of the set A (i.e. a word of type e) and word $\widetilde{ф}_1$ of type $\widetilde{ф}$, the condition $\widetilde{\mathfrak{Q}}^{\Phi}(e_1, \widetilde{ф}_1)$ is satisfied if and only if we can realize positive natural numbers l and $m_1, \cdots, m_l$, a point P_1 of the space $K_{m_1}, \cdots$, a point P_l of the space K_{m_l}, and real duplexes $\alpha_1, \cdots, \alpha_l$ such that

$$\widetilde{ф}_1 \underset{f}{=} \alpha_1 \cdot P_1 + \cdots + \alpha_l \cdot P_l, \quad |\alpha_1| + \cdots + |\alpha_l| = 1,$$

and

$$\mathfrak{N}^f(m_1 \square i_1 \square P_1) \leqslant 2^{-i_1} \& \cdots \& \mathfrak{N}^f(m_l \square i_l \square P_l) \leqslant 2^{-i_l},$$

where

$$i_j \rightleftharpoons \{e_1\}(m_j) \quad (j = 1, \ldots, l).$$

i) A two-parameter formula $\mathfrak{Q}^{\Phi}$, defined such that, for any words e_1 of type e and $ф_1$ of type $ф$, if $ф_1 \overline{\underline{\circ}}\, m \tau P$, then

$$\mathfrak{Q}^{\Phi}(e_1, ф_1) \equiv \widetilde{\mathfrak{Q}}^{\Phi}(e_1, P).$$

We now form the following two lists:

(25) $$Ч_0,\ \widetilde{\mathfrak{P}}^{\Phi},\ \mathfrak{E}^f,\ \underset{f}{+},\ \underset{f}{\cdot},\ \mathfrak{O}^f,\ Ч_0,\ A,\ \widetilde{\mathfrak{Q}}^{\Phi},$$

(26) $$A_r,\ \mathfrak{P}^{\Phi},\ \mathfrak{E}^{\Phi},\ \underset{\Phi}{+},\ \underset{\Phi}{\cdot},\ \mathfrak{O}^{\Phi},\ Ч_0,\ A,\ \mathfrak{Q}^{\Phi}.$$

These two lists define constructive locally convex spaces. The space (25) is the F-inductive limit, and the space (26) the FR-inductive limit, of the expanding sequence (24) of enumerably normed spaces. We denote the spaces (25) and (26) by $\widetilde{K}$ and K respectively.

It is obvious by definition that for any point P of the space $\widetilde{K}$ we can realize a natural number $m \geqq 1$ such that $m \tau P$ is a point of K. On the other hand, the following theorem holds.

1.3.1. *There cannot exist an algorithm which constructs for each point P of the space K a natural number $m \geqq 1$ such that $m \tau P$ is a point of $\widetilde{K}$.*

Proof. This theorem is proved in the same way as was used in the proof of Theorem 3.3.3 of Chapter IV. Therefore we will not present a detailed proof, but shall give only the outline. Suppose that $\mathscr{A}$ is an algorithm of the type (н $\dot{\to}$ н) whose applicability problem is undecidable, and G an algorithm applicable to any natural numbers of the type $l \square n$ and such that $G(l \square n) \doteq \Lambda$ if and only if the process of application of the algorithm $\mathscr{A}$ to the number l stops at exactly n steps.

We construct an algorithm $\mathfrak{B}$ of the type (ннд → д) and an algorithm $\mathfrak{C}$ of the type (нд → д) such that, for any natural numbers l and n and any duplex x,

$$\mathfrak{B}(l \square n \square x) \simeq \begin{cases} 0, & \text{if } G(l \square n) \neq \Lambda; \\ \chi(x-n), & \text{if } G(l \square n) \doteq \Lambda, \end{cases}$$

$$\mathfrak{C}(l \square x) = \sum_{i=0}^{\zeta(x)} \mathfrak{B}(l \square i \square x),$$

where χ is the function defined in Example 1 of §1.1, and ζ is an algorithm of the type (д → н) having the property $\forall x(\zeta(x) \geqq |x|)$.

It is possible to verify that for any l the algorithm ($\mathfrak{H} \uparrow l \square$) is an infinitely differentiable function. Moreover, one can construct a sequence $\mathfrak{D}$ of words of type f such that

$$\forall lx (\underbrace{\mathfrak{D}(l)^{(0)}}(x) = \mathfrak{C}(l \square x)).$$

Using this construction, one successively proves the following propositions:

1) For any word P of type f, if P is a term of the sequence $\mathfrak{D}$, then

$$\neg\neg \exists m \forall x (|x| \geqslant m \supset \underbrace{P^{(0)}}(x) = 0). \tag{27}$$

2) There cannot exist an algorithm transforming each word P of type f satisfying the condition (27) into a natural number m such that

$$\forall x (|x| \geqslant m \supset \underbrace{P^{(0)}}(x) = 0).$$

Finally, using proposition 2), one proves Theorem 1.3.1.

In concluding this subsection we introduce an algorithm for differentiation of words of type Ф, which we shall use in the sequel. This is the algorithm

D^{Φ}, and it is constructed in such a way that, for any natural number k and any word Φ_1 of type Φ, if $\Phi_1 \overset{\circ}{=} m\tau f_1$, then

$$D^{\Phi}(k \square \Phi_1) \mathrel{\overset{\cdot}{=}} m\tau D^f(k \square f_1).$$

For any k the algorithm $(D^{\Phi} \uparrow k \square)$ is an operator of type $(\Phi \to \Phi)$. $D^{\Phi}(k \square \Phi_1)$ is called the *derivative of order k of the word* Φ_1 *of type* Φ.

It is possible to construct other operations as well, for example multiplication, convolution, and linear substitution of a variable, over words of type Φ. However, we shall not use these operations in the following exposition, and so we shall not dwell on their construction.

1.4. Now we consider bounded sets in the spaces $\widetilde{K}$ and K. We prove the following theorem.

1.4.1. *It is possible to construct a bounded set* $\mathfrak{A}$ *in the space* $\widetilde{K}$ *having the property that there cannot exist a natural number m such that* $\mathfrak{A}$ *is contained in the space* K_m.

PROOF. We construct and then fix a word Z of type f such that

$$\neg(Z \underset{f}{=} \mathfrak{O}^f) \,\&\, \forall x (x \leqslant 0 \vee x \geqslant 1 \supset \underbrace{Z^{(0)}}(x) = 0).$$

Such a word Z may be constructed using Example 2 in §1.1 and the algorithm L^{Φ}. It is possible to construct a sequence λ of positive natural numbers such that

$$\forall k (\mathfrak{N}^f(1 \square k \square Z) \leqslant \lambda(k)). \tag{28}$$

Now we construct a set $\mathfrak{A}$ of words of type $\widetilde{\Phi}$ as follows:

$$\widetilde{\Phi}_1 \in \mathfrak{A} \rightleftharpoons \exists mk\, [\forall i (i \leqslant m \,\&!\, \mathfrak{U}(i \square m) \supset \mathfrak{U}(i \square m) \leqslant k) \,\&\, (\widetilde{\Phi}_1 \in K_m) \\ \,\&\, \mathfrak{N}^f(m \square k \square \widetilde{\Phi}_1) \leqslant 2^{-k} \,\&\, \forall j (\mathfrak{N}^f(m \square j \square \widetilde{\Phi}_1) \leqslant \lambda(j))],$$

where $\mathfrak{U}$ is a universal arithmetic algorithm.

1) We shall prove that the set $\mathfrak{A}$ is bounded in $\widetilde{K}$, i.e. that

$$\forall e_1 \exists a (a > 0 \,\&\, \forall \widetilde{\Phi}_1 (\widetilde{\Phi}_1 \in \mathfrak{A} \supset \widetilde{\mathfrak{Q}}^{\Phi}(e_1, a \cdot \widetilde{\Phi}_1))). \tag{29}$$

Suppose given a word e_1 of type e. Define a number a as follows:

$$a \rightleftharpoons 2^{-\mathfrak{U}(e_1 \square e_1)} \cdot \frac{1}{\lambda(\mathfrak{U}(e_1 \square e_1))}.$$

Obviously $a > 0$. It remains to prove that

$$\forall \widetilde{\Phi} (\widetilde{\Phi}_1 \in \mathfrak{A} \supset \widetilde{\mathfrak{Q}}^{\Phi}(e_1, a \cdot \widetilde{\Phi}_1)). \tag{30}$$

Suppose that $\widetilde{\Phi}_1$ is a point of K lying in $\mathfrak{A}$. Then we can realize natural numbers m and k such that

$$\forall i (i \leqslant m \,\&!\, \mathfrak{U}(i \square m) \supset \mathfrak{U}(i \square m) \leqslant k) \,\&\, (\widetilde{\Phi}_1 \in K_m) \,\&\, \mathfrak{N}^f(m \square k \square \widetilde{\Phi}_1) \\ \leqslant 2^{-k} \,\&\, \forall j (\mathfrak{N}^f(m \square j \square \widetilde{\Phi}_1) \leqslant \lambda(j)) \tag{31}$$

are realizable.

a) Suppose that $m \leqq e_1$. Then $\tilde{\Phi}_1 \in K_m$ implies that $\tilde{\Phi}_1 \in K_{e_1}$. Write $l \leftrightharpoons \{e_1\}(e_1)$, i.e. $l \leftrightharpoons \mathfrak{U}(e_1 \square e_1)$. From (31) we find that

$$\mathfrak{N}^f(e_1 \square l \square a \cdot \tilde{\Phi}_1) = 2^{-l} \cdot \frac{1}{\lambda(l)} \cdot \mathfrak{N}^f(e_1 \square l \square \tilde{\Phi}_1)$$
$$= 2^{-l} \cdot \frac{1}{\lambda(l)} \cdot \mathfrak{N}^f(m \square l \square \tilde{\Phi}_1) \leqslant 2^{-l}.$$

Hence it follows that $\mathfrak{Q}^\Phi(e_1, a \cdot \tilde{\Phi}_1)$.

b) Suppose that $m \geqq e_1$. Then it follows from (31) that $\{e_1\}(m) \leqq k$. Write $l \leftrightharpoons \{e_1\}(m)$. Obviously $l \leqq k$. Hence from (21) we get

$$\mathfrak{N}^f(m \square l \square \tilde{\Phi}_1) \leqslant \mathfrak{N}^f(m \square k \square \tilde{\Phi}_1) \leqslant 2^{-k} \leqslant 2^{-l}.$$

Hence it follows that

$$\mathfrak{N}^f(m \square l \square a \cdot \tilde{\Phi}_1) \leqslant 2^{-l}.$$

We again obtain $\mathfrak{Q}^\Phi(e_1, a \cdot \tilde{\Phi}_1)$.

Obviously $(m \leqq e_1 \vee m \geqq e_1)$. Therefore from what has been proved we find that $\mathfrak{Q}^\Phi(e_1, a \cdot \tilde{\Phi}_1)$. Formula (30), and thus formula (29), are proved.

2) Now we shall prove that $\mathfrak{A}$ is not contained in any of the spaces K_m, i.e. that

(32) $$\neg \exists m \forall \tilde{\Phi}_1 (\tilde{\Phi}_1 \in \mathfrak{A} \tilde{\Phi}_1 \in K_m).$$

Suppose that m is a natural number $\geqq 1$. Write $n \leftrightharpoons m+1$. We shall prove that

$$\exists k \forall i (i \leqslant n \,\&!\, \mathfrak{U}(i \square n) \supset \mathfrak{U}(i \square n) \leqslant k) \supset \exists \tilde{\Phi}_1 (\tilde{\Phi}_1 \in \mathfrak{A} \,\&\, \neg(\tilde{\Phi}_1 \in K_m)).$$

(33)

Indeed, suppose that we can realize a natural number k such that

(34) $$\forall i (i \leqslant n \,\&!\, \mathfrak{U}(i \square n) \supset \mathfrak{U}(i \square n) \leqslant k).$$

We construct words Y_1 and Y of type f as follows:

$$Y_1 \rightleftharpoons L^\Phi(1 \square - m \square Z) \text{ and } Y \rightleftharpoons 2^{-k} \cdot \frac{1}{\lambda(k)} \cdot Y_1.$$

From the properties of the word Z it follows that Y_1 has the properties

(35) $$\neg(Y_1 \underset{f}{=} \mathfrak{O}^f) \,\&\, \forall x (x \leqslant m \vee x \geqslant n \supset \underline{Y_1^{(0)}}(x) = 0),$$

(36) $$\forall j (\mathfrak{N}^f(n \square j \square Y_1) \leqslant \lambda(j)).$$

It follows from (35) and (36) that Y satisfies the following condition:

(37) $$(Y \in K_n) \,\&\, \forall j (\mathfrak{N}^f(n \square j \square Y) \leqslant \lambda(j)) \,\&\, \mathfrak{N}^f(n \square k \square Y) \leqslant 2^{-k}.$$

It follows from (34) and (37) that $Y \in \mathfrak{A}$. Moreover, it is obvious from (35) that $\neg(Y \in K_m)$. So formula (33) is proved.

In the proof of Theorem 1.2.2 we proved that

$$\neg\neg\exists k \forall i (i \leqslant n \,\&!\, \mathfrak{U}(i \square n) \supset \mathfrak{U}(i \square n) \leqslant k).$$

Therefore from (33) we conclude that

$$(38) \qquad \neg\neg\exists \tilde{\Phi}_1 (\tilde{\Phi}_1 \in \mathfrak{N} \,\&\, \neg(\tilde{\Phi}_1 \in K_m)).$$

Formula (38) is proved for any positive natural number m. Therefore we get

$$(39) \qquad \forall m \neg\neg\exists \tilde{\Phi}_1 (\tilde{\Phi}_1 \in \mathfrak{A} \,\&\, \neg(\tilde{\Phi}_1 \in K_m)).$$

From (39) there successively result the following formulas:

$$\neg\exists m \forall \tilde{\Phi}_1 \neg(\tilde{\Phi}_1 \in \mathfrak{A} \,\&\, \neg(\tilde{\Phi}_1 \in K_m)),$$

$$\neg\exists m \forall \tilde{\Phi}_1 (\tilde{\Phi}_1 \in \mathfrak{A} \supset \neg\neg(\tilde{\Phi}_1 \in K_m)),$$

$$\neg\exists m \forall \tilde{\Phi}_1 (\tilde{\Phi}_1 \in \mathfrak{A} \supset \tilde{\Phi}_1 \in K_m).$$

Here we note that the condition $(\tilde{\Phi}_1 \in K_m)$ is expressed by a normal formula. Thus (32) is proved. Theorem 1.4.1 is completely proved.

1.4.2. *It is possible to construct a bounded set $\mathfrak{B}$ in the space K having the property that there cannot exist a natural number m such that $\mathfrak{B}$ is contained in the space K_m.*[2)]

Proof. We construct a set $\mathfrak{B}$ such that for any word Φ_1 of type Φ

$$(\Phi_1 \in \mathfrak{B}) \equiv ([\Phi_1]_2^{\tau} \in \mathfrak{A}),$$

where $\mathfrak{A}$ is the set constructed in the proof of 1.4.1 (here $[\Phi_1]_2^{\tau}$ is the right wing of the natural imbedding of the sign τ in the word Φ_1). From the proof of Theorem 1.4.1 we deduce that $\mathfrak{B}$ has the property indicated in 1.4.2.

Remark. Theorems 1.4.1 and 1.4.2 show that a strictly expanding sequence of enumerably normed spaces is realizable such that both its F-strict inductive limit and its FR-strict inductive limit are not α-regular and hence not regular. Thus the constructive analogue of the theorem on the regularity of strict inductive limits of sequences of locally convex spaces in classical mathematics (see [**39**], or [**38**], Chapter III, or [**7**], Chapter XI) is refuted by an example.

1.5. In this subsection we consider the multinormability of the space K. In fact, we prove that K is multinormable.

We introduce the following two-parameter formulas:

$$\tilde{\mathfrak{Q}}_s(\mathrm{e}_1, \tilde{\Phi}_1) \rightleftharpoons \forall mjx (|x| \geqslant m - 1 \,\&\, j \leqslant \{\mathrm{e}_1\}(m) \supset |\tilde{\Phi}_1^{(j)}(x)| \leqslant 2^{-\{\mathrm{e}_1\}(m_i)});$$

$$\mathfrak{Q}_s(\mathrm{e}_1, \Phi_1) \rightleftharpoons \tilde{\mathfrak{Q}}_f(\mathrm{e}_1, [\Phi_1]_2^{\tau}).$$

In these formulas e_1 is a variable of genus e, $\tilde{\Phi}_1$ a variable of genus $\tilde{\Phi}$.

---

2) We say that the set $\mathfrak{B}$ of points of the space K is contained in K_m if the following condition is satisfied: for any element Φ_1 of the set $\mathfrak{B}$, if $\Phi_1 \overline{\underline{\circ}}\, n\tau f_1$, then f_1 is a point of K_m (see §3.3, Chapter II).

One may verify that the list $(\mathrm{Ч}_0, A, \widetilde{\mathfrak{Q}}_s)$ defines a locally convex topology on the linear space

(40) $$\mathrm{Ч}_0,\ \widetilde{\mathfrak{P}}^{\Phi},\ \mathfrak{C}^f,\ \underset{f}{+},\ \underset{f}{\cdot},\ \mathfrak{O}^f,$$

and the list $(\mathrm{Ч}_0, A, \mathfrak{Q}_s)$ defines a locally convex topology on the linear space

(41) $$\mathrm{A}_r,\ \widetilde{\mathfrak{P}}^{\Phi},\ \mathfrak{C}^{\Phi},\ \underset{\Phi}{+},\ \underset{\Phi}{\cdot},\ \mathfrak{O}^{\Phi}.$$

Now we prove the following theorem.

1.5.1. a) *The two locally convex topologies* $(\mathrm{Ч}_0, A, \widetilde{\mathfrak{Q}}^{\Phi})$ *and* $(\mathrm{Ч}_0, A, \widetilde{\mathfrak{Q}}_s)$, *defined on the linear space* (40), *are equivalent to one another.*

b) *The two locally convex topologies* $(\mathrm{Ч}_0, A, \mathfrak{Q}^{\Phi})$ *and* $(\mathrm{Ч}_0, A, \mathfrak{Q}_s)$, *defined on the linear space* (41), *are equivalent to one another.*

Proof. It suffices for us to prove part a) of the theorem. To this end we have to prove the following two formulas:

(42) $$\forall \mathrm{e}_1 \exists \mathrm{e}_2 \forall \widetilde{\phi}_1 (\widetilde{\mathfrak{Q}}^{\Phi}(\mathrm{e}_2,\ \widetilde{\phi}_1) \supset \widetilde{\mathfrak{Q}}_s(\mathrm{e}_1,\ \widetilde{\phi}_1));$$

(43) $$\forall \mathrm{e}_1 \dot{\exists} \mathrm{e}_2 \forall \widetilde{\phi}_1 (\widetilde{\mathfrak{Q}}_s(\mathrm{e}_2,\ \widetilde{\phi}_1) \supset \widetilde{\mathfrak{Q}}^{\Phi}(\mathrm{e}_1,\ \widetilde{\phi}_1)).$$

Let us prove (42). First we show that

(44) $$\forall m\, \mathrm{e}_1 \exists i\, \forall \phi_1^m (\mathfrak{N}^f(m \square i \square \phi_1^m) \leqslant 2^{-i} \supset \widetilde{\mathfrak{Q}}_s(\mathrm{e}_1,\ \phi_1^m)).$$

Suppose given a natural number $m \geqq 1$ and a word e_1 of type e. We write

$$i \leftrightharpoons \max(\{\mathrm{e}_1\}(1) \square \cdots \square \{\mathrm{e}_1\}(m)).$$

Suppose that P is a word of type ϕ^m such that $\mathfrak{N}^f(m \square i \square P) \leqq 2^{-i}$. Then we have

$$\forall jx(|x| \leqslant m \,\&\, j \leqslant i \supset |\underset{\llcorner\!\lrcorner}{P^{(j)}}(x)| \leqslant 2^{-i});$$

$$\forall jx(|x| \geqslant m \supset |\underset{\llcorner\!\lrcorner}{P^{(j)}}(x)| = 0).$$

From these formulas it follows that $\widetilde{\mathfrak{Q}}_s(\mathrm{e}_1, P)$. Hence (44) is proved.

Formula (44) shows that for any $m \geqq 1$ the identity mapping of the space K_m into the space (40) with the topology $(\mathrm{Ч}_0, A, \widetilde{\mathfrak{Q}}_s)$ is continuous. Therefore, in view of Theorem 3.1.2, Chapter II, the topology $(\mathrm{Ч}_0, A, \widetilde{\mathfrak{Q}}_s)$ is majorized by the topology $(\mathrm{Ч}_0, A, \widetilde{\mathfrak{Q}}^{\Phi})$ of the F-inductive limit $\widetilde{K}$. (In Chapter II Theorem 3.1.2 was proved for the FR-inductive limit, but it is valid for the F-inductive limit as well.) Thus (42) is proved.

Now we prove (43). We construct and then fix a sequence Z of words of type f satisfying the condition

(45) $$\forall jx((|x| \leqslant j \supset \underset{\llcorner\!\lrcorner}{Z_j^{(0)}}(x) = 1) \,\&\, (|x|) \geqslant j+1 \supset \underset{\llcorner\!\lrcorner}{Z_j^{(0)}}(x) = 0)),$$

where $Z_j \leftrightharpoons Z(j)$ (see Example 4, §1.1). For any j, Z_j is a word of type ϕ^{j+1}.

After doing this we construct a sequence G of words of type f such that

$$G(1) = Z_0; \quad G(j) \underset{}{=} Z_{j-1} \underset{f}{\times} (\tilde{1} - Z_{j-2}) \underset{f}{\times} \cdots \underset{f}{\times} (\tilde{1} - Z_0), \quad \text{if} \quad j \geqslant 2.$$

Here by the sign $\tilde{1}$ we understand a word of type f having the property

$$\forall x (\underbrace{\tilde{1}^{(0)}}(x) = 1).$$

For any $j \geqq 1$ the word G_j ($G_j \leftrightharpoons G(j)$) is a word of type Φ^j, i.e. G_j belongs to the space K_j.

We can construct an algorithm B of the type $(\text{HH} \to \text{H})$ satisfying the condition

$$\forall jn (\mathfrak{N}^f (j \square n \square G_j) \leqslant 2^{B(j \square n)}).$$

Then we construct an algorithm E of the type $(\text{HH} \to \text{H})$ such that

$$\forall jn \left(E(j \square n) \simeq \mu i \left(\sum_{k=0}^{n} C_n^k \cdot 2^{B(j \square k)} < 2^i \right) \right).$$

Suppose that m is a natural number $\geqq 1$. We will show that each word P of the type Φ^m may be represented in the following form:

(46) $$P \underset{f}{=} P_1 + \cdots + P_{m+1},$$

where $P_j \leftrightharpoons P \underset{f}{\times} G_j$ $(j = 1, \cdots, m+1)$.

Indeed, suppose that P is a word of type Φ^m. We define at the same time words $R_1, \cdots, R_{m+1}$ and $Q_1, \cdots, Q_{m+2}$ as follows:

$$Q_1 \leftrightharpoons P; \quad Q_{j+1} \leftrightharpoons Q_j - R_j \ (1 \leqslant j \leqslant m+1), \quad R_j \leftrightharpoons Q_j \underset{f}{\times} Z_{j-1} (1 \leqslant j \leqslant m+1).$$

It is easy to verify that for $2 \leqq j \leqq m+2$ we have

(47) $$Q_j \underset{f}{=} P \underset{f}{\times} (\tilde{1} - Z_{j-2}) \underset{f}{\times} \cdots \underset{f}{\times} (\tilde{1} - Z_0),$$

so that for any j with $1 \leqq j \leqq m+2$ we have

(48) $$R_j \underset{f}{=} P \underset{f}{\times} G_j.$$

On the other hand, from the definition of the words Q_j we get

(49) $$P - Q_{m+2} \underset{f}{=} R_1 + \cdots + R_{m+1}.$$

It follows from (45) and (47) that $Q_{m+2} \underset{f}{=} \mathfrak{O}^f$. Therefore (46) follows from (48) and (49).

We note that for $2 \leqq j \leqq m+2$ we have

$$\forall x (|x| \leqslant j - 2 \supset \underbrace{Q_j^{(0)}}(x) = 0)$$

(this follows from (47)). Hence from $P_j \underset{f}{=} R_j \underset{f}{=} Q_j \times Z_{j-1}$ $(1 \leqq j \leqq m+1)$ it follows that

(50) $$\forall jx (2 \leqslant j \leqslant m+1 \ \& \ |x| \leqslant j-2 \supset \underbrace{P_j^{(0)}}(x) = 0).$$

It is not difficult to verify that from the fact that $P_j = P \times G_j$ it follows that

(51) $$\forall jnx\,(1 \leqslant j \leqslant m+1 \supset \max_{0 \leqslant i \leqslant n} |\underline{P_j^{(i)}}(x)| < 2^{E(j\square n)} \cdot \max_{0 \leqslant i \leqslant n} |\underline{P^{(i)}}(x)|),$$

where E is the algorithm defined above.

Now we turn to the proof of formula (43). Suppose given a word e_1 of type e. It is possible to construct a word e_2 of type e such that

(52) $$\begin{aligned}&\{\mathrm{e}_2\}(1) = \max(n_1 + 1 + E(1\square n_1)\square n_2 + 2 + E(2\square n_2));\\ &\{\mathrm{e}_2\}(j) = n_{j+1} + j + 1 + E(j+1\square n_{j+1}), \quad \text{if } j \geqslant 2,\end{aligned}$$

where $n_i \leftrightharpoons \{\mathrm{e}_i\}(i)$. We will show that

(53) $$\forall \tilde{\Phi}_1 (\tilde{\mathfrak{Q}}_s(\mathrm{e}_2, \tilde{\Phi}_1) \supset \tilde{\mathfrak{Q}}^{\Phi}(\mathrm{e}_1, \tilde{\Phi}_1)).$$

Suppose that P is a word of type $\tilde{\Phi}$ such that $\tilde{\mathfrak{Q}}_s(\mathrm{e}_2, P)$. Then

(54) $$\forall kix\,(|x| \geqslant k-1 \,\&\, i \leqslant \{\mathrm{e}_2\}(k) \supset |\underline{P^{(i)}}(x)| \leqslant 2^{-\{\mathrm{e}_2\}(k)}).$$

Since P is a word of type $\tilde{\Phi}$, we can realize a natural number $m \geqq 1$ such that $P \in K_m$. Suppose that P is representable in the form (46). We note that if $1 \leqq j \leqq m+1$, then P_j belongs to K_j. From (54) we find, for $k = 1$,

(55) $$\forall ix\,(i \leqslant \{\mathrm{e}_2\}(1) \supset |\underline{P^{(i)}}(x)| \leqslant 2^{-\{\mathrm{e}_2\}(1)}).$$

From (55), (52) and (61) it follows that

(56) $$\mathfrak{N}^f(1\square n_1\square P_1) \leqslant 2^{-(n_1+1)} \,\&\, \mathfrak{N}^f(2\square n_2\square P_2) \leqslant 2^{-(n_2+2)}.$$

Suppose that j is a natural number > 2. From (54) we find that, for $k = j-1$,

$$\forall ix\,(|x| \geqslant j-2 \,\&\, i \leqslant \{\mathrm{e}_2\}(j-1) \supset |\underline{P^{(i)}}(x)| \leqslant 2^{-\{\mathrm{e}_2\}(j-1)}).$$

Hence

(57) $$\forall ix\,(|x| \geqslant j-2 \,\&\, i \leqslant n_j \supset |\underline{P^{(i)}}(x)| \leqslant 2^{-(n_j+j+E(j\square n_j))}).$$

It follows from (57), (51) and (50) that for $2 < j \leqq m+1$ one has

$$\forall x\,(\max_{0 \leqslant i \leqslant n_j} |P_j^{(i)}(x)| \leqslant 2^{-(n_j+j)}).$$

Therefore

(58) $$\forall j\,(2 < j \leqslant m+1 \supset \mathfrak{N}^f(j\square n_j\square P_j) \leqslant 2^{-(n_j+j)}).$$

Write $X_j \leftrightharpoons 2^f \cdot P_j$ $(1 \leqq j \leqq m+1)$ and $X_{m+2} \leftrightharpoons \mathfrak{O}^f$. Then from (46), (56) and (58) we find that

$$P \underset{f}{=} 2^{-1} \cdot X_1 + \cdots + 2^{-(m+1)} \cdot X_{m+1} + 2^{-(m+1)} \cdot X_{m+2},$$

$$2^{-1} + \cdots + 2^{-(m+1)} + 2^{-(m+1)} = 1,$$

$$\forall j\,(1 \leqslant j \leqslant m+2 \supset \mathfrak{N}^f(i\square n_j\square X_j) \leqslant 2^{-n_j}).$$

This means that $\tilde{\mathfrak{Q}}^{\Phi}(\mathrm{e}_1, P)$. Formula (53) and also (43) are proved. Thus Theorem 1.5.1 is proved.

1.5.2. *The space K is multinormable.*

Proof. We may construct an algorithm $\mathfrak{N}^s$ of the type ($\mathrm{e}\Phi \to$ д) such that, for any words e_1 of type e and Φ_1 of type Φ, if $\Phi_1 \overline{\underline{\circ}}\, m\, \tau f_1$, then

$$\mathfrak{N}^s(\mathrm{e}_1 \square \Phi_1) = \max_{1 \leqslant j \leqslant m} \{2^{\{\mathrm{e}_1\}(j)} \cdot \max_{0 \leqslant i \leqslant \{\mathrm{e}_1\}(j)} [\sup_{j-1 \leqslant |x| \leqslant m} |\underline{f_1^{(i)}}(x)|]\},$$

where the expression in square brackets denotes the largest of the least upper bounds of the function φ defined by the formula

$$\forall x (\varphi(x) \simeq |\underline{f_1^{(i)}}(x)|),$$

in the segments $-m\Delta - (j-1)$ and $j - 1\Delta m$.

Such an algorithm $\mathfrak{N}^s$ may be constructed because, for any word f_1 of type f and for any natural number i, $f_1^{(i)}$ is the complete cipher of an almost uniformly continuous function.

It is easy to verify that $\mathfrak{N}^s$ defines a multinorm in the space (41). Moreover, $\mathfrak{N}^s$ has the following property:

$$\forall \mathrm{e}_1 \mathrm{e}_2 \exists \mathrm{e}_3 \forall \Phi_1 (\mathfrak{N}^s(\mathrm{e}_3 \square \Phi_1) \geqslant \max(\mathfrak{N}^s(\mathrm{e}_1 \square 2 \cdot \Phi_1) \square \mathfrak{N}^s(\mathrm{e}_2 \square 2 \cdot \Phi_1))).$$

Therefore the locally convex topology corresponding to the multinorm $\mathfrak{N}^s$ may be defined by the list $(\text{Ч}_0, A, \mathfrak{Q}^*)$, where

$$\mathfrak{Q}\ (\mathrm{e}_1,\ \Phi_1) \equiv (\mathfrak{N}^s(\mathrm{e}_1 \square \Phi_1) \leqslant 1).$$

From the definition of $\mathfrak{Q}_s$ and $\mathfrak{Q}^*$ it follows that

$$\forall \mathrm{e}_1 \Phi_1 (\mathfrak{Q}_s(\mathrm{e}_1,\ \Phi_1) \equiv \mathfrak{Q}^*(\mathrm{e}_1,\ \Phi_1)).$$

Thus the topologies $(\text{Ч}_0, A, \mathfrak{Q}_s)$ and $(\text{Ч}_0, A, \mathfrak{Q}^*)$ are equivalent to one another. Hence we conclude from Theorem 1.5.1 that the topologies $(\text{Ч}_0, A, \mathfrak{Q}^{\Phi})$ and $(\text{Ч}_0, A, \mathfrak{Q}^*)$ are also equivalent to one another, i.e. that the space K is multinormable. The theorem is proved.

§2. Constructive linear operators and functionals in the enumerably normed space K_M

2.1. We fix a positive natural number M and consider the enumerably normed space K_M defined by the list

$$(1) \qquad \text{Ч}_0,\ \mathfrak{P}_K^M,\ \mathfrak{C}^f,\ \underset{f}{+},\ \underset{f}{\cdot},\ \mathfrak{D}^f,\ (\mathfrak{N}^f \uparrow M \square)$$

(see §1.2). In this subsection we consider linear operators mapping K_M into locally convex spaces.

Now, basing ourselves on Theorem 1.2.2, we prove the following theorem.

2.1.1. *It is possible to determine a constructive locally convex space* $\mathfrak{M}$ *and a constructive linear operator* λ, *mapping* K_M *into* $\mathfrak{M}$, *such that* λ *is bounded and* п*-continuous, but not continuous.*[3)]

Proof. We introduce the following formula:

$$\mathfrak{J}^0 \rightleftharpoons (\langle c_{0,1}\rangle_0 \in (\mathrm{e} \to \mathrm{H})).$$

The formula $\mathfrak{J}^0$ defines a set of words in Ч_0. This is the set of transcriptions of all possible algorithms of the type $(\mathrm{e} \to \mathrm{H})$. We note that the set of all words of type e, i.e. the set of all "indices" of general recursive functions, is normal. Therefore $\mathfrak{J}^0$ is a normal formula. We adjoin $\mathfrak{J}^0$ to the list Σ of formulas of the language $\mathfrak{A}_\Sigma$ and denote by ι the subordinate generic letter assigned to $\mathfrak{J}^0$ in this language.

It is possible to construct a two-parameter formula $\mathfrak{Q}^0$ such that, for any words ι_1 of type ι and Ф_1^M of type Ф^M, the condition $\mathfrak{Q}^0(\iota_1, \text{Ф}_1^M)$ is satisfied if and only if we can realize a natural number m, words $\mathrm{e}_1, \cdots, \mathrm{e}_m$ of type e, words $P_1, \cdots, P_m$ of type Ф^M and real duplexes $\alpha_1, \cdots, \alpha_m$ for which

$$\text{Ф}_1^M \underset{f}{=} \alpha_1 \cdot P_1 + \cdots + \alpha_m \cdot P_m, \quad |\alpha_1| + \cdots + |\alpha_m| = 1$$

and

$$\forall n \left(\mathfrak{N}^f (M \square n \square P_1) \leqslant \frac{\{\mathrm{e}_1\}(n) + 1}{\langle \iota_1 \rangle (\mathrm{e}_1) + 1} \right) \& \cdots$$
$$\cdots \& \forall n \left(\mathfrak{N}^f (M \square n \square P_m) \leqslant \frac{\{\mathrm{e}_m\}(n) + 1}{\langle \iota_1 \rangle (e_m) + 1} \right).$$

Now consider the following list:

(2) $$\text{Ч}_0,\ \mathfrak{P}_K^M,\ \mathfrak{C}^f,\ \underset{f}{+},\ \underset{f}{\cdot},\ \mathfrak{D}^f,\ \text{Ч}_0,\ \mathfrak{J}^0,\ \mathfrak{Q}^0.$$

In order to show that this list defines a locally convex space, we have to verify that $(\text{Ч}_0, \mathfrak{J}^0, \mathfrak{Q}^0)$ satisfies the topological axioms III1–III5 of §1.1, Chapter I. For simplicity of exposition we will not dwell on this verification.

We denote the locally convex space (2) by $\mathfrak{M}$. It is clear that the spaces K_M and $\mathfrak{M}$ differ only in the topologies defined in them.

Denote by λ the identity mapping of K_M into $\mathfrak{M}$. Evidently λ is a linear operator mapping K_M into $\mathfrak{M}$.

1) *The operator* λ *is bounded*. In order to prove this, we need to prove that any set bounded in K_M is bounded also in $\mathfrak{M}$. Suppose that $\mathfrak{B}$ is a bounded set in K_M. Then we can realize a word e_1 of type e such that

(3) $$\forall \text{Ф}_1^M (\text{Ф}_1^M \in \mathfrak{B} \supset \forall n (\mathfrak{N}^f (M \square n \square \text{Ф}_1^M) \leqslant \{\mathrm{e}_1\}(n) + 1)).$$

We shall prove that $\mathfrak{B}$ is bounded in $\mathfrak{M}$, i.e. that

(4) $$\forall \iota_1 \exists a (a > 0 \& \forall \text{Ф}_1^M (\text{Ф}_1^M \in \mathfrak{B} \supset \mathfrak{Q}^0 (\iota_1, a \cdot \text{Ф}_1^M))).$$

[3)] For the definition of the concepts of "bounded operator" and of "п-continuous operator", see §2.1, Chapter III.

Suppose that ι_1 is a word of type ι, i.e. an algorithm of the type $(\mathrm{e} \to \mathrm{H})$. Choose $a = 1 : (\langle \iota_1 \rangle (\mathrm{e}_1) + 1)$. Clearly $a > 0$. Suppose that Φ_1^M is a word of type Φ^M such that $\Phi_1^M \in \mathfrak{B}$. Then

$$\forall n (\mathfrak{N}^f (M \square n \square \Phi_1^M) \leqslant \{\mathrm{e}_1\} (n) + 1),$$

so that

$$\forall n \left(\mathfrak{N}^f (M \square n \square a \cdot \Phi_1^M) \leqslant \frac{\{\mathrm{e}_1\} (n) + 1}{\langle \iota \rangle (\mathrm{e}_1) + 1} \right),$$

from which it follows that $\mathfrak{Q}^0(\iota_1, a \cdot \Phi_1^M)$. Thus (4) is proved.

2) *The operator* λ *is* Π*-continuous.* For this it suffices to prove that every sequence φ of words of type Φ^M converging to $\mathfrak{O}^f$ in K^M converges to $\mathfrak{O}^f$ in $\mathfrak{M}$ as well.

Suppose that φ is a sequence of words of type Φ^M converging to $\mathfrak{O}^f$ in K^M. Then

(5) $$\forall i \exists m \forall n (n \geqslant m \supset \mathfrak{N}^f (M \square i \square \varphi(n)) \leqslant 2^{-i}).$$

We need to prove that the sequence φ converges to $\mathfrak{O}^f$ in $\mathfrak{M}$, i.e. that

(6) $$\forall \iota_1 \exists m \forall n (n \geqslant m \supset \mathfrak{Q}^0 (\iota_1, \varphi(n))).$$

Suppose that ι_1 is of type ι, i.e. the transcription of an algorithm of the type $(\mathrm{e} \to \mathrm{H})$. First we prove that a word e_1 of type e is realizable for which the following formula is valid:

(7) $$\exists m \forall n \left(n \geqslant m \supset \forall k \left(\mathfrak{N}^f (M \square k \square \varphi(n)) \leqslant \frac{\{\mathrm{e}_1\} (k) + 1}{\langle \iota \rangle (\mathrm{e}_1) + 1} \right) \right).$$

In view of the assumption (5) it is possible to construct an increasing sequence γ of natural numbers such that

(8) $$\forall i n (n \geqslant \gamma(i) \supset \mathfrak{N}^f (M \square i \square \varphi(n)) \leqslant 2^{-i}).$$

Then we may construct a sequence ψ of words of the type Φ^M such that, for any natural number n,

a) if $n < \gamma(1)$, then $\psi(n) \doteq \varphi(n)$;

b) if $\gamma(i) \leqq n < \gamma(i+1)$, then $\psi(n) \doteq 2^i \varphi(n)$, where i is any natural number $\geqq 1$.

From (8) and from the definition of the sequence ψ it follows that

(9) $$\forall i n (\gamma(i) \leqslant n < \gamma(i+1) \supset \mathfrak{N}^f (M \square i \square \psi(n)) \leqslant 1).$$

We recall that $\mathfrak{N}^f$ has the following property:

(10) $$\forall i j \Phi_1^M (i \leqslant j \supset \mathfrak{N}^f (M \square i \square \Phi_1^M) \leqslant \mathfrak{N}^f (M \square j \square \Phi_1^M)).$$

It follows from (9) and (10) that

(11) $$\forall i n (n \geqslant \gamma(i) \supset \mathfrak{N}^f (M \square i \square \psi(n)) \leqslant 1).$$

Now we may construct a word e_1 of type e such that for any natural number i

(12) $\max(\mathfrak{N}^f(M\square i\square\psi(0))\square\ldots\square\mathfrak{N}^f(M\square i\square\psi(\gamma(i)-1)))\leqslant\{\mathrm{e}_1\}(i).$

From (11) and (12) we obtain

$$\forall in(\mathfrak{N}^f(M\square i\square\psi(n))\leqslant\{\mathrm{e}_1\}(i)+1). \tag{13}$$

Now we prove that formula (7) holds for that word e_1. We find a natural number l such that $\langle\iota_1\rangle(\mathrm{e}_1)+1\leqq 2^l$, and then we write $m\leftrightharpoons\gamma(l)$. We will show that

$$\forall n(n\geqslant m\supset\forall k((\langle\iota_1\rangle(\mathrm{e}_1)+1)\cdot\mathfrak{N}^f(M\square k\square\varphi(n))\leqslant\{\mathrm{e}_1\}(k)+1)). \tag{14}$$

Suppose that n is a natural number $\geqq m$. Then from the definition of the sequence ψ it follows that

$$\forall k(2^l\cdot\mathfrak{N}^f(M\square k\square\varphi(n))\leqslant\mathfrak{N}^f(M\square k\square\psi(n))).$$

Hence, using (13), we find that

$$\forall k(2^l\cdot\mathfrak{N}^f(M\square k\square\varphi(n))\leqslant\{\mathrm{e}_1\}(k)+1),$$

so that

$$\forall k((\langle\iota_1\rangle(\mathrm{e}_1)+1)\cdot\mathfrak{N}^f(M\square k\square\varphi(n))\leqslant\{\mathrm{e}_1\}(k)+1).$$

Thus formula (14), and with it (7), are proved. By the definition of $\mathfrak{Q}^0$, for any word Φ_1^M of type Φ^M, if

$$\forall k\left(\mathfrak{N}^f(M\square k\square\Phi_1^M)\leqslant\frac{\{\mathrm{e}_1\}(k)+1}{\langle\iota_1\rangle(\mathrm{e}_1)+1}\right),$$

then $\mathfrak{Q}^0(\iota_1,\Phi_1^M)$. Therefore it follows from (7) that

$$\exists m\forall n(n\geqslant m\supset\mathfrak{Q}^0(\iota_1,\varphi(n))).$$

Thus (6) is proved. The sequence φ converges to zero in the space $\mathfrak{M}$.

3) *The operator* λ *is not continuous.* In this case it suffices for us to prove the following formula:

$$\exists\iota_1\neg\exists i\forall\Phi_1^M(\mathfrak{N}^f(M\square i\square\Phi_1^M)\leqslant 2^{-i}\supset\mathfrak{Q}^0(\iota_1,\Phi_1^M)). \tag{15}$$

We construct an algorithm ξ of the type $(\mathrm{e}\to\mathrm{H})$ such that

$$\forall\mathrm{e}_1(\xi(\mathrm{e}_1)\simeq\max(\{\mathrm{e}_1\}(0)\square\ldots\square\{\mathrm{e}_1\}(\mathrm{e}_1))). \tag{16}$$

Then we let $\iota_1\leftrightharpoons\mathcal{E}\,\xi\,\mathfrak{Z}$. It is clear that ι_1 is a word of type ι. Let us prove that

$$\forall\Phi_1^M(\mathfrak{Q}^0(\iota_1,\Phi_1^M)\supset(\Phi_1^M\in\mathfrak{A})), \tag{17}$$

where $\mathfrak{A}$ is the set of points of the space K_M defined in the proof of Theorem 1.2.2. Suppose that Φ_1^M is a word of type Φ^M such that $\mathfrak{Q}^0(\iota_1,\Phi_1^M)$. Then we can realize a natural number m, words $\mathrm{e}_1,\cdots,\mathrm{e}_m$ of type e, words $P_1,\cdots,P_m$ of type Φ^M, and real duplexes $\alpha_1,\cdots,\alpha_m$ such that

$$\Phi_1^M=\alpha_1\cdot P_1+\cdots+\alpha_m\cdot P_m;\quad|\alpha_1|+\cdots+|\alpha_m|=1; \tag{18}$$

$$\forall n\left(\mathfrak{N}^f(M\square n\square P_1)\leqslant\frac{\{e_1\}(n)+1}{\xi(e_1)+1}\right)\&\cdots \tag{19}$$
$$\ldots\&\forall n\left(\mathfrak{N}^f(M\square n\square P_m)\leqslant\frac{\{e_m\}(n)+1}{\xi(e_m)+1}\right).$$

We need to prove that $(\Phi_1^M\in\mathfrak{A})$, i.e. that

$$\forall nk((\forall j(j\leqslant n\&!\mathfrak{U}(j\square n)\supset\mathfrak{U}(j\square n)\leqslant k)\supset\mathfrak{N}^f(M\square n\square\Phi_1^M)\leqslant k+1). \tag{20}$$

Suppose that n and k are natural numbers such that

$$\forall j(j\leqslant n\&!\mathfrak{U}(j\square n)\supset\mathfrak{U}(j\square n)\leqslant k). \tag{21}$$

It follows from (16), (19) and (21) that, for any i with $1\leqq i\leqq m$:
a) if $e_i\leqq n$, then $\{e_i\}(n)=\mathfrak{U}(e_i\square n)\leqq k$, so that

$$\mathfrak{N}^f(M\square n\square P_i)\leqslant k+1;$$

b) if $e_i\geqq n$, then $\{e_i\}(n)\leqq\xi(i)$, so that

$$\mathfrak{N}^f(M\square n\square P_i)\leqslant 1.$$

Thus for any i with $1\leqq i\leqq m$ we have

$$\mathfrak{N}^f(M\square n\square P_i)\leqslant k+1. \tag{22}$$

From (18) and (22) it follows that

$$\mathfrak{N}^f(M\square n\square\Phi_1^M)\leqslant k+1.$$

Thus formula (20), and with it (17), are proved.
Suppose that

$$\exists i\forall\Phi_1^M(\mathfrak{N}^f(M\square i\square\Phi_1^M)\leqslant 2^{-i}\supset\mathfrak{Q}^0(\iota,\Phi_1^M)). \tag{23}$$

Then it follows from (23) and (17) that

$$\exists i\forall\Phi_1^M(\mathfrak{N}^f(M\square i\square\Phi_1^M)\leqslant 2^{-i}\supset(\Phi_1^M\in\mathfrak{A})). \tag{24}$$

However, as we proved in Theorem 1.2.2, formula (24) is not valid; see formula (11) in the proof of 1.2.2. Therefore (23) is not true. This means that (15) is proved. The operator is not continuous. Theorem 2.1.1 is completely proved.

REMARK 1. One may prove that the topology of the space $\mathfrak{M}$ majorizes the topology of K_M. Therefore every set bounded in $\mathfrak{M}$ is bounded in K_M as well. On the other hand, (15) shows that the topology of K_M does not majorize the topology of $\mathfrak{M}$. Thus K_M and $\mathfrak{M}$ have the same bounded sets, but their topologies are not equivalent.

2.1.2. *Suppose that $\mathfrak{M}$ is any constructive metric space. Then every operator mapping the space K_M into $\mathfrak{M}$ is continuous.*

We say that the operator λ mapping K_M into $\mathfrak{M}$ is continuous if the following condition is satisfied:

$$\forall \phi_1^M n \exists i \forall \phi_2^M (\mathfrak{N}^f (M \square i \square \phi_2^M - \phi_1^M) \leqslant 2^{-i} \supset \rho(\lambda(\phi_2^M) \square \lambda(\phi_1^M)) \leqslant 2^{-n}),$$

where ρ is the metric function of $\mathfrak{M}$.

PROOF. K_M is an enumerably normed space, and so it is metrizable (Theorem 4.2.1, Chapter I). In view of Theorem 1.2.1, K_M is complete and separable. Thus we may regard K_M as a complete separable constructive metric space. As is known, on the basis of the theorem of G. S. Ceĭtin [**27**] one finds that every algorithmic operator mapping a complete separable constructive metric space into any constructive metric space is continuous[4] (see also Theorem 3 of [**49**]). Thus Theorem 2.1.2 is proved.

The following corollary is evident.

2.1.3. *Suppose that* $\mathfrak{M}$ *is any constructive normed* (*or enumerably normed*) *space. Then every linear operator mapping the space* K_M *into* $\mathfrak{M}$ *is continuous.*

REMARK 2. As we know, in classical mathematics, with the classical understanding of terms and inferences, one proves that a linear operator mapping a countably normed space into any locally convex space is continuous if and only if it is bounded or countably continuous. For this see for example [**7**], Chapter XI. Theorem 2.1.1 shows that the constructive analogue of this classical theorem is refuted by an example. On the other hand, in classical mathematics, using the axiom of choice, one proves that there exist linear discontinuous functionals in the space $\mathscr{D}_m$ of infinitely differentiable functions, vanishing outside the segment $[-m, m]$, where m is any positive natural number. See for example [**52**], page 84 of the Russian translation (French p. 80).[5] Theorem 2.1.3 shows that this assertion of classical mathematics does not carry over to constructive mathematics. Thus Theorems 2.1.1–2.1.3 represent points of essential divergence between classical and constructive mathematics.

2.2. In this subsection we consider continuous linear functionals in the space K_M and in the space adjoint to it. The following is an obvious consequence of Theorem 2.1.3.

2.2.1. *Every linear functional in the space* K_M *is continuous.*

Consider the set $\mathfrak{P}_{л}^M$ defined by the formula

$$\mathfrak{P}_{л}^M \leftrightarrows (\langle c_{0,1} \rangle_3 \in \underline{\text{lin f}}_{K_M}),$$

---

[4] Here we have in mind operators defined everywhere. Such an operator is said to be continuous if it is continuous at all points.

[5] In [**52**] it was stated that there exist linear discontinuous functionals in $\mathscr{D}$. Hence, by the definition of the concept of convergence in the space $\mathscr{D}$ introduced in [**52**], it is possible to prove that there exist linear discontinuous functionals in the space $\mathscr{D}_m$ (convergence in $\mathscr{D}_m$ is uniform convergence of derivatives of all orders).

where ($\lambda \in \underline{\text{lin f}}_{K_M}$) denotes the condition "λ is a linear functional in K_M". The set $\mathfrak{P}^M_{\text{л}}$ is normal. It is the set of transcriptions, relative to the alphabet Ч^{sa}_3, of all possible linear functionals in K_M.

In view of Theorem 2.2.1, we have

$$\text{(25)} \qquad \forall X k \exists i \forall \phi^M_1 (\mathfrak{N}^f (M \square i \square \phi^M_1) \leqslant 2^{-i} \supset |\langle X \rangle_3 (\phi^M_1)| \leqslant 2^{-k}),$$

where X is a variable for elements of the set $\mathfrak{P}^M_{\text{л}}$.

By the principles of the constructive interpretation of mathematical inferences it follows that we can realize an algorithm $\mathfrak{Z}$ which transforms each element X of the set $\mathfrak{P}^M_{\text{л}}$ into the transcription $\mathfrak{Z}(X)$ of some algorithm of the type $(\text{н} \to \text{н})$ satisfying the condition

$$\forall k \phi^M_1 (\mathfrak{N}^f (M \square \langle \mathfrak{Z}(X) \rangle_0 (k) \square \phi^M_1) \leqslant 2^{-\langle \mathfrak{Z}(X) \rangle_0 (k)} \supset |\langle X \rangle_3 (\phi^M_1)| \leqslant 2^{-k}).$$

In other words, $\mathfrak{Z}$ *transforms the transcription of any linear functional in* K_M *into the transcription of its regulator of continuity.*

Now we shall construct the space adjoint to K_M. For convenience in the further exposition we construct a sequence of spaces adjoint to the spaces of the sequence defined by the list (24) in §1.3.

We shall use the alphabet $\text{A}_\pi = \text{Ч}_0 \cup \{ \otimes \}$ introduced in §2.2 of Chapter III.

It is possible to construct in the language Я_Σ a two-parameter formula $\mathfrak{P}_{\text{c}}$, whose parameters are a variable m for positive natural numbers and a variable $t_{\pi,1}$ for words in A_π, such that, for any natural number $M \geqq 1$ and any word P in A_π, the formula $\mathfrak{P}_{\text{c}}(M, P)$ (here $\mathfrak{P}_{\text{c}}(M, P) \leftrightharpoons F^{m,\, t_{\pi,1}}_{M,\, P} \llcorner \mathfrak{P}_{\text{c}} \lrcorner$) expresses the condition "P is a complete cipher of a continuous linear functional in the space K_M" (see the definition of the complete cipher of a continuous linear functional in §2.2, Chapter III). We adjoin this formula $\mathfrak{P}_{\text{c}}$ to the list $\triangle$ of formulas of the language $\text{Я}^{\triangle}_\Sigma$ and denote by щ the pregeneric letter assigned to the formula $\mathfrak{P}_{\text{c}}$ in that language.

Suppose that M is a natural number $\geqq 1$. Write

$$\mathfrak{P}^M_{\text{c}} \leftrightharpoons F^m_M \llcorner \mathfrak{P}_{\text{c}} \lrcorner.$$

Words of the type щ^M, i.e. elements of the set $\mathfrak{P}^M_{\text{c}}$, are complete ciphers of continuous linear functionals in K_M.

Suppose that θ is any object term. Here $\underline{\theta}_\pi$ and $\overline{\theta}^\pi$ will be understood as follows:

$$\theta_\pi \leftrightharpoons \langle \overline{\text{lef}}_\pi(\theta) \rangle_3; \quad \theta^\pi \leftrightharpoons \langle \overline{\text{rig}}_\pi(\theta) \rangle_0$$

(see §2.2, Chapter III).

We introduce a formula $\mathfrak{E}_{\text{c}}$ as follows:

$$\mathfrak{E}_{\text{c}} \leftrightharpoons \forall f_1 (\forall x (|x| \geqslant m \supset f^{(0)}_1 (x) = 0) \supset (\underline{t_{\pi,1}}_\pi (f_1) = \underline{t_{\pi,2}}_\pi (f_1))).$$

This is a three-parameter formula of the language Я_Σ whose parameters are the variables m, $t_{\pi,1}$ and $t_{\pi,2}$. We shall use the notation

$$\mathfrak{E}^M_{\text{c}} \leftrightharpoons F^m_M \llcorner \mathfrak{E}_{\text{c}} \lrcorner; \quad (P \underset{\text{c},\, M}{=} Q) \leftrightharpoons F^{m,\, t_{\pi,1},\, t_{\pi,2}}_{M,\, P,\, Q} \llcorner \mathfrak{E}_{\text{c}} \lrcorner,$$

where M, P and Q are arbitrary object terms. Obviously, for any natural number $M \geqq 1$ and any words P and Q of type щ^M,

$$(P \underset{c,\,M}{=} Q) \equiv \forall \phi_1^M (\underset{\llcorner\pi}{P} (\phi_1^M) = \underset{\llcorner\pi}{Q} (\phi_1^M)).$$

It is possible to construct algorithms $\underset{c}{+}$ and $\underset{c}{\cdot}$ such that for any natural number $M \geqq 1$ the algorithm $(\underset{c}{+} \uparrow M\square)$ defines the sum of complete ciphers of continuous linear functionals in K_M and the algorithm $(\underset{c}{\cdot} \uparrow M\square)$ defines the multiplication of such complete ciphers by real duplexes. We will write

$$\underset{c,\,M}{+} \leftrightharpoons (\underset{c}{+} \uparrow M\square), \qquad \underset{c,\,M}{\cdot} \leftrightharpoons (\underset{c}{\cdot} \uparrow M\square).$$

Suppose that U is the transcription of an algorithm transforming any word of type f into the number 0, and V the transcription of an algorithm transforming any natural number into the number 0. Then $\mathfrak{O}^c \leftrightharpoons U \otimes V$. $\mathfrak{O}^c$ is the complete cipher of the zero functional in the space K_M.

We introduce further the following formula:

$$\mathfrak{L}_c \leftrightharpoons \forall f_1 (\forall x (|x| \geqslant m \supset \underset{\llcorner\,\lrcorner}{f_1^{(0)}}(x) = 0) \,\&\, \forall n (\mathfrak{N}^f (m\square n\square f_1) \leqslant \{e_1\}(n) + 1) \supset |\underset{\llcorner\pi}{t_{\pi,1}} (f_1)| \leqslant 1).$$

This is a three-parameter formula of the language $Я_\Sigma$ whose parameters are the variables m, e_1 and $t_{\pi,1}$. Here e_1 is a variable of genus e, i.e. a variable for elements of the set A (the set of "indices" of all general recursive functions). We will write

$$\mathfrak{L}_c^M \leftrightharpoons F_M^m \llcorner \mathfrak{L}_c \lrcorner;\ \mathfrak{L}_c(M, U, P)\ [\text{or}\ \ \mathfrak{L}_c^M (U, P)] \leftrightharpoons F_{M,U,P}^{m,e_1,t_{\pi,1}} \llcorner \mathfrak{L}_c \lrcorner,$$

where M, U and P are arbitrary object terms. Obviously, for any natural number $M \geqq 1$ and any words e_1 of type e and P of type щ^M,

$$\mathfrak{L}_c(M, e_1, P) \equiv \forall \phi_1^M (R^M (e_1, \phi_1^M \supset)|\underset{\llcorner\,\lrcorner}{P_\pi} (\phi_1^M)| \leqslant 1),$$

where R^M is the formula defined in §1.2.

Collecting together the objects just constructed, we obtain the list

$$(26) \qquad Ч_0,\ N,\ A_\pi,\ \mathfrak{P}_c,\ \mathfrak{E}_c,\ \underset{c}{+},\ \underset{c}{\cdot},\ \mathfrak{O}^c,\ Ч_0,\ A,\ \mathfrak{L}_c.$$

Here N denotes the set of all positive natural numbers. This list defines an N-family of locally convex spaces. This N-family of spaces will be considered in §3 of the present chapter.

Now we turn to the investigation of the space K_M, where M is a fixed positive number. The list

$$(27) \qquad A_\pi,\ \mathfrak{P}_c^M,\ \mathfrak{E}_c^M,\ \underset{c,\,M}{+},\ \underset{c,\,M}{\cdot},\ \mathfrak{O}^c,\ Ч_0,\ A,\ \mathfrak{L}_c^M$$

represents a locally convex space adjoint to K_M (see §2.4 of Chapter III). We denote the space (27) by K_M^c.

REMARK 1. Using the algorithm $\mathfrak{Z}$ constructed above, we may construct an algorithm which transforms the transcription of any linear functional in K_M into the complete cipher of that functional. Therefore as the points of the space adjoint to K_M we may adopt the transcriptions of linear functionals in K_M, rather than the complete ciphers of continuous linear functionals in K_M. In the make-up of the complete cipher of a continuous linear functional the transcription of the regulator of continuity of the functional constitutes "superfluous information", which can be gotten from the transcription of the functional itself. However, for convenience of exposition we shall admit this "superfluous information", and, in accordance with the general definition of adjoint space, take as the points of the space adjoint to K_M the complete ciphers of continuous linear functionals in K_M.

2.2.2. *The space K_M^c adjoint to K_M is complete.*

PROOF. Suppose that $\mathfrak{X}$ is a two-parameter formula defined as follows:

$$\mathfrak{X}(e_1, e_2) \rightleftharpoons \forall ш_c^M (\mathfrak{L}_c^M(e_2, ш_1^M) \supset \mathfrak{L}_c^M(e_1, ш_1^M)).$$

In place of the notation $\mathfrak{X}(e_1, e_2)$ we shall write $(e_1 \prec e_2)$. The list $(A, \mathfrak{X})$ defines a directed set of indices of neighborhoods of the origin in the space K_M^c.[6] For the proof of the theorem we need to construct an $(A, \mathfrak{X})$-algorithm of passing to the limit in K_M^c. Suppose that ψ is the index of the alphabet $A_\pi \cup \{\odot\}$. The alphabet A_ψ is intended for the construction of FR-$(A, \mathfrak{X})$-constructs in the space K_M^c. We write

$$\underline{\theta_\psi} \leftrightharpoons \langle \overline{\mathrm{lef}_\psi}(\theta)\rangle_\pi, \quad \overline{\theta_\psi} \leftrightharpoons \langle \overline{\mathrm{rig}_\psi}(\theta)\rangle_0,$$

where θ is any object term ($\overline{\mathrm{lef}_\psi}$ and $\overline{\mathrm{rig}_\psi}$ are defined analogously to §3.1, Chapter I).

First we construct auxiliary algorithms ζ of the type $(Ф^M \to e)$, and η of the type $(\text{н}e \to e)$, satisfying the following conditions:

$$\forall Ф_1^M\, n\, (\mathfrak{N}^f(M \square n \square Ф_1^M) \leqslant \{\zeta(Ф_1^M)\}(n) + 1); \tag{28}$$

$$\forall i\, e_1\, n (\{\eta(i \square e_1)\}(n) \simeq 2^i(\{e_1\}(n) + 1) - 1). \tag{29}$$

Suppose that $t_{\psi,1}$ is any FR-$(A, \mathfrak{X})$-construct in K_M^c. Put

$$B \rightleftharpoons \underline{t_{\psi,1}}_\psi, \quad D \rightleftharpoons \overline{t_{\psi,1}}^\psi.$$

We construct in the alphabet A_ψ^{sa} algorithms α and β such that, for any FR-$(A, \mathfrak{X})$-construct $t_{\psi,1}$ in K_M^c and any word $Ф_1^M$ of type $Ф^M$,

$$\alpha(t_{\psi,1} \square Ф_1^M) \overset{+}{\in} c_0; \quad \beta(t_{\psi,1} \square Ф_1^M) \overset{+}{\subset} c_0;$$

$$\forall e_1 (\langle \alpha(t_{\psi,1} \square Ф_1^M)\rangle_3 (e_1) \simeq \underline{B(e_1)}_\pi (Ф_1^M));$$

[6] Here we are applying the concepts and notation introduced in §3 of Chapter I.

$$\forall i (\langle\beta(t_{\Phi,1}\square\Phi_1^M)\rangle_0(i) \simeq D(\eta(i\square\zeta(\Phi_1^M)))).$$

Suppose that $t_{\psi,1}$ is an FR-$(A,\mathfrak{X})$-construct in K_M^c and Φ_1^M a word of type Φ^M. We shall show that the word

(30) $$\alpha(t_{\Phi,1}\square\Phi_1^M)\odot\beta(t_{\psi,1}\square\Phi_1^M)$$

is an FR-$(A,\mathfrak{X})$-construct in the space R of real duplexes. For this it suffices for us to prove the following formula:

(31) $$\forall i\, e_2 e_3 (D(\eta(i\square\zeta(\Phi_1^M)))\prec e_2, e_3 \supset |\underline{B(e_2)}_\pi(\Phi_1^M) - \underline{B(e_3)}_\pi(\Phi_1^M)| \leqslant 2^{-i}).$$

We shall prove (31). Since $t_{\psi,1}$ is an FR-$(A,\mathfrak{X})$-construct in K_M^c, we have

$$\forall e_1 e_2 e_3 (D(e_1)\prec e_2, e_3 \supset \mathfrak{Q}_c^M(e_1, B(e_2) \underset{c,M}{-} B(e_3))),$$

i.e.

$$\forall e_1 e_2 e_3 (D(e_1)\prec e_2, e_3$$

(32) $$\supset \forall \Phi_2^M (R^M(e_1, \Phi_2^M) \supset |\underline{B(e_2)}_\pi(\Phi_2^M) - \underline{B(e_3)}_\pi(\Phi_2^M)| \leqslant 1)).$$

Suppose that i is a natural number, and e_2 and e_3 words of type e, such that

$$D(\eta(i\square\zeta(\Phi_1^M)))\prec e_2, e_3.$$

Then from (32) we get

(33) $$\forall \Phi_2^M (R^M(\eta(i\square\zeta(\Phi_1^M)), \Phi_2^M) \supset |\underline{B(e_2)}_\pi(\Phi_2^M) - \underline{B(e_3)}_\pi(\Phi_2^M)| \leqslant 1).$$

We note that in view of (28) we have

$$\forall n (\mathfrak{N}^f(M\square n\square\Phi_1^M) \leqslant \{\zeta(\Phi_1^M)\}(n)+1).$$

Hence, and from (29), it follows that

$$\forall n (\mathfrak{N}^f(M\square n\square 2^i\cdot\Phi_1^M) \leqslant \{\eta(i\square\zeta(\Phi_1^M))\}(n)+1),$$

i.e. that

$$R^M(\eta(i\square\zeta(\Phi_1^M)), 2^i\cdot\Phi_1^M).$$

Hence in view of (33) we get

$$|\underline{B(e_2)}_\pi(2^i\cdot\Phi_1^M) - \underline{B(e_3)}_\pi(2^i\cdot\Phi_1^M)| \leqslant 1,$$

so that

$$|\underline{B(e_2)}_\pi(\Phi_1^M) - \underline{B(e_3)}_\pi(\Phi_1^M)| \leqslant 2^{-i}.$$

Thus formula (31) is proved.

Suppose that lim is an $(A,\mathfrak{X})$-algorithm of passing to the limit in the space R of real duplexes. Such an algorithm is realizable because the space R is complete; see Theorem 3.1.3 of Chapter I. Now we construct an algorithm

И in the alphabet A_{ψ}^{sa} such that for any FR-$(A, \mathfrak{X})$-construct $t_{\psi,1}$ in the space K_M^c

$\text{И}(t_{\psi,1}) \overset{+}{\in} c_0$ and $\forall \Phi_1^M (\langle \text{И}(t_{\psi,1})\rangle_3 (\Phi_1^M) \simeq \lim (\alpha (t_{\psi,1} \square \Phi_1^M) \odot \beta (t_{\psi,1} \square \Phi_1^M)))$.

From the definitions of the algorithms α, β and lim it is not difficult to see that, for any FR-$(A, \mathfrak{X})$-construct $t_{\psi,1}$ in K_M^c, $\langle \text{И}(t_{\psi,1})\rangle_3$ is a linear functional in the space K_M.

Finally, we construct an algorithm $\mathscr{L}$ as follows:

$$\forall t_{\psi,1} (\mathscr{L}(t_{\psi,1}) \simeq \text{И}(t_{\psi,1}) \otimes \mathfrak{Z}(\text{И}(t_{\psi,1}))),$$

where $\mathfrak{Z}$ is the algorithm defined at the beginning of this subsection.

It is clear that, for any FR-$(A, \mathfrak{X})$-construct $t_{\psi,1}$ in K_M^c, $\mathscr{L}(t_{\psi,1})$ is the complete cipher of a continuous linear functional in K_M, i.e. a point of the space K_M^c.

We shall show that $\mathscr{L}$ is an $(A, \mathfrak{X})$-algorithm of passage to the limit in the space K_M^c. To this end it suffices for us to show that, for any FR-$(A, \mathfrak{X})$-construct $t_{\psi,1}$ in K_M^c,

$$\forall e_1 \exists e_2 \forall e_3 (e_2 \prec e_3 \supset \Omega_c^M (e_1, B(e_3) \underset{c,M}{-} \mathscr{L}(t_{\psi,1}))). \tag{34}$$

Suppose given a word e_1 of type e. Put $U \leftrightharpoons D(\eta(1 \square e_1))$. Then, using formula (32), one can prove that

$$\forall e_3 e_4 \left(U \prec e_3, e_4 \supset \forall \Phi_1^M \left(R^M (e_1, \Phi_1^M) \supset |\underbrace{B(e_3)}_{\pi} (\Phi_1^M) - \underbrace{B(e_4)}_{\pi} (\Phi_1^M)| \leqslant \frac{1}{2}\right)\right). \tag{35}$$

We shall show that

$$\forall e_3\ U \prec e_3 \supset \forall \Phi_1^M (R^M (e_1, \Phi_1^M) \supset |\underbrace{B(e_3)}_{\pi} (\Phi_1^M) - \langle \text{И}(t_{\psi,1})\rangle_3 (\Phi_1^M)| \leqslant 1). \tag{36}$$

Suppose that e_3 is a word of type e such that $U \prec e_3$, and Φ_1^M a word of type Φ^M such that $R^M(e_1, \Phi_1^M)$. We note that $\langle \text{И}(t_{\psi,1})\rangle_3(\Phi_1^M)$ is the limit of an $(A, \mathfrak{X})$-sequence of real duplexes $\langle \alpha(t_{\psi,1} \square \Phi_1^M)\rangle_3$, so that we can realize a word V of type e such that

$$\forall e_4 \left(V \prec e_4 \supset |\underbrace{B(e_4)}_{\pi} (\Phi_1^M) - \langle \text{И}(t_{\psi,1})\rangle_3 (\Phi_1^M)| \leqslant \frac{1}{2}\right). \tag{37}$$

It is possible to construct a word W of type e such that $U, V \prec W$. Then, in view of formulas (35) and (37), it follows from the hypotheses $U \prec e_3$ and $R^M(e_1, \Phi_1^M)$ that

$$|\underbrace{B(e_3)}_{\pi} (\Phi_1^M) - \underbrace{B(W)}_{\pi} (\Phi_1^M)| \leqslant \frac{1}{2},$$

$$|\underbrace{B(W)}_{\pi} (\Phi_1^M) - \langle \text{И}(t_{\psi,1})\rangle_3 (\Phi_1^M)| \leqslant \frac{1}{2},$$

so that we get

$$|\underline{B(e_3)}_\pi(\text{ф}_1^M) - \langle И(t_{\psi,1})\rangle_3(\text{ф}_1^M)| \leqslant 1.$$

Thus formula (36), and with it (34), are proved. Theorem 2.2.2 is completely proved.

REMARK 2. In classical mathematics one also proves that for any $m \geqq 1$ the space adjoint to $\mathscr{D}_m$ is complete ($\mathscr{D}_m$ is the classical space of all infinitely differentiable functions vanishing outside the segment $[-m, m]$, equipped with the topology of uniform convergence of derivatives of all orders). However, there is an essential difference between the classical proof of this assertion and the constructive proof of Theorem 2.2.2. This difference consists in the following: in the classical proof the continuity of the limit functional is established on the basis of the theorem which states that every bounded linear functional in a countably normed space is continuous (in the literature, as far as the author knows, the proof of this theorem is not constructive), while in the constructive proof of Theorem 2.2.2 the continuity of the limit functional is established on the basis of Theorem 2.2.1 (as we have noted, the assertion corresponding to Theorem 2.2.1 does not hold in classical mathematics).

2.3. Now we consider bounded sets in the space K_M^c. Suppose that $\mathfrak{B}$ is a set of points of K_M^c. By definition $\mathfrak{B}$ is a bounded set in K_M^c if

(38) $$\forall e_1 \exists a (a > 0 \,\&\, \forall \text{ш}_1^M ((\text{ш}_1^M \in \mathfrak{B}) \supset \mathfrak{Q}_c^M(e_1, a \cdot \text{ш}_1^M))).$$

Formula (38) is equivalent to the following formula:

(39) $$\forall e_1 \exists a (a > 0 \,\&\, \forall \text{ш}_1^M \text{ф}_1^M ((\text{ш}_1^M \in \mathfrak{B}) \,\&\, R^M(e_1, \text{ф}_1^M) \supset |\underline{\text{ш}_1^M}_\pi(\text{ф}_1^M)| \leqslant a)).$$

Formula (39) means that the set $\mathfrak{B}$ is bounded in any bounded set of the space K_M. Hence we obtain the following.

2.3.1. *Suppose that $\mathfrak{B}$ is a set of points of the space K_M^c. The set $\mathfrak{B}$ is a bounded set in K_M^c if and only if $\mathfrak{B}$ is bounded in any bounded set of K_M.*

Let i be a natural number. It is not hard to see that

$$\forall e_1 \exists a (a > 0 \,\&\, \forall \text{ф}_1^M (R^M(e_1, \text{ф}_1^M) \supset \mathfrak{N}^f(M \square i \square a \cdot \text{ф}_1^M) \leqslant 2^{-i}))$$

[we recall that

$$R^M(e_1, \text{ф}_1^M) \equiv \forall n (\mathfrak{N}^f(M \square n \square \text{ф}_1^M) \leq \{e_1\}(n) + 1))].$$

Therefore, if the set $\mathfrak{B}$ of points of K_M^c satisfies the condition

$$\exists a (a > 0 \,\&\, \forall \text{ш}_1^M \text{ф}_1^M ((\text{ш}_1^M \in \mathfrak{B}) \,\&\, \mathfrak{N}^f(M \square i \square \text{ф}_1^M) \leqslant 2^{-i} \supset |\underline{\text{ш}_1^M}_\pi(\text{ф}_1^M)| \leqslant a)),$$

then it satisfies (39) as well. We therefore obtain the following from 2.3.1.

2.3.2. *Suppose that $\mathfrak{B}$ is a set of points of the space K_M^c. If $\mathfrak{B}$ is bounded in some neighborhood of the origin of the space K_M, then $\mathfrak{B}$ is a bounded set in K_M^c.*

Now we prove the following theorem.

2.3.3. *It is possible to construct a bounded set* $\mathfrak{B}$ *in the space* K_M^c *which is not bounded in any neighborhood of the origin in* K_M.

Proof. Consider a set $\mathfrak{B}$ defined as follows:

$$(\text{ш}_1^M \in \mathfrak{B}) \leftrightharpoons \exists nk(\forall i(i \leqslant n \& !\mathfrak{U}(i\square n) \supset \mathfrak{U}(i\square n) \leqslant k) \& \forall \Phi_1^M (\underset{\pi}{\underline{\text{ш}_1^M}}(\Phi_1^M) = \frac{1}{k+1}\Phi_1^{M(n)}(0))),$$ [7]

where $\mathfrak{U}$ is a universal arithmetic algorithm.

We shall show that the set $\mathfrak{B}$ satisfies condition (39). Suppose given a word e_1 of type e. Put

$$a \leftrightharpoons \max(\mathfrak{U}(e_1 \square 0) \square \cdots \square \mathfrak{U}(e_1 \square e_1)) + 1.$$

Suppose that ш_1^M is a point of the space K_M^c such that $(\text{ш}_1^M \in \mathfrak{B})$. Then we can realize natural numbers n and k such that

$$\forall i(i \leqslant n \& !\, \mathfrak{U}(i\square n) \supset \mathfrak{U}(i\square n) \leqslant k) \tag{40}$$

and

$$\forall \Phi_2^M \left(\underset{\pi}{\underline{\text{ш}_1^M}}(\Phi_2^M) = \frac{1}{k+1} \cdot \Phi_2^{M(n)}(0)\right). \tag{41}$$

Suppose that Φ_1^M is a word of type Φ^M such that $R^M(e_1, \Phi_1^M)$. Then we have

$$\forall i (\mathfrak{N}^f(M \square i \square \Phi_1^M) \leqslant \mathfrak{U}(e_1 \square i) + 1). \tag{42}$$

From (41) and (42) it follows that

$$|\underset{\pi}{\underline{\text{ш}_1^M}}(\Phi_1^M)| \leqslant \frac{1}{k+1} \cdot (\mathfrak{U}(e_1 \square n) + 1). \tag{43}$$

On the other hand, from (40) and from the definition of the number a it follows that

$$(n \leqslant e_1 \supset \mathfrak{U}(e_1 \square n) + 1 \leqslant a) \& (n \geqslant e_1 \supset \mathfrak{U}(e_1 \square n) \leqslant k). \tag{44}$$

It follows from (43) and (44) that

$$|\underset{\pi}{\underline{\text{ш}_1^M}}(\Phi_1^M)| \leqslant a.$$

Thus we have proved that the set $\mathfrak{B}$ satisfies condition (39), i.e. $\mathfrak{B}$ is bounded in any bounded set of K_M. Therefore $\mathfrak{B}$ is a bounded set in K_M^c.

It remains to be proved that $\mathfrak{B}$ is not bounded in any neighborhood of the origin in K_M, i.e. that[8]

$$\neg \exists ih \forall \text{ш}_1^M \Phi_1^M (\text{ш}_1^M \in \mathfrak{B} \& \mathfrak{N}^f(M \square i \square \Phi_1^M) \leqslant 2^{-i} \supset |\underset{\pi}{\underline{\text{ш}_1^M}}(\Phi_1^M)| \leqslant h). \tag{45}$$

---

[7] We recall that if P is a word of type f, then $P^{(n)} \leftrightharpoons \langle P \rangle_4(n)$.

[8] In (45) and (46), h denotes a variable for positive natural numbers.

In order to prove this, it suffices for us to prove the following formula:

(46) $\forall ih \neg\neg \exists ш_1^M \phi_1^M (ш_1^M \in \mathfrak{B} \& \mathfrak{N}^f(M \square i \square \phi_1^M) \leqslant 2^{-i} \& |\underset{\pi}{\underline{ш_1^M}}(\phi_1^M)| > h)$

The fact that (46) implies (45) may be proved within the limits of constructive logic. Now suppose given natural numbers i and h, with $h > 0$. Write $n \leftrightharpoons 2i + 2 + h$. We shall prove that

(47)
$$\begin{aligned} &\exists k \forall j (j \leqslant n \&!\ \mathfrak{U}(j \square n) \supset \mathfrak{U}(j \square n) \leqslant k) \\ &\supset \exists ш_1^M \phi_1^M (ш_1^M \in \mathfrak{B} \& \mathfrak{N}^f (M \square i \square \phi_1^M) \leqslant 2^{-i} \& |\underset{\pi}{\underline{ш_1^M}}(\phi_1^M)| > h). \end{aligned}$$

Suppose that we can realize a natural number k such that

(48) $$\forall j (j \leqslant n \&!\ \mathfrak{U}(j \square n) \supset \mathfrak{U}(j \square n) \leqslant k).$$

Write $\alpha \leftrightharpoons \max(k \square 2)$. We can construct a word Z of type ϕ^M such that

(49) $$\mathfrak{N}^f (M \square i \square Z) \leqslant 2^{-i} \& \mathscr{N}(M \square Z^{(n)}) \geqslant 2^{-i} \cdot \alpha^{n-i}$$

For this see the proof of Theorem 1.2.2. That word Z may be constructed in such a way that the following further condition is satisfied:

(50) $$\forall x \left(|x| \geqslant \frac{M}{2} \supset \underline{Z^{(0)}}(x) = 0\right).$$

From the second term of the conjunction (49) it follows that

(51) $$\mathscr{N}(M \square Z) \geqslant 2^{-i} \cdot \alpha^{(i+2+h)} > \alpha^2 \cdot h > (k+1) \cdot h.$$

The real duplex $\mathscr{N}(M \square Z^{(n)})$ is the least upper bound of the uniformly continuous function f on the segment defined by the formula

$$\forall x (f(x) \simeq |\underline{Z^{(n)}}(x)|).$$

Therefore it follows from (51) that we can realize a real duplex β such that

(52) $$|\beta| \leqslant \frac{M}{2} \& |\underline{Z^{(n)}}(\beta)| > (k+1) \cdot h.$$

Write $Y \leftrightharpoons L^f(1 \square \beta \square Z)$. Then it follows from (50) that Y is a word of type f satisfying the condition

$$\forall x (|x| \geqslant M \supset \underline{Y^{(0)}}(x) = 0).$$

This means that Y is a word of type ϕ^M. It follows from (49) and (52) that

$$\mathfrak{N}^f (M \square i \square Y) \leqslant 2^{-i} \& |\underline{Y^{(n)}}(0)| > (k+1) \cdot h.$$

Now we may define a word F of type $ш^M$ such that

$$\forall \phi_1^M (\underline{F_\pi}(\phi_1^M) = \frac{1}{k+1} \underline{\phi_1^{M(n)}}(0)).$$

For this word F it is obvious that

$$|\underset{\pi}{\underline{F}}(Y)| > h.$$

Thus we have proved that, for any natural numbers i and h with $h > 0$, if we can realize a natural number k satisfying the condition (48), where $n \leftrightharpoons 2i + 2 + h$, then we can also realize words F of type щ^M and Y of type Φ^M such that

$$\exists nk\Big(\forall j(j \leqslant n \,\&\, !\,\mathfrak{U}(j\,\square\, n) \supset \mathfrak{U}(j\,\square\, n) \leqslant k) \,\&\, \forall \phi_1^M\Big(\underset{\pi}{\underline{F}}(\phi_1^M) = \frac{1}{k+1} \cdot \underline{\phi_1^{M(n)}}(0)\Big)\Big),$$

$$\mathfrak{N}^f(M \,\square\, i \,\square\, Y) \leqslant 2^{-i} \text{and } |\underset{\pi}{\underline{F}}(Y)| > h.$$

In other words, formula (47) is proved for arbitrary natural numbers i and h with $h > 0$. In view of formula (12) and the proof of Theorem 1.2.2 we have

$$\forall ih \neg\neg \exists k \forall j (j \leqslant n \,\&\, !\, \mathfrak{U}(j\,\square\, n) \supset \mathfrak{U}(j\,\square\, n) \leqslant k) \tag{53}$$

with $n \leftrightharpoons 2i + 2 + h$. Finally, (46) is obtained from (53) and (47). Theorem 2.3.3 is completely proved.

REMARK. As we know, in classical mathematics one proves that every bounded set in a space which is adjoint to an arbitrary countably normed space is bounded in some neighborhood of the origin of that countably normed space: see for example [2], Chapter I. Theorem 2.2.3 shows that the constructive analogue of that theorem is refuted by an example.

2.4. We introduce the following formula:

$$\mathfrak{K} \leftrightharpoons (\langle c_{0,1}\rangle_3 \in (\mathrm{e} \to \mathrm{p})) \,\&\, \forall \mathrm{e}_1 (\langle c_{0,1}\rangle_3(\mathrm{e}_1) > 0).$$

$\mathfrak{K}$ is a one-parameter formula whose parameter is the variable $c_{0,1}$. We adjoin the formula $\mathfrak{K}$ to the list Σ of formulas of the language $Я_\Sigma$ and denote by κ the subordinate generic letter assigned to $\mathfrak{K}$ in this language. Thus words of type κ are the transcriptions of algorithms transforming any word of type e into a positive rational number.

We further introduce a two-parameter formula $\mathfrak{R}_c^M$ such that, for any words κ_1 of type κ and щ_1^M of type щ^M,

$$\mathfrak{R}_c^M(\kappa_1, \text{щ}_1^M) \equiv \forall \mathrm{e}_1 \mathfrak{L}_c^M(\mathrm{e}_1, \langle \kappa_1\rangle_3(\mathrm{e}_1) \cdot \text{щ}_1^M).$$

The list $(\mathfrak{K}, \mathfrak{R}_c^M)$ defines a *normal fundamental system of bounded sets in the space* K_M^c (Theorem 4.1.1, Chapter I).

Now we construct a formula $\mathfrak{Q}_1$ such that, for any words κ_1 of type κ and ϕ_1^M of type Φ^M,

$$\mathfrak{Q}_1(\kappa_1, \phi_1^M) \equiv \forall \text{щ}_1^M (\mathfrak{R}_c^M(\kappa_1, \text{щ}_1^M) \supset |\underset{\pi}{\underline{\text{щ}_1^M}}(\phi_1^M)| \leqslant 1).$$

The list $(\text{Ч}_0, \mathfrak{K}, \mathfrak{Q}_1)$ defines the *strong topology* in the space K_M (see the definition of strong topology in §2.5, Chapter III).

2.4.1. *In the space K_M the strong topology majorizes the original topology.*

PROOF. We need to prove the following theorem:

$$\forall i \exists \varkappa_1 \forall \text{ф}_1^M (\mathfrak{Q}_1(\lambda, \text{ф}_1^M) \supset \mathfrak{N}^f(M \square i \square \text{ф}_1^M) \leqslant 2^{-i}). \tag{54}$$

Suppose given a natural number i. We construct a word κ_1 of type κ such that

$$\forall \mathrm{e}_1 (\langle \varkappa_1 \rangle (\mathrm{e}_1) \simeq 2^{-i} : (\max \{\mathrm{e}_1\}(0) \square \cdots \square \{\mathrm{e}_1\}(i)) + 1).$$

Suppose that ф_1^M is a word of type ф^M such that $\mathfrak{Q}_1(\kappa_1, \text{ф}_1^M)$. Then

$$\forall \text{щ}_1^M (\mathfrak{R}_c^M (\varkappa_1, \text{щ}_1^M) \supset |\underline{\text{щ}_1^M}_\pi (\text{ф}_1^M)| \leqslant 1). \tag{55}$$

Suppose that j is a natural number $\leqq i$. We construct an algorithm φ of the type $(\text{д} \to \text{щ}^M)$ such that

$$\forall x \text{ф}_1^M (\underline{\varphi(x)}_\pi (\text{ф}_1^M) = \underline{\text{ф}_1^{M(j)}}(x)). \tag{56}$$

It is not hard to prove that

$$\forall x \mathrm{e}_1 \text{ф}_2^M (\forall n (\mathfrak{N}^f(M \square n \square \text{ф}_2^M) \leqslant \{\mathrm{e}_1\}(n) + 1) \supset \langle \varkappa_1 \rangle (\mathrm{e}_1) \cdot 2^i \cdot |\underline{\varphi(x)}_\pi (\text{ф}_2^M)| \leqslant 1).$$

This means that

$$\forall x \mathfrak{R}_c^M (\varkappa_1, 2^i \cdot \varphi(x)). \tag{57}$$

It follows from (55) and (57) that

$$\forall x (2^i \cdot |\underline{\varphi(x)}_\pi (\text{ф}_1^M)| \leqslant 1).$$

Therefore from (56) we find that

$$\forall x (|\underline{\text{ф}_1^{M(j)}}(x)| \leqslant 2^{-i}),$$

i.e. $\mathcal{N}(M \square \text{ф}_1^{M(j)}) \leqq 2^{-i}$. This is true for any $j \leqq i$. Therefore we find that $\mathfrak{N}^f(M \square i \square \text{ф}_1^M) \leqq 2^{-i}$. Thus formula (54), and with it Theorem 2.4.1, are proved.

REMARK 1. As was proved in Chapter I (Theorem 4.4.1), in the most general case of a constructive locally convex space the strong topology does not necessarily majorize the original topology. In Theorem 4.4.1, Chapter I, a case was indicated where the strong topology does not majorize the original topology; on the other hand, Theorem 2.4.1 indicates a case where the strong topology majorizes the original topology.

In conclusion of this section we prove the following theorem.

2.4.2. *In the space K_M the initial topology does not majorize the strong topology.*

PROOF. We shall prove the following formula:

$$\exists \varkappa_1 \neg \exists i \forall \text{ф}_1^M (\mathfrak{N}^f(M \square i \square \text{ф}_1^M) \leqslant 2^{-i} \supset \mathfrak{Q}_1(\varkappa_1, \text{ф}_1^M)). \tag{58}$$

Suppose that κ_1 is a word of type κ satisfying the condition

$$\forall \mathrm{e}_1 (\langle \varkappa_1 \rangle_3 (\mathrm{e}_1) \simeq 1 : (\max (\mathfrak{U}(\mathrm{e}_1 \square 0) \square \cdots \square \mathfrak{U}(\mathrm{e}_1 \square \mathrm{e}_1)) + 1)),$$

where $\mathfrak{U}$ is an arithmetic universal algorithm.

Suppose that $\mathfrak{B}$ is a set of points of the space K_M^c, as defined in the proof of Theorem 2.3.3. In the proof of Theorem 2.3.3 we have already proved that $\mathfrak{B}$ satisfies the condition

$$\forall \mathrm{e}_1 \text{ш}_1^M \text{ф}_1^M (\text{ш}_1^M \in \mathfrak{B} \,\&\, R^M(\mathrm{e}_1, \text{ф}_1^M) \supset |\underline{\text{ш}_1^M}_{\pi}(\text{ф}_1^M)| \leqslant 1 : \langle \varkappa_1 \rangle(\mathrm{e}_1)).$$

This condition may be expressed as follows:

$$\forall \text{ш}_1^M (\text{ш}_1^M \in \mathfrak{B} \supset \forall \mathrm{e}_1 (\forall \text{ф}_1^M (R^M(\mathrm{e}_1, \text{ф}_1^M) \supset |\langle \varkappa_1 \rangle(\mathrm{e}_1) \cdot \underline{\text{ш}_1^M}_{\pi}(\text{ф}_1^M)| \leqslant 1))).$$

This means that

$$\forall \text{ш}_1^M (\text{ш}_1^M \in \mathfrak{B} \supset \mathfrak{R}_c^M(\varkappa_1, \text{ш}_1^M)). \tag{59}$$

From formula (45) (in the proof of 2.3.3) it follows that

$$\neg \exists i \,\forall \text{ф}_1^M (\mathfrak{N}^f(M \,\square\, i \,\square\, \text{ф}_1^M) \leqslant 2^{-i} \supset \forall \text{ш}_1^M (\text{ш}_1^M \in \mathfrak{B} \supset |\underline{\text{ш}_1^M}_{\pi}(\text{ф}_1^M)| \leqslant 1)). \tag{60}$$

It follows from (59) and (60) that

$$\neg \exists i \,\forall \text{ф}_1^M (\mathfrak{N}^f(M \,\square\, i \,\square\, \text{ф}_1^M) \leqslant 2^{-i} \supset \forall \text{ш}_1^M (\mathfrak{R}_c^M(\varkappa_1, \text{ш}_1^M) \supset |\underline{\text{ш}_1^M}_{\pi}(\text{ф}_1^M)| \leqslant 1)),$$

i.e. that

$$\neg \exists i \,\forall \text{ф}_1^M (\mathfrak{N}^f(M \,\square\, i \,\square\, \text{ф}_1^M) \leqslant 2^{-i} \supset \mathfrak{Q}_1(\varkappa_1, \text{ф}_1^M)).$$

Formula (59), and with it Theorem 2.4.2, are proved.

REMARK 2. In classical mathematics one proves that in any countably normed space the strong and original topologies are equivalent to one another (see for example [2], Chapter I). The constructive analogue of this theorem, as is shown by Theorem 2.4.2, is refuted by an example.

§3. Constructive continuous linear functionals in the space K

3.1. Now we consider linear functionals in the space K constructed in §1.3. By definition, the space K is the FR-inductive limit of an expanding sequence of spaces K_m. More precisely, it is the FR-inductive limit of the sequence of spaces defined by the list (24) in §1.3. The points of the space K, i.e. words of type Φ, are words of the type $m \tau f_1$, where m is a positive natural number and f_1 is a word of type f satisfying the condition

$$\forall x (|x| \geqslant m \supset \underline{f_1}^{(0)}(x) = 0).$$

First we prove the following theorem.

3.1.1. *Every linear functional in the space K is continuous.*

PROOF. Let λ be a linear functional in K. For the proof of continuity it suffices for us to prove that

$$\forall k \,\exists \mathrm{e}_1 \forall \text{ф}_1 (\mathfrak{Q}^{\Phi}(\mathrm{e}_1, \text{ф}_1) \supset |\lambda(\text{ф}_1)| \leqslant 2^{-k}). \tag{1}$$

We note that, for any positive natural number M, the algorithm $(\lambda \uparrow M\tau)$

represents a linear functional in K_M. In view of Theorem 2.2.1, for any $M \geqq 1$ the functional $(\lambda \uparrow M_T)$ is continuous in K_M. Therefore

$$\forall km \exists i \forall \phi_1^m (\mathfrak{N}^f (m \square i \square \phi_1^m) \leqslant 2^{-i} \supset |\lambda (m\tau \phi_1^m)| \leqslant 2^{-k}). \tag{2}$$

Suppose given a natural number k. It follows from (2) that we can realize a word e_1 of type e such that

$$\forall m \phi_1^m (\mathfrak{N}^f (m \square \{e_1\}(m) \square \phi_1^m) \leqslant 2^{-\{e_1\}(m)} \supset |\lambda (m\tau \phi_1^m)| \leqslant 2^{-k}). \tag{3}$$

Suppose that ϕ_1 is a word of type ϕ such that $\mathfrak{Q}^\Phi(e_1, \phi_1)$. Suppose that $\phi_1 \underline{\circ} M_T P$, where M is a natural number $\geqq 1$ and P a word of type ϕ^M. Then we can realize natural numbers l and $m_1, \cdots, m_l$, a point P_1 of the space $K_{m_1}, \cdots$, a point P_l of the space K_{m_l}, and real duplexes $\alpha_1, \cdots, \alpha_l$ such that

$$P \underset{f}{=} \alpha_1 \cdot P_1 + \cdots + \alpha_l \cdot P_l, \quad |\alpha_1| + \cdots + |\alpha_l| = 1, \tag{4}$$

$$\begin{aligned} &\mathfrak{N}^f (m_1 \square \{e_1\}(m_1) \square P_1) \leqslant 2^{-\{e_1\}(m_1)} \,\& \\ &\cdots \& \,\mathfrak{N}^f (m_l \square \{e_1\}(m_l) \square P_l) \leqslant 2^{-\{e_1\}(m_l)}. \end{aligned} \tag{5}$$

It follows from (3) and (5) that

$$|\lambda (m_1 \tau P_1)| \leqslant 2^{-k} \& \cdots \& |\lambda (m_l \tau P_l)| \leqslant 2^{-k}. \tag{6}$$

Then from (4) and (6) we find that $|\lambda(\phi_1)| \leqq 2^{-k}$. Hence formula (1), and with it Theorem 3.1.1, are proved.

Remark. Theorem 3.1.1 is also valid for the F-inductive limit $\widetilde{K}$. In other words, every linear functional in the space $\widetilde{K}$ is continuous.

Now, following the general method described in §2.4, Chapter III, we construct a locally convex space adjoint to the space K. We construct the following objects:

a) A one-parameter formula $\mathfrak{P}^{щ}$, defining the set of all complete ciphers of continuous linear functionals in K. To the formula $\mathfrak{P}^{щ}$ we assign a subordinate generic letter, which we denote by щ. Thus *words of type* щ (i.e. elements of the set $\mathfrak{P}^{щ}$) are words in the alphabet A_* $(= Ч_0 \cup \{\otimes\})$ of the type $P \otimes Q$, where P is the transcription of a continuous linear functional in K and Q is the transcription of a regulator of continuity.[9]

b) A two-parameter formula $\mathfrak{E}^{щ}$ defining the relation of equality of words of type щ.

c) The algorithm $\underset{щ}{+}$ for addition of words of type щ.

d) The algorithm $\underset{щ}{\cdot}$ for multiplication of words of type щ by real duplexes.

e) The zero element $\mathfrak{O}^{щ}$ of type щ.

The objects $\mathfrak{E}^{щ}$, $\underset{щ}{+}$, $\underset{щ}{\cdot}$ and $\mathfrak{O}^{щ}$ are constructed by means of the method indicated in §4.3, Chapter I.

[9] For words of type щ one may make the same remark as was made for points of the space K_M^c (see Remark 1, §2.2).

We denote by η the index of the alphabet $\text{A}_r \cup \text{Ч}_3$, i.e. $\text{A}_\eta = \text{A}_r \cup \text{Ч}_3$ (recall that $\text{A}_r = \text{Ч}_0 \cup \{\tau\}$).

Suppose that θ is a word of type щ. The notation $\underline{\theta}_\tau$ and $\overline{\theta}^\tau$ is understood as follows:

$$\underline{\theta}_\tau \leftrightharpoons \langle \overline{\text{lef}_\tau(\theta)} \rangle_\eta; \quad \overline{\theta}^\tau \leftrightharpoons \langle \overline{\text{rig}_\tau(\theta)} \rangle_0.$$

Now we construct a fundamental system of bounded sets in the space K. Suppose that $\mathfrak{K}$ is the set defined in §2.4. We construct a two-parameter formula $\mathfrak{R}^\Phi$ so that for any word $\varkappa_1$ of type κ (i.e. any element of $\mathfrak{K}$) and any word ϕ_1 of type ϕ (i.e. any point of K)

$$\mathfrak{R}^\Phi(\varkappa_1, \phi_1) \equiv \forall e_1 \mathfrak{Q}^\Phi(e_1, \langle \varkappa_1 \rangle_3 (e_1) \underset{\Phi}{\cdot} \phi_1).$$

In view of Theorem 1.1.1, Chapter III, the list $(\text{Ч}_0, \mathfrak{K}, \mathfrak{R}^\Phi)$ defines a normal fundamental system of bounded sets in K.

Finally, we construct a formula $\mathfrak{Q}^{\text{ш}}$ such that for any words κ_1 of type κ and щ$_1$ of type щ,

$$\mathfrak{Q}^{\text{ш}}(\varkappa_1, \text{щ}_1) \equiv \forall \phi_1 (\mathfrak{R}^\Phi(\varkappa_1, \phi_1) \supset |\text{щ}_{1\pi}(\phi_1)| \leqslant 1).$$

By the definition of adjoint space, the list

$$\text{(7)} \qquad \text{A}_\pi, \mathfrak{P}^{\text{щ}}, \mathfrak{E}^{\text{щ}}, \underset{\text{щ}}{+}, \underset{\text{щ}}{\cdot}, \mathfrak{D}^{\text{ш}}, \text{Ч}_0, \mathfrak{K}, \mathfrak{Q}^{\text{щ}}$$

defines a locally convex space, the space adjoint to K. This space (7) will be denoted by $K^{\text{ш}}$.

3.2. Now we shall construct yet another locally convex topology on the linear space:

$$\text{(8)} \qquad \text{A}_\pi, \mathfrak{P}^{\text{щ}}, \mathfrak{E}^{\text{щ}}, \underset{\text{щ}}{+}, \underset{\text{щ}}{\cdot}, \mathfrak{D}^{\text{ш}}.$$

First we construct a normal formula $\mathfrak{S}$, defining the set of all possible words in the alphabet A_r of the form $m \tau e_1$, where m is a positive natural number and e_1 is a word of type e. We assign to this formula $\mathfrak{S}$ a subordinated generic letter, which we shall denote by υ. Then we construct a two-parameter formula $\mathfrak{R}^{\text{сп}}$ such that, for any word υ_1 of type υ and word ϕ_1 of type ϕ, if $\upsilon_1 \,\underline{\circ}\, m \tau e_1$ and $\phi \,\underline{\circ}\, M \tau f_1$, then

$$\mathfrak{R}^{\text{сп}}(\upsilon_1, \phi_1) \equiv \forall x (|x| \geqslant m \supset \underline{f_1^{(0)}}(x) = 0) \,\&\, \forall n (\mathfrak{R}^f(m \Box n \Box f_1) \leqslant \{e_1\}(n) + 1).$$

It is not hard to prove that the list $(\text{A}_r, \mathfrak{S}, \mathfrak{R}^{\text{сп}})$ defines a normal system of bounded sets in K. Furthermore, this system of bounded sets satisfies the following conditions:

$$\text{(9)} \qquad \forall \upsilon_1 \upsilon_2 \exists \upsilon_3 \forall \phi_1 (\mathfrak{R}^{\text{сп}}(\upsilon_1, \phi_1) \vee \mathfrak{R}^{\text{сп}}(\upsilon_2, \phi_2) \supset \mathfrak{R}^{\text{сп}}(\upsilon_3, \phi_3));$$

$$\text{(10)} \qquad \forall \upsilon_1 \exists \upsilon_2 \forall \phi_1 (\mathfrak{R}^{\text{сп}}(\upsilon_1, \phi_1) \supset \mathfrak{R}^{\text{сп}}(\upsilon_2, 2 \cdot \phi_1)).$$

Next we construct a two-parameter formula $\mathfrak{Q}^{\text{сп}}$ such that, for any word υ_1 of type υ and any word щ$_1$ of type щ,

$$\mathfrak{Q}^{\text{сп}}(\mathfrak{v}_1, \text{ш}_1) \equiv \forall \text{ф}_1 (\mathfrak{R}^{\text{сп}}(\mathfrak{v}_1, \text{ф}_1) \supset |\underline{\text{ш}_1}_\pi(\text{ф}_1)| \leqslant 1).$$

Taking into account properties (9) and (10), we can prove that the list $(\mathrm{A}_r, \mathfrak{S}, \mathfrak{Q}^{\text{сп}})$ defines a locally convex topology on the linear space (8). Thus the list

$$(11) \qquad \mathrm{A}_\tau, \mathfrak{P}^{\text{ш}}, \mathfrak{C}^{\text{ш}}, \underset{\text{ш}}{+}, \underset{\text{ш}}{\cdot}, \mathfrak{D}^{\text{ш}}, \mathrm{A}_r, \mathfrak{S}, \mathfrak{Q}^{\text{сп}}$$

defines a locally convex space. We shall denote this space by $K^{\text{сп}}$.

3.2.1. *The two locally convex topologies* $(\text{Ч}_0, \mathfrak{K}, \mathfrak{Q}^{\text{ш}})$ *and* $(\mathrm{A}_r, \mathfrak{S}, \mathfrak{Q}^{\text{сп}})$ *defined on the linear space* (8) *are not equivalent to one another, and the first majorizes the second.*

Proof. For the proof of the theorem it suffices for us to prove the following two formulas:

$$(12) \qquad \forall \mathfrak{v}_1 \exists \varkappa_1 \forall \text{ш}_1 (\mathfrak{Q}^{\text{ш}}(\varkappa_1, \text{ш}_1) \supset \mathfrak{Q}^{\text{сп}}(\mathfrak{v}_1, \text{ш}_1));$$

$$(13) \qquad \neg \cdot \forall \varkappa_1 \exists \mathfrak{v}_1 \forall \text{ш}_1 (\mathfrak{Q}^{\text{сп}}(\mathfrak{v}_1, \text{ш}_1) \supset \mathfrak{Q}^{\text{ш}}(\varkappa_1, \text{ш}_1)).$$

Formula (12) is verified without difficulty. It follows from the simpler formula

$$\forall \mathfrak{v}_1 \exists \varkappa_1 \forall \text{ф}_1 (\mathfrak{R}^{\text{сп}}(\mathfrak{v}_1, \text{ф}_1) \supset \mathfrak{R}^{\Phi}(\varkappa_1, \text{ф}_1)).$$

And the last formula follows from the fact that the list $(\mathrm{A}_r, \mathfrak{S}, \mathfrak{R}^{\text{сп}})$ defines a system of bounded sets in K.

Now we shall prove formula (13). To this end we prove the following stronger formula:

$$(14) \qquad \exists \varkappa_1 \forall \mathfrak{v}_1 \neg \forall \text{ш}_1 (\mathfrak{Q}^{\text{сп}}(\mathfrak{v}_1, \text{ш}_1) \supset \mathfrak{Q}^{\text{ш}}(\varkappa_1, \text{ш}_1)).$$

Consider a set $\mathfrak{A}_1$ of words of type ф defined in such a way that for any word ф_1 of type ф

$$(\text{ф}_1 \in \mathfrak{A}_1) \equiv ([\text{ф}_1]_2^\tau \in \mathfrak{A}),$$

where $\mathfrak{A}$ is the set of words of the space $\widetilde{K}$ defined in the proof of Theorem 1.4.1.

We construct a word κ_1 of type κ such that

$$\forall \mathrm{e}_1(\langle \varkappa_1 \rangle_3 (\mathrm{e}_1) \simeq 2^{-\mathfrak{u}(\mathrm{e}_1 \square \mathrm{e}_1)} : \lambda(\mathfrak{U}(\mathrm{e}_1 \square \mathrm{e}_1))).$$

Here $\mathfrak{U}$ and λ are understood in the same way as in the proof of Theorem 1.4.1.

Suppose that υ_1 is any word of type υ and $\upsilon_1 \,\overline{\underline{\circ}}\, m\,\tau\, \mathrm{e}_1$. We have to prove that

$$(15) \qquad \neg \forall \text{ш}_1 (\mathfrak{Q}^{\text{сп}}(\mathfrak{v}_1, \text{ш}_1) \supset \mathfrak{Q}^{\text{ш}}(\varkappa_1, \text{ш}_1)).$$

From formula (30) and the proof of Theorem 1.4.1 it follows that

$$(16) \qquad \forall \text{ф}_1 (\text{ф}_1 \in \mathfrak{A}_1 \supset \mathfrak{R}^{\Phi}(\varkappa_1, \text{ф}_1)).$$

We shall show that

(17) $\exists \text{ф}_1(\text{ф}_1 \in \mathfrak{A}_1 \,\&\, \neg([\text{ф}_1]_2^\tau \in K_m)) \supset \exists \text{щ}_1(\mathfrak{Q}^{\text{сп}}(\upsilon_1, \text{щ}_1) \,\&\, \neg \mathfrak{Q}^{\text{щ}}(\varkappa_1, \text{щ}_1))$.

Suppose that we can realize a word Z of type ф such that $Z \in \mathfrak{A}_1$ and $\neg(Y \in K_m)$, where $Y \leftrightarrows [Z]_2^\tau$. Then it is possible to find a real duplex α and a natural number h such that

(18) $$|\alpha| > m \,\&\, |\underline{Y^{(0)}}(\alpha)| > 2^{-h}.$$

It is possible to construct a word щ_1 of type щ such that, for any word ф_1 of type ф,

(19) $$\underline{\text{щ}_1}_{\pi}(\text{ф}_1) \simeq 2^h \cdot f_1^{(0)}(\alpha);$$

where

$$f_1 \leftrightarrows [\text{ф}_1]_2^\tau.$$

Clearly, if $\mathfrak{R}^{\text{сп}}(\upsilon_1, \text{ф}_1)$, then

$$\forall x(|x| \geqslant m \supset \underline{f_1^{(0)}}(x) = 0) \qquad (f_1 \leftrightarrows [\text{ф}_1]_2^\tau).$$

Therefore it follows from (18) and (19) that $\underline{\text{щ}_1}_{\pi}(\text{ф}_1) = 0$. So we find that

$$\forall \text{ф}_1(\mathfrak{R}^{\text{сп}}(\upsilon_1, \text{ф}_1) \supset |\underline{\text{щ}_1}_{\pi}(\text{ф}_1)| \leqslant 1),$$

i.e. $\mathfrak{Q}^{\text{сп}}(\upsilon_1, \text{щ}_1)$.

On the other hand, from $Z \in \mathfrak{A}_1$ and from (16) it follows that $\mathfrak{R}^{\Phi}(\varkappa_1, Z)$; from (18) and (19) it follows that $|\underline{\text{щ}_1}_{\pi}(Z)| > 1$. Hence we get

$$\exists \text{ф}_1(\mathfrak{R}^{\Phi}(\varkappa_1, \text{ф}_1) \,\&\, |\underline{\text{щ}_1}_{\pi}(\text{ф}_1)| > 1),$$

from which it follows that $\neg \mathfrak{Q}^{\text{щ}}(\varkappa_1, \text{щ}_1)$. Thus formula (17) is proved.

We note that from formula (39) and the proof of Theorem 1.4.1 we may deduce that

$$\neg\neg \exists \text{ф}_1(\text{ф}_1 \in \mathfrak{A} \,\&\, \neg([\text{ф}_1]_2^\tau \in K_m)).$$

Therefore from (17) we find that

(20) $$\neg\neg \exists \text{щ}_1(\mathfrak{Q}^{\text{сп}}(\upsilon_1, \text{щ}_1) \,\&\, \neg \mathfrak{Q}^{\text{щ}}(\varkappa_1, \text{щ}_1)).$$

It is not hard to see that

$$\forall \text{щ}_1(\mathfrak{Q}^{\text{сп}}(\upsilon_1, \text{щ}_1) \supset \mathfrak{Q}^{\text{щ}}(\varkappa_1, \text{щ}_1)) \supset \neg \exists \text{щ}_1(\mathfrak{Q}^{\text{сп}}(\upsilon_1, \text{щ}_1) \,\&\, \neg \mathfrak{Q}^{\text{щ}}(\varkappa_1, \text{щ}_1)).$$

Thus (15) follows from (20). Formula (15), and with it (14), are proved. The theorem is proved.

Now we consider the N-family of locally convex spaces defined by the list (25) in §2.2, i.e. the list

(21) $$\text{Ч}_0,\ N,\ \text{А}_\pi,\ \mathfrak{P}_{\text{с}},\ \mathfrak{E}_{\text{с}},\ \underset{\text{с}}{+},\ \underset{\text{с}}{\cdot},\ \mathfrak{D}^{\text{с}},\ \text{Ч}_0,\ A,\ \mathfrak{Q}_{\text{с}}.$$

For each natural number $M \geqq 1$ the space $K_M^{\text{с}}$ of this N-family is adjoint to the space K_M.

We note that for any $M \geqq 1$ each complete cipher of a continuous linear functional in K_{M+1} is also the complete cipher of a continuous linear functional in K_M. Denote by ω the algorithm defined as follows:

$$\forall m l (\omega (m \square l \square t_{\pi, 1}) \doteq t_{\pi, 1}).$$

For any m and l, if $1 \leqq l \leqq m$, then $(\omega \uparrow m \square l \square)$ is the identity mapping of K_m^c into K_l^c. It is easy to see that this mapping is linear and continuous.

Thus *the list* (21), *considered along with the algorithm* ω, *represents the projective spectrum of constructive locally convex spaces.* (Here N denotes the directed set of positive natural numbers.)

We may construct the projective limit of this projective spectrum using the method presented in §2.2, Chapter II. We construct successively the following objects:

a) The set $\mathfrak{P}^*$ of words $c_{0,1}$ in Ч_0 defined by the condition

$$\forall m (\langle c_{0,1} \rangle_\pi (m) \in \mathfrak{P}_c^m) \,\&\, \forall m l (1 \leqslant l \leqslant m \supset \langle c_{0,1} \rangle_\pi (l) \underset{c,\, l}{=} \langle c_{0,1} \rangle_\pi (m)).$$

The word P in Ч_0 belongs to the set $\mathfrak{P}^*$ if and only if P is the transcription of a sequence of words in A_r such that, for any $m \geqq 1$, $\langle P_\pi \rangle (m)$ is the complete cipher of a continuous linear functional in K_m and, if $1 \leqq l \leqq m$, then $\langle P \rangle_\pi (m)$ is equal to $\langle P \rangle_\pi (l)$ in the space K_l^c.

b) A two-parameter formula $\mathfrak{E}^*$ defined so that, for any elements P and Q of the set $\mathfrak{P}^*$,

$$\mathfrak{E}^* (P,\ Q) \equiv \forall m (\langle P \rangle_\pi (m) \underset{c,\, m}{=} \langle Q \rangle_\pi (m)).$$

c) An addition algorithm $\underset{\theta}{+}$ for elements of the set $\mathfrak{P}^*$.

d) An algorithm $\underset{\theta}{\cdot}$ for multiplication of elements of $\mathfrak{P}^*$ by real duplexes.[10)]

e) The transcription $\mathfrak{O}^*$ of an algorithm transforming any natural number into the word $\mathfrak{O}^c$ ($\mathfrak{O}^c$ is the zero element in each space K_m^c).

f) The alphabet $A_r = \text{Ч}_0 \cup \{\tau\}$, and the set $\mathfrak{S}$ of words of the form $m \tau e_1$, where m is a positive natural number, e_1 a word of type e, and A_r and $\mathfrak{S}$ have already been defined above.

g) A two-parameter formula $\mathfrak{Q}^*$ defined such that for any elements v_1 of $\mathfrak{S}$ and θ_1 of $\mathfrak{P}^*$, if $v_1 \underline{\circ} m \tau e_1$, then

$$\mathfrak{Q}^* (v_1,\ \theta_1) \equiv \mathfrak{Q}_c (m,\ e_1,\ \langle \theta_1 \rangle_\pi (m)).$$

The list

$$\text{Ч}_0,\ \mathfrak{P}^*,\ \mathfrak{E}^*,\ \underset{\theta}{+},\ \underset{\theta}{\cdot},\ \mathfrak{O}^*,\ A_r,\ \mathfrak{S},\ \mathfrak{Q}^* \tag{22}$$

defines a locally convex space. This space is the projective limit of the projective spectrum (21). We denote this space by $K^{\text{пр}}$.

---

10) The algorithms $\underset{\theta}{+}$ and $\underset{\theta}{\cdot}$ are constructed according to the general method in §2.2, Chapter II.

It is possible to construct an algorithm λ_1 transforming any point θ_1 of the space $K^{\text{пр}}$ into a point $\lambda_1(\theta_1)$ of the space $K^{\text{ш}}$ such that, for any word ϕ_1 of type ϕ, if $\phi_1 \overline{\underline{\circ}}\, m \tau f_1$, then

$$\underbrace{\lambda_1(\theta_1)\ \ (\phi_1)}_{\pi} \simeq \underbrace{\langle\theta_1\rangle_\pi\ (m)\ \ (f_1)}_{\pi}.$$

It is not hard to verify that λ_1 is a linear and one-to-one mapping of the space $K^{\text{пр}}$ into the space $K^{\text{ш}}$ (or into the space $K^{\text{сп}}$). We shall call λ_1 the *canonical mapping* of $K^{\text{пр}}$ into $K^{\text{ш}}$ $(K^{\text{сп}})$.

It is also possible to construct an algorithm λ_2 which transforms each point ш_1 of the space $K^{\text{ш}}$ into a point $\lambda_2(\text{ш}_1)$ of the space $K^{\text{пр}}$ such that, for any natural number $m \geqq 1$ and any point ϕ_1^M of K_m,

$$\underbrace{\langle \lambda_2\ (\text{ш}_1)\rangle_\pi (m)\ \ (\phi_1^m)}_{\pi} \simeq \underbrace{\text{ш}_1}_{\pi}\ (m\tau\phi_1^m).$$

One may prove also that λ_2 is a linear mapping of $K^{\text{ш}}$ $(K^{\text{сп}})$ into $K^{\text{пр}}$, while λ_2 is the operator inverse to λ_1.

Suppose that υ_1 is an arbitrary word of type υ and $\upsilon_1 \overline{\underline{\circ}}\, m \tau e_1$. From the definitions of the formulas $\mathfrak{Q}^{\text{сп}}$ and $\mathfrak{Q}^*$ it follows that, for any points ш_1 of the space $K^{\text{ш}}$ $(K^{\text{сп}})$ and any point θ_1 of $K^{\text{пр}}$,

$$\mathfrak{Q}^{\text{сп}}(\upsilon_1, \text{ш}_1) \equiv \forall f_1 (\forall x(|x| \geqslant m \supset f_1^0(x) = 0) \,\&\, \forall n (\mathfrak{N}^f(m \square n \square f_1) \leqslant \{e_1\}(n) + 1) \supset |\underbrace{\text{ш}_1}_{\pi}\ (m\tau f_1)| \leqslant 1);$$

$$\mathfrak{Q}^*(\upsilon_1, \theta_1) \equiv \forall f_1 (\forall x(|x| \geqslant m \supset \underbrace{f_1^{(0)}}(x) = 0) \,\&\, \forall n (\mathfrak{N}^f(m \square n \square f_1) \leqslant \{e_1\}(n) + 1) \supset |\underbrace{\langle\theta_1\rangle_\pi\ (m)}_{\pi}\ (f)| \leqslant 1).$$

Therefore we get

(23) $$\forall \upsilon_1\theta_1 (\mathfrak{Q}^*(\upsilon_1, \theta_1) \equiv \mathfrak{Q}^{\text{сп}}(\upsilon_1, \lambda_1, (\theta_1)));$$

(24) $$\forall \upsilon_1\text{ш}_1 (\mathfrak{Q}^{\text{сп}}(\upsilon_1, \text{ш}_1) \equiv \mathfrak{Q}^*(\upsilon_1, \lambda_2(\text{ш}_1))),$$

where θ_1 is a variable for points of $K^{\text{пр}}$ and ш_1 is a variable for points of $K^{\text{ш}}(K^{\text{сп}})$.

Moreover, in view of Theorem 3.2.1 (see formulas (12) and (13) in the proof of that theorem) it follows from (24) and from the obvious formula[11)]

$$\forall \text{ш}_1 (\lambda_1(\lambda_2(\text{ш}_1)) \underset{\text{ш}}{=} \text{ш}_1)$$

that

(25) $$\forall \upsilon_1 \exists \varkappa_1 \forall \text{ш}_1 (\mathfrak{Q}^{\text{ш}}(\varkappa_1, \text{ш}_1) \supset \mathfrak{Q}^*(\upsilon_1, \lambda_2(\text{ш}_1)));$$

(26) $$\neg \forall \varkappa_1 \exists \upsilon_1 \forall \theta_1 (\mathfrak{Q}^*(\upsilon_1, \theta_1) \supset \mathfrak{Q}^{\text{ш}}(\varkappa_1, \lambda_1(\theta_1))).$$

Summing up the above results, we obtain

[11)] Here $\underset{\text{ш}}{=}$ denotes equality of points of the space $K^{\text{ш}}$.

3.2.2. a) *The canonical mapping* λ_1 *is an isomorphism of the space* $K^{\text{пр}}$ *onto the space* $K^{\text{сп}}$.

c) *The canonical mapping* λ_1 *is not an isomorphism of the space* $K^{\text{сп}}$ *onto the space* $K^{\text{ш}}$, *while the mapping* λ_1 *is not continuous, and the mapping* λ_2 *inverse to* λ_1 *is continuous.*

In view of Theorem 2.2.2, for any $M \geqq 1$ the space $K_M^{\text{с}}$ is complete. It is not hard to show that $K_M^{\text{с}}$ is T'-separated. Therefore, in view of Theorem 2.2.3, Chapter II, we conclude that the space $K^{\text{пр}}$ is complete. In view of Theorem 3.2.2 the spaces $K^{\text{пр}}$ and $K^{\text{сп}}$ are isomorphic to one another. Thus the space $K^{\text{сп}}$ is also complete (see 3.1.5, Chapter I). So we get

3.2.3. *The space* $K^{\text{сп}}$ *is complete.*

REMARK. Part b) of Theorem 3.2.2 shows that within the limits of constructive mathematics it is possible to determine a strictly increasing sequence of constructive locally convex spaces such that the adjoint to the strict inductive limit is not isomorphic (by means of the canonical mapping) to the projective limit of the sequence of adjoint spaces. This represents a point of essential divergence between constructive and classical mathematics.

3.3. By definition, words of type щ are complete ciphers of continuous linear functionals in the space K. Thus it is possible to regard the concept of a "word of type щ" as the constructive analogue of the concept of "generalized function" defined by L. Schwartz in classical mathematics ([**51**]; see also [**1,2**]).

The spaces $K^{\text{ш}}$ and $K^{\text{сп}}$ are the constructive analogues of the space of generalized functions in the sense of Schwartz in classical mathematics.

It is possible to define the operation of differentiation and other operations over words of type щ and then "carry over" into constructive mathematics certain portions of the classical theory of generalized functions developed by Schwartz and others.

In concluding this chapter we shall consider briefly the question of the connection between the concept of the "complete cipher of a continuous linear functional in the space K" and the concept of "constructive generalized function" defined in Chapter IV.

First of all we construct an algorithm α of the type (н → н), an algorithm β of the type (dн → н) and an algorithm γ of the type (фн → н) such that, for any natural numbers m and k with $m \geqq 1$, any word d_1 of type d and any word ф$_1$ of type ф,[12]

$$\alpha(m) = \mu j\,(2m \leqslant 2^j);$$

$$\mathcal{N}(m \,\square\, F(U \,\square\, m \,\square\, d_1)) \leqslant 2^{\beta(d_1 \square m)};$$

$$\mathfrak{N}^f(l \,\square\, k \,\square\, f_1) \leqslant 2^{\gamma(\text{ф}_1 \square k)},$$

[12] Here we use the notation introduced in Chapter IV for constructive generalized functions.

where

$$U \leftrightharpoons \mathrm{Or}(d_1),\ l \leftrightharpoons [\Phi_1]_1^{\tau},\ f_1 \leftrightharpoons [\Phi_1]_2^{\tau}\ (\text{т. е. } \Phi_1 \overline{\overline{\circ}}\, l\tau f_1).$$

Then we may construct algorithms ξ and ζ such that, for any words d_1 of type d and Φ_1 of type Φ,

$$\xi(d_1 \Box \Phi_1) \overset{+}{\in} c_0;\ \zeta(d_1 \Box \Phi_1) \overset{+}{\in} c_0;$$

$$\forall n (\langle \xi(d_1 \Box \Phi_1)\rangle_3(n) \cong \int_{-l}^{l} (\underset{\llcorner d}{P}(n) \underset{\text{ш}}{\times} f_1^0));$$

$$\forall i (\langle \zeta(d_1 \Box \Phi_1)\rangle_0(i) \simeq \overline{P}^{d}\ (\max(l \Box i + \alpha(l) \dot{+} \gamma(\Phi_1 \Box \{U\}(l))))),$$

where U, P, l and f_1 are defined by the following relations:

(27) $$d_1 \overline{\overline{\circ}}\, U\tau P; \qquad \Phi_1 \overline{\overline{\circ}}\, l\tau f_1.$$

Using the rule of integration by parts, we may show that for any natural number n

(28) $$\langle \xi(d_1 \Box \Phi_1)\rangle_3(x) = (-1)^k \cdot \int_{-l}^{l} (\Phi(U \Box l \Box \underset{\llcorner d}{P}(n)) \underset{\text{ш}}{\times} f_1^{(k)}),$$

where $k \leftrightharpoons \{U\}(l)$.

Obviously, for any words d_1 of type d and Φ_1 of type Φ, $\langle \xi(d_1 \Box \Phi_1)\rangle_3$ defines a sequence of real duplexes. Moreover, using (28), we can prove that

$$\forall inp (\langle \zeta(d_1 \Box \Phi_1)\rangle_0(i) \leqslant n,\ p \supset |\langle \xi(d_1 \Box \Phi_1)\rangle_3(n) - \langle \xi(d_1 \Box \Phi_1)\rangle_3(p)| \leqslant 2^{-i}),$$

i.e. $\langle \zeta(d_1 \Box \Phi_1)\rangle_0$ is a regulator of convergence in itself of the sequence $\langle \xi(d_1 \Box \Phi_1)\rangle_3$.

Suppose that lim is the algorithm for passing to the limit in the space of real duplexes. We construct an algorithm $\mathfrak{B}$ such that, for any word d_1 of type d, $\mathfrak{B}(d_1)$ is the transcription of an algorithm of the type $(\Phi \to \text{д})$ relative to the alphabet $A_\text{ч} = \text{Ч}_3 \cup \{\tau\}$ such that

(29) $$\forall \Phi_1 (\langle \mathfrak{B}(d_1)\rangle_\text{ч}(\Phi_1) \simeq \lim(\xi(d_1 \Box \Phi_1) \odot \zeta(d_1 \Box \Phi_1))).$$

From the definition it is not difficult to see that for any word d_1 of type d the algorithm $\langle \mathfrak{B}(d_1)\rangle_\text{ч}$ represents a linear functional in the space K.

Suppose that $d_1 \overline{\overline{\circ}}\, U\tau P$ and $\Phi_1 \overline{\overline{\circ}}\, l\tau f_1$. From the definition of the algorithm we have

$$\Phi(U \Box l \Box \underset{\llcorner \Phi}{P}(n)) \xrightarrow[n\to\infty]{C_m} F(U \Box l \Box d_1).$$

Hence from formulas (28) and (29) we may deduce that

$$\langle \mathfrak{B}(d_1)\rangle_\text{ч}(\Phi_1) \doteq (-1)^k \cdot \int_{-l}^{l} (F(U \Box l \Box d_1) \underset{\text{ш}}{\times} f_1^{(k)}),$$

where $k \leftrightharpoons \{U\}(l)$.

Now we may construct an algorithm $\mathfrak{C}$ such that, for any word d_1 of type d, $\mathfrak{C}(d_1)$ is the transcription of an algorithm of type $(\mathrm{H} \rightarrow \mathrm{e})$ such that

$$\forall im (\{\langle \mathfrak{C}(d_1) \rangle_0 (i)\}(m) \simeq \max (\{\mathrm{Or}(d_1)\}(m) \square i + \alpha(m) + \beta(d_1 \square m))).$$

It is not difficult to prove the following formula:

$$\forall d_1 i \phi_1 (\mathfrak{Q}^{\Phi}(\langle \mathfrak{C}(d_1) \rangle_0 (i), \phi_1) \supset |\langle \mathfrak{B}(d_1) \rangle_\eta (\phi_1)| \leqslant 2^{-i}).$$

This means that $\langle \mathfrak{C}(d_1) \rangle_0$ is a regulator of continuity of the linear functional $\langle \mathfrak{B}(d_1) \rangle_\eta$ in K.

We construct finally an algorithm Ψ as follows:

$$\forall d_1 (\Psi(d_1) \simeq \mathfrak{B}(d_1) \otimes \mathfrak{C}(d_1)).$$

It is possible to prove that the algorithm Ψ is a continuous linear mapping of the space $\mathfrak{D}$ of constructive generalized functions into the space $K^{\text{ш}}$ adjoint to the space K.

For any word d_1 of type d, $\Psi(d_1)$ may be regarded as the *representative* of the constructive generalized function d_1 in the space $K^{\text{ш}}$. In this sense one may suppose that every constructive generalized function "is" the complete cipher of a continuous linear functional in the space K.

The following question remains open. Is it true that for any complete cipher ш_1 of a continuous functional in K one can realize a constructive generalized function d_1 such that $\Psi(d_1) \underset{\text{ш}}{=} \text{ш}_1$? We note that in classical mathematics the question is resolved in the affirmative. In fact, in any finite interval every generalized function, as defined in the sense of Schwartz, may be represented in the form of the "derivative" of some order of a continuous function.

It seems to the author that within the limits of constructive mathematics the answer to the above question will be negative. However, he has not yet succeeded in proving this.

Appendix

TWO LEMMAS ON CONSTRUCTIVE REAL NUMBERS

Suppose that $\mathscr{A}$ is a nonempty set of words in some alphabet (we recall that a set as a rule is given by means of a one-parameter formula in a definite logico-mathematical language). The subordinate generic letter assigned to this set $\mathscr{A}$ will be denoted by ν.

Lemma 1. *If $\mathscr{A}$ is an algorithmically enumerable set, then, for any algorithm λ of the type $(\nu \to \text{Д})$,*

$$\neg \forall \nu_1 (\lambda(\nu_1) = 0) \supset \exists \nu_1 (\lambda(\nu_1) \neq 0);$$
$$\neg \forall \nu_1 (\lambda(\nu_1) \geqslant 0) \supset \exists \nu_1 (\lambda(\nu_1) < 0);$$
$$\neg \forall \nu_1 (\lambda(\nu_1) \leqslant 0) \supset \exists \nu_1 (\lambda(\nu_1) > 0),$$

where ν_1 is a variable of genus ν.

Proof. Suppose that λ is an algorithm of the indicated type. We shall prove the first of the formulas noted above. Suppose that

$$\neg \forall \nu_1 (\lambda(\nu_1) = 0). \tag{1}$$

Using formula (a) in Theorem 4.8.1 of [**26**], we may deduce that formula (1) is equivalent to the following formula:

$$\neg \forall \nu_1 k (|\underline{\lambda(\nu_1)}\,\overline{(\lambda(\nu_1)}(k+1))| \leqslant 2^{-(k+1)}). \tag{2}$$

The formula inside the outside parentheses of (2) expresses an algorithmically verifiable condition. The characteristic sets of the variables ν_1 and k are nonempty enumerable sets. Therefore, by the principle of constructive choice, it follows from (2) that

$$\exists \nu_1 k (|\underline{\lambda(\nu_1)}\,\overline{(\lambda(\nu_1)}(k+1))| > 2^{-(k+1)}).$$

In view of Theorem 4.8.1 of [**6**] we therefore find that

$$\exists \nu_1 (\lambda(\nu_1) \neq 0),$$

as we were required to prove. The remaining portions of the lemma are proved analogously.

Lemma 2. *It is possible to construct a set $\mathscr{A}$ containing at least one element and having the following property: it is not true that for any algorithm λ of the type $(\nu \to \text{Д})$ one can have*

$$\neg \forall \nu_1 (\lambda(\nu_1) = 0) \supset \exists \nu_1 (\lambda(\nu_1) \neq 0). \tag{3}$$

Proof. Suppose that π_1 is a universal partially-recursive rational-valued function for functions of one variable, and π_3 a universal function for functions of three variables (see [12], §1). Suppose that S_2^1 is a primitive recursive function such that

$$\forall lnkm\,(\pi_3(l, n, k, m) \simeq \pi_1(S_2^1(l, n, k), m)).$$

Suppose that t is a primitive recursive function of two variables such that it is not possible to construct a general recursive function ν satisfying the following condition: $v(n) = 0$ if and only if $t(m, n) \neq 0$ for every m. It is possible to construct a rational-valued partially-recursive function χ such that

$$\chi(n, k, m) \simeq 2 \cdot \overline{\mathrm{sg}}\left(\prod_{i \leqslant m} t(n, i + k)\right).$$

By the theorem on universal functions ([12], §1.6.1) we can realize a natural number N such that, for any n, k and m,

$$\chi(n, k, m) \simeq \pi_3(N, n, k, m).$$

Now we define a set $\mathscr{A}$ of natural numbers as follows:

$$l \in \mathcal{A} \leftrightharpoons \forall mk(!\pi_1(l, k) \,\&\, !\,\pi_1(l, m) \,\&\, (k \leqslant m \supset |\pi_1(l, k) - \pi_1(l, m)| \leqslant 2^{-k}).$$

This set $\mathscr{A}$ is constructed in the language of recursive functions. However, it may also be constructed in an appropriate logico-mathematical language of the type described in [24]. The transition from recursive functions to the corresponding normal algorithms can always be realized because of the equivalence of these concepts ([3]).

Suppose that ν is a subordinate generic letter assigned to the set $\mathscr{A}$.

It is possible to construct a two-parameter formula $\mathfrak{A}$ such that, for any element ν_1 of the set $\mathscr{A}$ and any natural number n,

$$\mathfrak{A}(\nu_1, n) \equiv \exists k\,(\nu_1 = S_2^1(N, n, k)).$$

The set $\mathscr{A}$ and the formula $\mathfrak{A}$ were constructed in the proof of Theorem 4 of [5]. Obviously the set $\mathscr{A}$ contains at least one element. In [5] it was proved that the formula

(4) $$\forall n\nu_1\,(\mathfrak{A}(\nu_1, n) \vee \neg\,\mathfrak{A}(\nu_1, n))$$

is constructively true, but that the formula

(5) $$\forall n\,(\neg\neg\,\exists \nu_1 \mathfrak{A}(\nu_1, n) \supset \exists \nu_1 \mathfrak{A}(\nu_1, n))$$

is not constructively true.

Now we construct an algorithm $\mathfrak{B}$ of the type (н$\nu \to$ д) as follows:

$$\mathfrak{B}(n \square \nu_1) \simeq \begin{cases} 1, & \text{if } \mathfrak{A}(\nu_1, n), \\ 0, & \text{if } \neg\,\mathfrak{A}(\nu_1, n). \end{cases}$$

In view of (4) it is obvious that $(\mathfrak{B} \uparrow n \square)$ is an algorithm of the type ($\nu \to$ д) for any natural number n.

We suppose that formula (3) holds for any algorithm λ of the type ($\nu \to$ д). Then the following formula is constructively true:

$$\forall n (\neg \forall \nu_1 (\mathfrak{B}(n \square \nu_1) = 0) \supset \exists \nu_1 (\mathfrak{B}(n \square \nu_1) \neq 0)).$$

Since

$$\mathfrak{B}(n \square \nu_1) = 0 \equiv \neg \mathfrak{A}(\nu_1, n) \quad \text{and} \quad \mathfrak{B}(n \square \nu_1) \neq 0 \equiv \mathfrak{A}(\nu_1, n),$$

it follows that the formula

$$\forall n (\neg \forall \nu_1 \neg \mathfrak{A}(\nu_1, n) \supset \exists \nu_1 \mathfrak{A}(\nu_1, n))$$

is true, so that (5) holds. But (5) is not constructively true, so that it is not true that (3) holds for any algorithm λ of the type ($\nu \to$ д). Lemma 2 is proved.

REMARK 1. The set $\mathscr{A}$ constructed in the proof of Lemma 2 is expressed by a normal formula.

REMARK 2. Lemma 2 remains valid if in its formulation formula (3) is replaced by one of the following formulas:

$$\neg \forall \nu_1 (\lambda(\nu_1) \geqslant 0) \supset \exists \nu_1 (\lambda(\nu_1) < 0);$$

$$\neg \forall \nu_1 (\lambda(\nu_1) \leqslant 0) \supset \exists \nu_1 (\lambda(\nu_1) > 0);$$

$$\neg \forall \nu_1 (\lambda(\nu_1) \neq 0) \supset \exists \nu_1 (\lambda(\nu_1) = 0);$$

$$\neg \forall \nu_1 (\lambda(\nu_1) < 0) \supset \exists \nu_1 (\lambda(\nu_1) \geqslant 0);$$

$$\neg \forall \nu_1 (\lambda(\nu_1) > 0) \supset \exists \nu_1 (\lambda(\nu_1) \leqslant 0.$$

For each of these cases the lemma is proved by means of the appropriate change in the definition of the algorithm $\mathfrak{B}$.

BIBLIOGRAPHY

[1] I. M. Gel'fand and G. E. Šilov, *Generalized functions*. Vol. 1: *Operations on them*, Fizmatgiz, Moscow, 1959; English transl., Academic Press, New York, 1964. MR **20** #4182; **29** #3869.

[2] ———, *Generalized functions*. Vol. 2: *Spaces of fundamental functions*, Fizmatgiz, Moscow, 1958; English transl., Academic Press; Gordon and Breach, New York, 1968. MR **21** #5142a; **37** #5693.

[3] V. K. Detlovs, *The equivalence of normal algorithms and recursive functions*, Trudy Mat. Inst. Steklov. **52** (1958), 75-139; English transl., Amer. Math. Soc. Transl. (2) **23** (1963), 15-81. MR **21** #5.

[4] I. D. Zaslavskiĭ, *Some properties of constructive real numbers and constructive functions*, Trudy Mat. Inst. Steklov. **67** (1962), 385-457; English transl., Amer. Math. Soc. Transl. (2) **57** (1966), 1-84. MR **27** #3519.

[5] ———, *On some differences between basic and subordinate variables in logico-mathematical languages*, Mathematical problems of cybernetics and computational techniques, Izdat. Akad. Nauk Armjan. SSR, Erevan, 1963, pp. 13-29. (Russian) RŽMat **1963** #12A66.

[6] I. D. Zaslavskiĭ and G. S. Ceĭtin, *On singular coverings and related properties of constructive functions*, Trudy Mat. Inst. Steklov. **67** (1962), 458-502; English transl., Amer. Math. Soc. Transl. (2) **98** (1971), 41-90. MR **27** #2408.

[7] L. V. Kantorovič and G. P. Akilov, *Functional analysis in normed spaces*, Fizmatgiz, Moscow, 1959; English transl., Internat. Series of Monographs in Pure and Appl. Math., vol. 46, Macmillan, New York, 1964. MR **22** #9837; **35** #4699.

[8] B. A. Kušner, *On the existence of unbounded analytic constructive functions*, Dokl. Akad. Nauk SSSR **160** (1965), 29-31 = Soviet Math. Dokl. **6** (1965), 26-28. MR **31** #55.

[9] B. M. Makarov, *Inductive limits of normed spaces*, Dokl. Akad. Nauk SSSR **119** (1958), 1092-1094. (Russian) MR **20** #5412.

[10] ———, *Some pathological properties of inductive limits of B-spaces*, Uspehi Mat. Nauk **18** (1963), no. 3 (111), 171-178. (Russian) MR **27** #2839.

[11] A. A. Markov, *Theory of algorithms*, Trudy Mat. Inst. Steklov. **42** (1954); English transl., Israel Program for Scientific Translations, Jerusalem, 1961. MR **17**, 1038; **24** #A2527.

[12] ———, *On constructive functions*, Trudy Mat. Inst. Steklov. **52** (1958), 315-348; English transl., Amer. Math. Soc. Transl. (2) **29** (1963), 163-195. MR **20** #5131; **27** #3527.

[13] ———, *Constructive mathematics*, Trudy Mat. Inst. Steklov. **67**(1962), 8-14; English transl., Amer. Math. Soc. Transl. (2) **98** (1971), 1-10. MR **27** #3528.

[14] ———, *On a principle of constructive mathematical logic*, Proc. Third All-Union Math. Congress, vol. II, Izdat. Akad. Nauk SSSR, Moscow, 1956, pp. 146-147. (Russian) RŽMat **1958** #946.

[15] G. E. Minc, *The differentiability predicate and the differentiation operator in constructive mathematical analysis*, Dokl. Akad. Nauk SSSR **147** (1962), 1032-1034 = Soviet Math. Dokl. **3** (1962), 1774-1776. MR **26** #6055.

[16]* ———, *On predicate and operator variants of the formation of the theories of constructive mathematics*, Trudy Mat. Inst. Steklov. **72** (1964), 383-436; English transl., Amer. Math. Soc. Transl. (2) **100** (1971), 1-68. MR **33** #2543.

[17]* ———, *Corrigenda and addenda to the article* "On predicate and operator variants of the formation of the theories of constructive mathematics", Trudy Mat. Inst. Steklov. **93** (1967), 257-258 = Proc. Steklov Inst. Math. **93** (1967), 327. MR **37** #35.

---

* *Editor's note*. The corrections made in [**17**] were incorporated into the English translation of [**16**].

[18] B. S. Mitjagin, *Approximate dimension and bases in nuclear spaces,* Uspehi Mat. Nauk **16** (1961), no. 4 (100), 63-132 = Russian Math. Surveys **16** (1961), no. 4, 59-127. MR **27** #2837.

[19] A. D. Myškis and A. Ja. Lepin, *On the definition of generalized functions,* Mat. Sb. **43** (**85**) (1957), 323-348. (Russian) MR **21** #293.

[20] V. P. Orevkov, *On certain types of continuity of constructive operators,* Trudy Mat. Inst. Steklov. **93** (1967), 164-186 = Proc. Steklov Inst. Math. **93** (1967), 211-240. MR **37** #1242.

[21] È. S. Orlovskiĭ, *Some questions in the theory of algorithms,* Trudy Mat. Inst. Steklov. **52** (1958), 140-171; English transl., Amer. Math. Soc. Transl. (2) **94** (1970), 1-36. MR **20** #6357; **42** #1622.

[22] D. A. Raĭkov, *On two classes of locally convex spaces which are important in applications,* Voronež. Gos. Univ. Trudy Sem. Funkcional. Anal. **1957**, no. 5, 22-34. (Russian) MR **21** #6526.

[23] A. O. Slisenko, *On certain algorithmic problems connected with arithmetic operations on duplexes,* Trudy Mat. Inst. Steklov. **72** (1964), 488-523; English transl., Amer. Math. Soc. Transl. (2) **100** (1972), 135-183. MR **34** #2460.

[24] N. A. Šanin, *On the constructive interpretation of mathematical judgements,* Trudy Mat. Inst. Steklov. **52** (1958), 226-311; English transl., Amer. Math. Soc. Transl. (2) **23** (1963), 109-189. MR **21** #2.

[25] ———, *Über einen Algorithmus zur konstruktiven Dechiffrierung mathematischer Urteile,* Z. Math. Logik Grundlagen Math. **4** (1958), 293-303. (Russian) MR **21** #5566.

[26] ———, *Constructive real numbers and constructive function spaces,* Trudy Mat. Inst. Steklov. **67** (1962), 15-294; English transl., Transl. Math. Monographs, vol. 21, Amer. Math. Soc., Providence, R. I., 1968. MR **28** #30.

[27] G. S. Ceĭtin, *Algorithmic operators in constructive metric spaces,* Trudy Mat. Inst. Steklov. **67** (1962), 295-361; English transl., Amer. Math. Soc. Transl. (2) **64** (1967), 1-80. MR **27** #2406.

[28] ———, *Mean-value theorems in constructive analysis,* Trudy Mat. Inst. Steklov. **67** (1962), 362-384; English transl., Amer. Math. Soc. Transl. (2) **98** (1971), 11-40. MR **27** #2407.

[29] Phan Dinh Diêu [Fan Din′ Zieu], *Constructive locally convex linear topological spaces,* Dokl. Akad. Nauk SSSR **162** (1965), 766-769 = Soviet Math. Dokl. **6** (1965), 764-767. MR **33** #6337.

[30] ———, *The metrizability, normability and multinormability of constructive locally convex spaces,* Dokl. Akad. Nauk SSSR **162** (1965), 1011-1014 = Soviet Math. Dokl. **6** (1965), 814-817. MR **33** #6338.

[31] ———, *On spaces adjoint to constructive locally convex spaces,* Dokl. Akad. Nauk SSSR **116** (1966), 45-48 = Soviet Math. Dokl. **7** (1966), 43-46. MR **33** #1702.

[32] ———, *Constructive generalized functions,* Dokl. Akad. Nauk SSSR **174** (1967), 37-40 = Soviet Math. Dokl. **8** (1967), 598-601. MR **36** #43.

[33] ———, *Certain properties of constructive generalized functions,* Dokl. Akad. Nauk SSSR **174** (1967), 298-301 = Soviet Math. Dokl. **8** (1967), 643-646. MR **36** #44.

[34] ———, *On spaces of constructive infinitely differentiable functions and on functionals in them,* Dokl. Akad. Nauk SSSR **180** (1968), 799-802 = Soviet Math. Dokl. **9** (1968), 695-698. MR **38** #529.

[35] ———, *A language of constructive mathematics involving systems of sets,* Trudy Mat. Inst. Steklov. **93** (1967), 123-141 = Proc. Steklov Inst. Math. **93** (1967), 157-182. MR **36** #4994.

[36] ———, *On closed and open sets in constructive topological spaces,* Trudy Mat. Inst. Steklov. **93** (1967), 250-256 = Proc. Steklov Inst. Math. **93** (1967), 319-326. MR **37** #2580.

[37] N. Bourbaki, *Topologie générale,* Chaps. 1, 2, 4th ed., Actualités Sci. Indust., no. 1142, Hermann, Paris, 1965; English transl., Hermann, Paris; Addison-Wesley, Reading, Mass., 1966; Russian transl., "Nauka", Moscow, 1968. MR **39** #6237, 6238; **41** #984.

[38] ———, *Espaces vectoriels topologiques,* Chaps. $1-5$, fascicule de résultats, Actualités Sci. Indust., nos. 1189, 1229, 1230, Hermann, Paris, 1953, 1955; Russian transl., IL, Moscow, 1958. MR **14**, 880; **17**, 1109.

[39] J. Dieudonné and L. Schwartz, *La dualité dans les espaces (F) et (LF)*, Ann. Inst. Fourier (Grenoble) **1** (1949/50), 61-101; Russian transl., Matematika **2** (1958), no. 2, 77-107. MR **12**, 417.

[40] H. R. Fischer, *Limesräume*, Math. Ann. **137** (1959), 269-303. MR **22** #225.

[41] R. L. Goodstein, *Recursive analysis*, Studies in Logic and Foundations of Math., North-Holland, Amsterdam, 1961. MR **24** #A1821.

[42] ———, *The relatively exponential, logarithmic and circular functions in recursive function theory*, Acta Math. **92** (1954), 171-190. MR **16**, 783.

[43] S. C. Kleene, *Introduction to metamathematics*, Van Nostrand, Princeton, N. J., 1952; Russian transl., IL, Moscow, 1957. MR **14**, 525; **19**, 2.

[44] A. Kolmogoroff, *Zur Normierbarkeit eines allgemeinen topologischen linearen Raumes*, Studia Math. **5** (1934), 29-43.

[45] J. Korevaar, *Distributions defined from the point of view of applied mathematics*. I – V, Nederl. Akad. Wetensch. Pric. Ser. A. **58** = Indag. Math. **17** (1955), 368-378, 379-389, 483-493, 494-503, 663-674. MR **17**, 63; 354; 594.

[46] H.-J. Kowalsky, *Limesräume und Komplettierung*, Math. Nachr. **12** (1954), 301-340. MR **17**, 390.

[47] S. Łojasiewicz, *Sur la valeur et la limite d'une distribution en un point*, Studia Math. **16** (1957), 1-36. MR **19**, 433.

[48] J. Mikusiński and R. Sikorski, *The elementary theory of distributions*. I, II, Rozprawy Mat. **12** (1957); ibid. **25** (1961); Russian transl., IL, Moscow, 1959, 1963. MR **20** #1214; **23** #A4006.

[49] Y. N. Moschovakis, *Recursive metric spaces*, Fund. Math. **55**(1964), 215-238. MR **32** #45.

[50] J. von Neumann, *On complete topological spaces*, Trans. Amer. Math. Soc. **37** (1935), 1-20.

[51] L. Schwartz, *Théorie des distributions*. Tomes I, II, Actualités Sci. Indust., nos. 1091, 1122, Hermann, Paris, 1950, 1951. MR **12**, 31; 833.

[52] ———, *Méthodes mathématiques pour les sciences physiques*, Enseignement des Sciences, Hermann, Paris, 1961; English transl., Hermann, Paris, and Addison-Wesley, Reading, Mass., 1966; Russian transl., "Nauka", Moscow, 1965. MR **26** #919; **32** #6721; **34** #7309.

[53] J. Sebastião e Silva, *Su certe classi di spazi localemente convessi importanti per le applicazioni*, Rend. Mat. e Appl. (5) **14** (1955), 388-410; Russian transl., Matematika **1** (1957), no. 1, 60-77. MR **16**, 1122.

[54] E. Specker, *Nicht konstruktive beweisbare Sätze der Analysis*, J. Symbolic Logic **14** (1949), 145-158. MR **11**, 151.

[55] G. Temple and Z. Zieleźny, *On Fourier transforms of rapidly increasing distributions*, Studia Math. **23** (1963), 119-140. MR **28** #424.

[56] J. Wloka, *Limesräume und Distributionen*, Math. Ann. **152** (1963), 351-409. MR **28** #1486.

[57] Z. Zieleźny, *Sur la définition de Łojasiewicz de la valeur d'une distribution dans un point*, Bull. Acad. Polon. Sci. Cl. III **3** (1955), 519-520. MR **17**, 595.